Embedded Systems

AF785794

For further volumes:
http://www.springer.com/series/8563

Brian Bailey · Grant Martin

ESL Models and their Application

Electronic System Level Design and Verification in Practice

Springer

Brian Bailey
9100 SW 161st Ave
Beaverton OR 97007
USA
brian_bailey@acm.org

Grant Martin
Tensilica, Inc.
3255-6 Scott Boulevard
Santa Clara CA 95054-3013
USA
gmartin@tensilica.com

ISBN 978-1-4614-2541-0 e-ISBN 978-1-4419-0965-7
DOI 10.1007/978-1-4419-0965-7
Springer New York Dordrecht Heidelberg London

Printed on acid-free paper

Springer is part of Springer Science+Business Media (www.springer.com)

Preface

The book *ESL Design and Verification* (Elsevier 2007) [1] coincided with a point in time at which the authors believed that ESL had started to become something real and tangible. Before then it had been a collection of university projects, isolated pockets of tool development, and even patchier levels of adoption. Most of the tools that were created were based on proprietary languages, with little if any external interfaces or extensibility. When we started to write the book, we believed that ESL had turned the corner and that we were at the starting point of something big happening in the industry. That book laid the groundwork for the entire domain that is categorized as ESL and we provided our best assessments about where the technology was heading and the things that people should pay close attention to.

That was just a few years ago, and in that very short period of time, there has been a significant maturing and development in certain areas of the ESL flow. These developments have brought ESL from a technology to keep your eye on to one that you should be considering adopting if you want to have the highest levels of productivity, or improved power and/or performance in the products that you design, and to attain the highest possible levels of quality. At the same time, almost nothing from that first book has been made redundant or has been shown to be a wrong prediction of where the industry was heading. As such this book is not a replacement for the previous book, but one that expands on some of the concepts that were first introduced there.

It is also uncanny how a small development can have such a profound set of implications, especially when that development is not even very advanced. As people have often said in the past, when we look at the progress we have made in a year, we are usually disappointed. When we look at the progress we made over the period of 10 years, we are often amazed at how much was accomplished. In the past year, one such event happened that on the surface of it was not that significant, but it already showing the magnitude of the implications that it will have.

We are talking about the release of the OSCI TLM 2.0 specification, which creates a standardized way to connect models described at the untimed or approximately timed transaction level. While it is essentially able to connect any forms of models, it really only addresses the memory-mapped communications domain. The ramifications of its introduction have been huge. Instead of every vendor of system-level virtual platforms having their own proprietary languages, models, and tools, every major developer of these platforms is now beginning to standardize on the use of TLM 2.0 as the way in which to interconnect models or is planning to do so within their next development cycle. Models developed for one system will be able to work on another, meaning that the problem of model availability and true interoperability is now being solved. A concerted effort is going into making SystemC a usable simulation environment and is becoming the accepted interconnect language for a large part of the ESL flow.

This does not mean that the industry has agreed upon the general usefulness of SystemC as a modeling language, and many vendors still only use it where they want to create interoperability layers in their products. This may change as each of the pieces of a SystemC flow becomes more optimized. High-level synthesis tool vendors are still locked in a battle for the best input language, with C, C++, and SystemC being the three strong contenders. At the same time, the software community is looking for solutions that enable them to embed the concepts of distributed computing into their domain and to remove the unwanted dependencies created in languages such as C. We think we have a long way to go before we eventually settle on the ideal language(s) on which we can base the next few decades of evolution.

This book will take a more practical approach than the first book, in that it will only cover the areas of the flow that are seeing real adoption and where the level of maturity is such that significant gains can be achieved by those who adopt them. It will concentrate on the models that are the cornerstone of those flows and will provide concrete examples showing how they are created, how they are used, and how they get transformed.

In order to provide that level of detail, it becomes necessary to showcase specific languages, tools, and flows. The choice of tools is in most cases arbitrary and readers should not take away any impression that we are endorsing those tools – even if personally we consider them among the best available! In many cases, several competing tools exist, and you should evaluate all of them before making any purchase decision.

Who Should Read This Book

We hope that this book will be useful to many types of readers, including

- **The novice in ESL**, who may be a student in electronic design, EDA, software design, system design, or related fields of study; or may be a designer seeking to learn more about ESL modeling, design, and verification. This book will provide you with an in-depth understanding of the areas of ESL that are actively being adopted by the design community. For those just entering the domain, it is hoped that the taxonomy will provide a clear understanding of the attributes of models as they exist in this domain, something that is often obfuscated by marketing departments.
- **The experienced designer** who has some base in ESL design and verification may find that there are many aspects of the process with which they are not familiar. For example, someone who is familiar with the concepts of high-level

synthesis may not be aware of the developments in the virtual prototyping space. In addition, someone contemplating the adoption of virtual prototypes may be confused by the wide variation in capabilities offered by the vendors, not understanding that the platforms are directed toward completely different tasks and have thus made implementation choices that are optimized for those tasks. This book provides a good opportunity for specialists in all aspects of the design flow to fill in gaps in their experience and knowledge base and improve their understanding of the overall tradeoffs and issues.

- **The design manager** who wishes to understand particular parts of the ESL flow, to improve the capabilities of his or her design team, who wants to reduce the risk in their development projects, who wants to expand the scope of their design projects and identify key missing components of the team will have a lot to gain by reading this book from cover to cover. It is also suggested that they also read the earlier book *ESL Design and Verification* [1], which will provide them with a more complete overview of the entire ESL flow.
- **The researcher** in academia or in an industrial research lab may have an excellent view of their subject area, but not have a good view of how it fits into the overall ESL design space. In addition, they may not be fully aware of the state-of-the-art industrial practice in ESL: those methods that have been proven to work and are recommended and those methods that have not found favor with practical design teams. This book will give them considerable insight in particular into industrial practice. We hope as well that the taxonomy defined in this book will be adopted by the academic community and extended, or modified, with the advances that they are making. A common classification scheme will help to remove a lot of the ambiguity that has been so common in this field.
- **The educator** in academia or within an industrial company who needs a text on which they can base a class or course. Many such courses were based on the first book *ESL Design and Verification*, especially at the post-graduate level, and this book provides a lot more detailed material on which more extensive courses can be based. While it is not specifically written as a textbook, the authors are willing to work with educators to help them develop materials.

Structure of the Book

This book is constructed in two major sections:

- The first section concentrates on the languages and models that are becoming commonly used within ESL flows. We try to be open and honest about these, giving both the positives and the negatives associated with them. We have tried to keep personal bias out of this as much as possible, but in some cases it may be a little obvious what the authors' opinions are. Tools are introduced when it helps to demonstrate the concepts being discussed. In many cases dedicated books exist that will provide a lot more depth than is possible in this book but the information provided here should enable you to understand the concepts behinds the models. References are provided for those who wish to go into much more depth.

- The second section will showcase some of those languages and models in use. In order to do this, actual tools will be shown. These demonstrate system-level virtual prototype at various levels of abstraction and the kinds of analysis that they can be used for. We will look at optimizing processors for specific applications, at making hardware/software tradeoffs, and at synthesis of the hardware portions. These are meant to demonstrate the types of capabilities available in these classes of tools, but it should also be clear that in the space available it is impossible to talk about all of their capabilities and features.

The first section of the book will be especially useful to those who want to understand some of the concepts behind the models and languages that are in use today, particularly in the functional modeling and verification spaces. The second section may be more useful to those who want to understand particular aspects of the ESL flow in more detail and how tools are developing to meet the needs of various types of practitioners.

Chapter Listing

- *Chapter 1: Introduction:* This chapter presents the fundamentals of models and languages by taking a historical look at the development of some early models. It looks at how models can evolve over time and why several abstractions of those models are necessary. Languages are used to build those models and we briefly review some of the basics of language construction. The chapter briefly describes various model types that are in active use for the design and verification of systems within an ESL flow. The model taxonomy is used to show how some of the functional languages that are being used have different characteristics that make them better at certain kinds of tasks. Finally, the chapter provides a set of definitions that are used throughout the book.
- *Chapter 2: IP Meta Models for SoC Assembly and HW/SW Interfaces:* This chapter explores in detail one class of models that are called meta models. It looks at the efforts of the SPIRIT standards organization in their attempts to create an IP metadata model. The goal of this was to enable platform assembly in a somewhat intelligent manner by encapsulating knowledge about the interfaces. The most valuable part of this effort has led to the standardization of register definitions, which has enabled metadata to be shared between the hardware and software teams. This chapter explores this usage of metadata extensively.
- *Chapter 3: Functional Models:* This chapter looks at models that are the mainstay of most design and verification flows, namely functional models. Starting from high-level mathematical models it examines the models in active use in ESL flows, concentrating on the SystemC language and TLM 2.0. It looks at the main concepts of the SystemC language that enable computation, communications, and

synchronization. TLM 2.0 provides communications and timing capabilities that enable modeling at various levels of timing accuracy. The chapter also examines the rules that go along with TLM usage and the need for compliance checking in order to insure interoperability. Finally, the chapter takes a brief look at some declarative models.

- *Chapter 4: Testbench Models:* Testbenches are constructed from a number of models that work together with a variety of tools to produce a verification environment. This chapter looks at how those models work together and briefly surveys some of those models, such as comparison models, constraint models, coverage models, and verification plans. It looks at some of the problems created when these environments are constructed from multiple languages and some techniques that are being used to overcome them. Finally the chapter looks at verification IP and standardization efforts that are underway.
- *Chapter 5: Virtual Prototypes and Mixed Abstraction Modeling:* This is the first of several chapters that take an in-depth look at models, and how they are used within specific parts of an ESL flow. This chapter looks at a system-level virtual prototype of the hardware system that is specifically intended to be used for the development and testing of software. This chapter also demonstrates the transition that is going on in the industry: away from proprietary systems and interfaces toward open standards. The chapter uses the Innovator tool from Synopsys to demonstrate many of the capabilities that can be found in tools of this type.
- *Chapter 6: Processor-Centric Design: Processors, Multi-processors and Software:* Most embedded systems have one or more processors at their core. This chapter looks into the issues involved in processor-centric design using a commercial configurable and extensible processor IP as an example. The chapter uses the Xtensa product line from Tensilica to demonstrate the design of a JPEG decoder. Starting from performance analysis, a series of design possibilities are investigated which lead to some surprising tradeoffs being made between performance, area, and power. Finally the chapter looks at some of the issues associated with the design and debugging of multi-processor systems.
- *Chapter 7: Codesign Experiences Based on a Virtual Platform:* This chapter delves into the world of hardware–software codesign where hardware and software are considered to be equal partners. Using a JPEG decoder as an example, this chapter explores the entire flow from specification to implementation using a platform-based development approach based on SystemC. The methodology allows easy migration of modules between the two domains and explores a number of ways in which the design can be partitioned and mapped onto a number of platforms having different numbers of processors, communications channels, and memory architectures.
- *Chapter 8: Transaction-Level Platform Creation:* Not all system-level virtual prototypes concentrate on the same issues or are used for the same purpose. In this chapter we look at a transaction-level platform that is intended for making hardware architectural choices. This requires being able to quickly and easily modify the timing associated with computation and communications within the platform. This requires layering of the timing model onto a functional model without

requiring changes to the functional model, all while using open industry standards. This chapter looks at the way that the Vista product from Mentor Graphics overcomes these challenges.
- *Chapter 9: C/C++ Hardware Design for the Real World:* This chapter explores the domain of high-level synthesis of hardware from a C/C++ description. It explores the language choices that could be made and looks at some of the fundamentals associated with this level of synthesis. Unlike RTL synthesis, high-level synthesis is an exploration process and requires a lot of analysis and feedback tools as well as understanding how languages embed architectural constraints. Using the CatapultC synthesis tool from Mentor Graphics, an almost real-time example is provided for the development of a matrix inversion block, which is core to many complex communications systems.

Relationship to First Book

ESL Design and Verification was published in 2007 and contained a complete overview of the ESL domain and the state of tools and languages at that time. It developed many of the foundations and concepts that are extended in this second book and as such these two books are meant to be read as if they were a single larger work.

We have attempted to make this second book completely self-contained and this does mean that there are several places where concepts were introduced in *ESL Design and Verification* and need to be understood within the scope of this book. In those cases we have provided a summary of the concepts in this book, but have not attempted to bring the entirety of the section forward. Those wishing to fully understand those concepts should read the corresponding sections of *ESL Design and Verification.*

One piece of feedback we received on the first book was that it did not go into enough detail. We hope we have rectified that problem with this book. While it may still not provide enough detail for people to become experts in the use of the various tools, we hope that there will now be enough detail that they can understand the concepts well and be able to fill in some of the blanks for themselves.

Companion Web Site

We are always open to feedback, either good or bad. We like to hear about the problems you are facing and what you would like to appear in the next book (if there is a third). We would encourage you to go to the companion site for these books http://ElectronicSystemLevel.com where we will be providing additional corrections and in some cases additional material that did not make it into the book, such as complete code listings. We also have a discussion forum attached to that site and we would encourage you to ask the community for help or guidance for anything related to the ESL space.

Reference

1. Brian Bailey, Grant Martin, Andrew Piziali. ESL Design and Verification: A Prescription for Electronic System Level Methodology, Elsevier, 2007.

Acknowledgments

Brian Bailey would like to thank his wife, Jaime Bailey, for her support and patience while working on this book. He would also like to acknowledge the many colleagues over the years that have helped him to build the knowledge that has enabled him to put this book together.

Grant Martin, as always, would like to thank his wife Margaret Steele and his daughters Jennifer and Fiona, who continue to put up with his book-writing endeavors now as patiently as they did more than a decade ago when he first collaborated on putting a book together.

The authors would also like to thank several companies who made their tools, people, and time available to put together many of the examples shown in this book. They believed in us and our ability to portray their technology to the community. They have spent many hours helping us understand the technical aspects of their technology so that we could present it to you. Without their help and support, it would not have been possible to put this book together. We know that you can get marketing pitches from any of them, and we have tried to keep that to a minimum while trying to expose some of the details of their technologies. We hope that we have succeeded in that respect.

Thanks go to Mentor Graphics and in particular Dinesh Jain and Thomas Nagler for assistance with their CatapultC high-level synthesis solution and to Rami Rachamim and Yossi Veller for assistance with their Vista architectural platform development tool. Thanks also to Frank Schirrmeister and Filip Thoen at Synopsys for the information about their Innovator software development platform. Thanks to Tensilica for allowing Grant to showcase their tools and for the support necessary to work on this book (most of which was done on evenings and weekends). A special thanks to Guy Bois. Guy was one of the first educators to adopt the book *ESL Design and Verification* into his graduate teaching at École Polytechnique in Montréal. He had a few suggestions with regard to the first book and thus we invited him to help us correct those problems with this book.

We would also like to thank David Murray of Duolog and Andrea Kroll of JEDA Technologies for their help in the construction of this book. Without their help, this book would not be as complete as it is.

Contents

About the Authors

Brian Bailey: Brian Bailey is an independent consultant working in the spaces of functional verification and ESL. Before that he was with Mentor Graphics for 12 years, with his final position being the Chief Technologist for verification, Synopsys, Zycad, Ridge Computers and GenRad. He graduated from Brunel University in England with a first class honours degree in electrical and electronic engineering.

Brian is the co-editor of the book *Taxonomies for the Development and Verification of Digital Systems* (Springer, 2005), the executive editor and author for *The Functional Verification of Electronic Systems: An Overview from Various Points of View* (IEC Press, 2005) and *Intellectual Property for Electronic Systems: An Essential Introduction* (IEC Press, 2007) and co-author of the best-selling book *ESL Design and Verification: A Prescription for Electronic System Level Methodology* (Elsevier, 2007). He has published many technical papers, given keynote speeches at conferences, performed seminars around the world, and been both a contributor and a moderator of panels at all of the major conferences.

Brian established the functional verification track in the DesignCon conferences, which quickly grew to be one of the major tracks of the conference. He chairs the interfaces standards group within Accellera and has in the past chaired other standards groups in Accellera and VSIA.

Brian is primarily interested in the specification, simulation, analysis, and implementation of embedded systems and today is moving into the problems associated with, and solutions necessary for, multi-processor systems.

Grant Martin: Grant Martin is a Chief Scientist at Tensilica, Inc. in Santa Clara, California. Before that, Grant worked for Burroughs in Scotland for 6 years; Nortel/BNR in Canada for 10 years; and Cadence Design Systems for 9 years, eventually becoming a Cadence Fellow in their Labs. He received his bachelor's and master's degrees in mathematics (Combinatorics and Optimisation) from the University of Waterloo, Canada, in 1977 and 1978.

Grant is a co-author or co-editor of nine books dealing with SoC design, SystemC, UML, modeling, EDA for integrated circuits and system-level design, including the first book on SoC design published in Russian. His last book, *ESL Design and Verification: A Prescription for Electronic System Level Methodology*, written with Brian Bailey and Andrew Piziali, was published by Elsevier Morgan

Kaufmann in February 2007. He has also contributed invited chapters to many more books and technical papers and talks at many conferences and workshops around the world.

He was co-chair of the DAC Technical Program Committee for Methods for 2005 and 2006. His particular areas of interest include system-level design, IP-based design of system-on-chip, platform-based design, multi-processor design, and embedded software. Grant is a Senior Member of the IEEE.

About the Contributors

Dr. Guy Bois holds a B.Sc.A and Ph.D. in computer science from the University of Montreal. He is a professor in the Department of Computer Engineering at École Polytechnique de Montréal and director of the SoC Platform laboratory. His research interests include HW/SW codesign and coverification for embedded systems. He is a cofounder of Space Codesign Systems Inc. He was assisted by *Laurent Moss* who is pursuing a Ph.D. at École Polytechnique de Montréal, *Luc Filion* a cofounder of Space Codesign Systems Inc., and *Sebastien Fontaine* who is an R&D engineer at STMicroelectronics. Laurent, Luc and Sebastien received a B.Ing. and M.Sc.A in computer engineering from École Polytechnique de Montréal.

Dr. Andrea Kroll has a Ph.D. in EE from the University of Technology in Aachen Germany related to high-level model validation. She works at JEDA technology as their VP of marketing and business development. Prior to that she worked at Synopsys and CoWare as an application specialist for system-level tools including CoCentic System Studio and CoWare Processor Designer, before she became director of product marketing for Processor Designer at CoWare. Her main interest is driving ESL hardware and software solutions and standards like SystemC and TLM2.0 from the technology as well as the business side.

David Murray is CTO of Duolog Technologies, where he is technical champion for Socrates, a chip integration platform. His career has spanned IC design and verification, software development, and EDA solutions at leading semiconductor companies and has written and presented many papers across this wide range of topics, from using algorithmic IP to functional verification. He has previously worked in Toucan Technologies and Silicon and Software Systems and was a founder of Inoru, a company focused on emerging SoC automation technologies. In Duolog, his research interest includes ESL tools, SoC integration, and automation flows. David holds an honors BEng degree from the University of Limerick, Ireland.

Thomas Nagler was born in Poland and grew up in Africa before coming to the United States. He obtained a master's and bachelor's from Northeastern University in Boston, MA, in electrical and, computer engineering. Currently he is a technical marketing engineer at Mentor Graphics working on the CatapultC high-level synthesis solution. Before this, he was a principal design engineer working on ATE instrumentation at LTX.

Frank Schirrmeister is director of product management in the solutions group at Synopsys. His research interests include electronic system-level design and verification, particularly pre-silicon virtual platforms for software development, architecture exploration, and verification. Frank got his MSEE in electrical engineering from the Technical University of Berlin with a focus on microelectronics and computer science. He has been a member of IEEE and the Association of German Engineers (VDI) since 1992.

Yossi Veller has over 16 years of experience in EDA. He was previously the chief scientist at Summit Design and head of the HDL and TLM simulation groups at that company. Currently he is serving as Mentor Graphics ESL chief scientist based in Israel. Before starting his EDA career he led an ADA cross-compiler project. He holds an M.Sc. degree in computer science, B.Sc. degrees in mathematics and in electronic engineering, all from Beer Sheva University, Israel.

Chapter 1
Introduction

In this introductory chapter we will present the fundamentals of models, starting with an understanding about the importance of models and how they can change over time. We will discuss why multiple models may exist at the same time and why a single abstraction is not always the ideal situation. We will explore the basic building blocks of languages (models of computation) and show how language syntaxes can be developed from them. We will briefly talk about the kinds of models that exist, their purposes, and the relationships that may exist between them. The chapter will also expand on the model taxonomy that was developed in the book *ESL Design and Verification* [1] and use it to show how many of the common functional models fit into the taxonomy and where the existing languages have problems that may need to be addressed in future languages. The chapter concludes with a set of definitions for many of the terms used throughout this book.

1.1 A Definition of a Model

A very good operational definition of an engineering model is found in [2]. The author has an elegant description of the use of scale models to construct the dome of the cathedral in Milan in the 15th century. Selic continues "The main purpose of engineering models is to help us understand the interesting aspects of a complex system *before* going through the expense and effort of actually constructing it." He also points out their use in analyzing existing systems and to communicate design ideas. His summary of "characteristics of good engineering models" includes support for abstraction, expressed in an intuitive form for ease of comprehension, accurate in faithfully representing interesting aspects of the modeled system, useful for prediction about the real system based on the model, and significantly cheaper to construct or experiment with than the real system.

However, although a good operational description of what a model is for and what its characteristics are, it still does not define what a model is. Turning to that growing source, Wikipedia, we find an interesting capsule definition [3]: "*A model* is a pattern, plan, representation (especially in miniature), or description designed to show the main object or workings of an object, system, or concept." Wikipedia

B. Bailey, G. Martin, *ESL Models and their Application*, Embedded Systems,
DOI 10.1007/978-1-4419-0965-7_1, © Springer Science+Business Media, LLC 2010

defines many further model types under the general category "model," among which is Computer Model, defined as follows [4]:

> A computer simulation, a computer model or a *computational model* is a computer program, or network of computers, that attempts to simulate an abstract model of a particular system.

We could debate ad infinitum the right definition of model, but these are reasonable starting points. Let us talk some more about what models are for.

1.2 A Day in the Life of a Model

Models allow us to reason about things that do not exist yet, are intangible, or are not fully understood. A model helps us to understand something. That understanding may allow us to refine the model or make decisions based on our expectation of the fidelity and accuracy of the model.

As a first example, think about the models for the Universe. The traditional belief was that the Earth was at the center of the Universe and that everything rotated around it. This model was created about 400 BC and is attributed to the Greek philosopher Eudoxus of Cnidus. About 300 BC it was suggested that the Sun was actually at the center, but this was rejected as it was clearly wrong based on the evidence that they had to work with at the time. If the Sun was stationary, then the Earth would be moving and thus we would expect unsecured objects to fly around and we would feel the motion, just as we did when we were moving.

The model, namely the geocentric model, had been updated many times to attempt to explain the movement of celestial bodies and those models got more and more complex over time. Problems remained with the model as it failed to explain the irregular motion of the planets and had no way to explain retrograde motion at all. Ptolemy updated the model to account for many of these phenomena by introducing orbits that were not exactly at the center of the Earth and that moved in their own epicycles. See Figure 1.1. Problems still remained and other observed effects could not be predicted from the model.

While many people had postulated that the model was flawed and that the Earth was not the center of the solar system or Universe, Copernicus, in his 1543 book entitled *De revolutionibus orbium coelestium* (*On the Revolutions of the Celestial Spheres*), was the first recorded person who proposed a completely new model which became the starting point for modern-day astronomy [Figure 1.2.]

That model has been improved and updated over the course of the last few centuries. Of course it is still a model and has not been proven to be correct, but so far no evidence has been found to foster a doubt about the fundamentals of the models, even though there is still active discussion about certain aspects of it. In the words of Ptolemy (approximately AD 168), who unfortunately based his work on the geocentric model:

> the simplest model that gets the right numbers should be used.

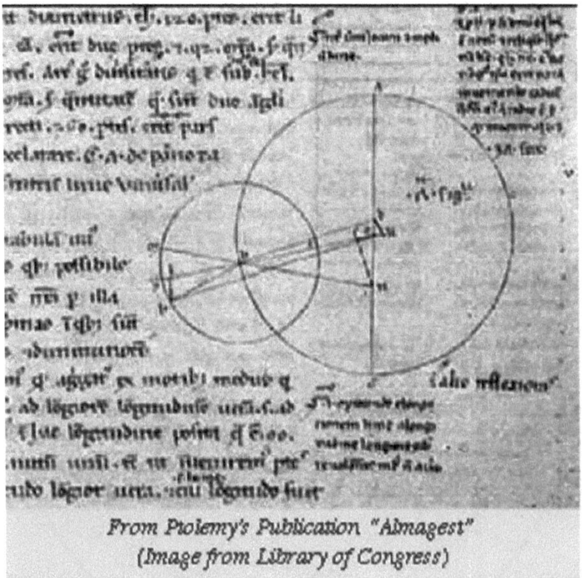

Fig. 1.1 Ptolemy's geocentric model

That statement holds true for today and is something that this book will come back to many times. Galileo then used his newly developed telescope to observe things in the solar system and found that the Ptolemaic model did not predict these effects while the Copernican model did. Using this model Kepler tested some other theories and found that they all supported the Copernican model. He devised a set of equations that related speed and distance to the Sun with a constant, and that this was true of all orbits except that the constant was different for each one. He suggested this attractive force to be magnetism, and it was not until Newton came along that he modified the thinking to be a relationship between mass, distance, and a universal constant.

This theory held true for about 200 years until Albert Einstein arrived at a time when we had the ability to take much more accurate measurements. He found that there were small discrepancies, especially with planets close to the Sun. He modified the simple formula to take into account objects that are moving at very high speed (close to the speed of light), or are moving in a region of very intense gravity. This led to the publication of his General Theory of Relativity in 1915.

So while general relativity is the best known model available to us today and has the lowest margin of error, it is clearly not always the right model to use. While it would be wrong to use the Ptolemaic model for anything because it is just plain wrong, the models of Copernicus are still good enough for most applications and for us mortals down here on Earth,

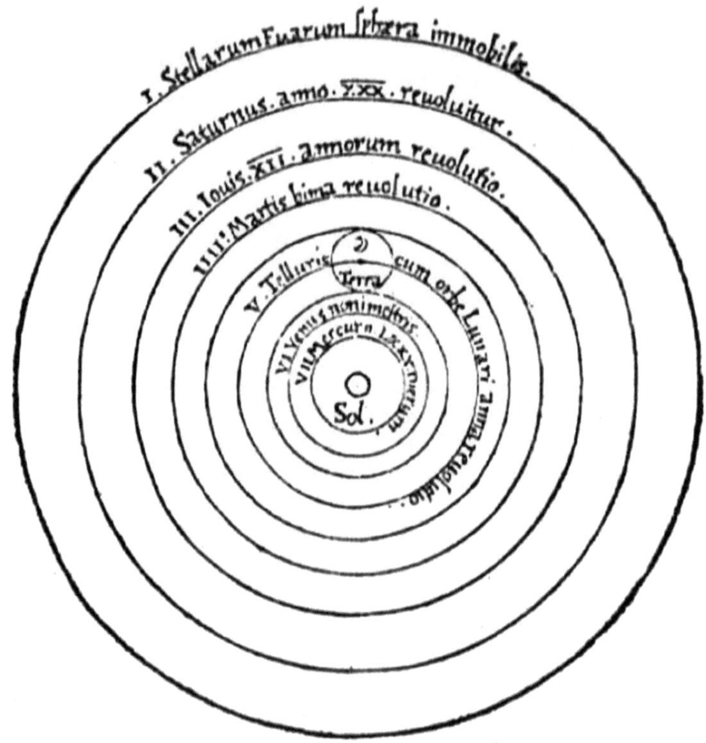

Fig. 1.2 Copernicus' heliocentric model

$$\mathrm{F_g} = \frac{\mathrm{Gm^1m^2}}{\mathrm{r^2}}$$

is much simpler and easier to use than

$$R_{\mu\nu} - \frac{1}{2}g_{\mu\nu}R + g_{\mu\nu}\Lambda = \frac{8\pi G}{C^4}T_{\mu\nu}$$

So the model with the greatest accuracy is not always the appropriate model. We have to make choices in the models that we use based on the operation that we are trying to perform.

As another example, consider models that are used to predict weather. The Farmers Almanac was a formula devised in 1792 by Robert B. Thomas. While the formula is still secret, it is based primarily on sunspot activity and relating patterns in that activity with weather patterns that were seen in the past based on similar profiles. Those predictions are still published to this day. It is interesting to note that their web site contains this statement:

> Modesty requires, however, that we add this caveat: It is obvious that neither we nor anyone else has as yet gained sufficient insight into the mysteries of the universe to predict weather long-range with anything resembling total accuracy.

Now that is a prediction that the authors of this book can support. But this was hardly the first attempt to predict the weather. As far back as 350 BC, Aristotle published his book Meteorology, which describes some of the basic cycles that make up part of our weather system, such as the cycle of evaporation and condensation of water in the atmosphere. The invention of the telegraph in 1837 enabled detailed surface data to be gathered from wide areas such that weather maps could be created and the ways in which weather changed could be studied. It was not until 1904 when it was first prognosticated that weather could be completely described by natural and physical laws. It was not long until fluid dynamics equations governing atmospheric flow were developed, but the problem was that only small numbers of these could be solved without the use of computers. It was estimated that 64,000 human computers would be necessary to calculate enough small areas to be able to predict global weather patterns. In 1948 John von Neumann developed ENIAC (see Figure 1.3) and recognized that weather forecasting was a natural for his computing machinery.

Fig. 1.3 von Neumann with his ENIAC computer

With the ability to collect and analyze data, more was learned about the atmosphere, and this led to the discovery of the chaotic nature of weather and that a number of possible outcomes were probable from any given set of initial conditions. The first regular weather prediction service used an IBM 701 in 1955 and issued twice daily updates. As more computer power became available and the models were improved over time, the forecasts became more accurate as shown in Figure 1.4.

Compute power has become a major factor in the complexity of the models that can be run. Not only are the models still being refined, but the available compute power limits the accuracy of the results that can be obtained. It is not economically viable to use the largest and most expensive computers available and for most forecasting, such as for the local news stations, something less accurate is adequate. Tracking and predicting the path of a hurricane is much more important. Back in

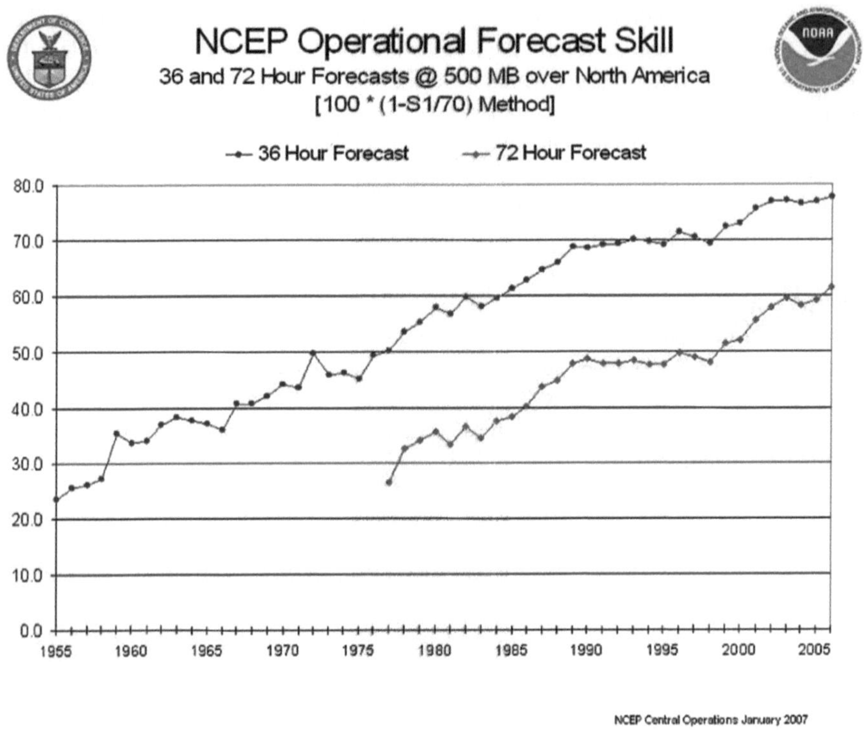

Fig. 1.4 This graph shows computer power and time vs. model accuracy as defined by the S1 score (a measure of the skill of the forecast) of 36- and 72-hour NCEP 500-millibar forecasts [5]

1985, forecasters could predict a hurricane's path within a 366 mile cone. Today, that cone has decreased to being 111 miles wide [6]. Many of the same lessons hold true with the use of models in the development of complex digital systems. We do not have unlimited compute-power available to us (well, a few companies seem to) and we must make rational decisions about what provides the most information for the cost we can bear. This is a classical design trade-off.

Models are a fundamental aspect of everything we do in the electronic design automation (EDA) industry. They enable us to abstract away the physics that makes transistors operate, they allow us to verify a design before it has been built, they allow us to increase design efficiency by having tools that can transform one model into another model that is closer to the final implementation. In short, the semiconductor development process is about the creation, application, and transformation of models.

Models are expensive to create and maintain, so we must limit the number of them that get used in the development process, often resulting in models being used for things that they were never intended for. If the model contains too little information it will result in inaccurate results, with error tolerances outside the range we feel comfortable with. But having too much information is equally bad. While we may then get very accurate results, the processing time necessary to get that result is not acceptable.

Engineers are thus continually playing a trade-off game between the time and expense of creating models that would enable them to perform a specific task and the value that those models would provide. Methodologies have been stalled or abandoned because of the model creation time, especially when much of a design is potentially coming from reuse. Is it worth creating new models for the parts of the design that are being reused? Would the time it takes for that modeling effort sink the current design by making it too late in the market, or would those models enable the design to have a much better power profile than any other product on the market and as a result blow the competition away on features such as battery life?

1.3 Types of Model

There are many types of models, including conceptual, physical, logical, mathematical, statistical, and visual. They can be models of the device that is to be built or the environment in which they are meant to operate. Most of them can exist at many different levels of abstraction. They all have inputs, perform some processing, and produce outputs. Many also contain objectives, constraints, and assumptions.

A conceptual model is a qualitative model that helps to highlight important connections between processes. This is probably the least understood of all model types as most people just equate this with an abstract model. Conceptual models exist primarily in the problem domain and not in the solution domain. Most of the time, the simplest model is the best, but since this is a model that is likely to evolve over time, it also needs to be built on a solid infrastructure so that modification can be made easily [7]. Conceptual models are often process flow diagrams, Petri Nets, event graphs, or some aspects of the Unified Modeling Language (UML), such as class diagrams, where classes represent the functions or concepts, and associations define the relationships that exist between them.

A physical model is most often something that can be created, observed, and manipulated. One or more aspects of the model have characteristics that are similar to the real system and thus allow concepts to be analyzed and understood.

A mathematical model enables equations to be analyzed. In our industry most of these will be numerical models although occasionally analytical models may exist as well. Statistical models allow variances to be modeled and to find means and variances.

Visual models help us to visualize how something will operate within the real world by providing some graphic or visual outputs, often with animations.

A logical model includes some aspects of the structure or form of the device that is to be built. This includes architectural models, transaction level models, and any model that includes time. These are thus implementation models and help us refine the ideas into a complete solution.

1.4 Models of Computation

A model of computation (MoC) is an attempt to distil the fundamental aspects of a system's ability to solve problems. The computer itself is a classic example of this. It provides a set of operations that it can perform, has the ability to store information in several ways such as registers, random access memory, and sequential memory, such as an old-fashioned tape store. Given these capabilities it becomes possible to predict how a piece of software will execute on that abstracted computer and may also be possible to estimate non-functional aspects of its execution, such as power, time or memory utilization.

A model of computation abstracts away many of the details of the machine so that it can be both fast and accurate in terms of the analysis which is to be performed. An example of this is the way in which various models of computation treat time. While it would be expected that this is a fundamental aspect of the system's capability, many models of computation assume that all operations happen in zero time. If we think about an algorithm, such as a differential equation, the model of computation tells us how to solve the equation. While order of execution may be important, time really has no meaning in this context unless we wish to actually run that algorithm on a computer or implement it in hardware. Thus time is a manifestation of an implementation, and if we are not close to an implementation, then this is an attribute of execution that can be ignored.

It should thus be seen that there is no one abstraction of a system and thus no one model of computation that is useful for all tasks in a typical ESL flow. Multiple models of computation will be required to progress from an idea to an implementation. At the same time, it is often found that a system needs multiple models of computation to describe separate pieces of the system. For example, a model of a computer would not be much help when attempting to solve a continuous system such as those found in analog components.

Models of computation may choose to describe many aspects of a system, such as time, concurrency, liveness, resource management as well as the basic functionality of a system. When a particular model of computation is chosen to describe a system, we are inherently assuming certain aspects of a solution. Consider for a moment the most fundamental model of computation in the software world. It is that of the von Neumann architecture. It defines the set of operations that can be performed by a processor and defines the availability of a memory to store a defined number of pieces of data that are accessible to the processor. In this way, all data is considered global, even though the program syntax itself may hide some of them from view in certain states. In addition it defines the relationship between the processor and the memory as a set of sequential operations.

Languages are ways to map the capabilities of a model of computation into something that can be described and communicated. These can either be textual or graphical in nature. They must have the ability to describe the memory storage capabilities, operations, and means of getting data into and out of a system, just as defined by the MoC. When described in this way, it is often thought that the language is the syntax and the model of computation underlying it is the semantics. We need both in order to fully describe something. Now the converse of this is not always true. There are languages that have been constructed that do not have a complete model of computation. These are sometimes referred to as ad hoc languages and generally result in problems such as non-deterministic behavior or ill-defined semantics that cause problems when implemented by multiple vendors and models are passed between the vendors resulting in different interpretations of the language. Perhaps languages such as Verilog and SystemVerilog come to mind here. These not only have all of the problems described, but also have problems with transformations because their semantics do not completely match the types of things that exist in the hardware world. We thus have to deal with problems such as differing functionality between simulation and synthesis tools that finish up requiring extra pragmas to resolve.

Returning to the example of von Neumann architecture, the languages encode these limitations, such as the sequential nature of the operations between processor and memory. This is why languages such as C, which were devised exactly for this model, have severe problems describing concurrency, or why data coherence becomes an issue. These are not conceived as possibilities by the fundamental model of computation.

It is clear that we need different models of computation to describe hardware systems and software systems even though they share many of the same needs, such as a way to express concurrency and time.

Even within the software domain, different models reflect different usage models. Consider a typical desktop computer running an application. The most important aspect of it is the transformation of data from one file to another. Speed and memory footprint are the two most important operational characteristics. Now if we contrast that with an embedded application, real time response may be more important than speed, power becomes an issue and the interaction of real things in the unpredictable physical world drive many of the implementation decisions. So where the Turing model can be used for a desktop program, embedded systems are classed as "defective" because they are non-terminating programs. Ironically if an embedded program terminates, it would be considered defective.

Time is a central aspect that differentiates many of the MoCs. In some, such as discrete event, all events can be arranged on a timeline with specific values attached to them. In other cases, while time advances, it is discrete as would be the case in a synchronous system. In others, time is a way in which causality is shown and is often referred to as partially ordered. This is why time is one of the fundamental axes of the taxonomy (see Section 1.9). It not only depicts abstraction, but also the underlying model of computation.

A good reference on models of computation and underlying theory can be found in [8].

1.5 Simplification

Models allow us to simplify a problem, often down to the essential elements necessary to perform a particular function, computation, or type of analysis. The primary means of simplification are abstraction, elimination, and structure.

1.5.1 Abstraction

This can exist in many ways. In the software realm: consider that the processor directly implies the lowest level of software abstraction, that is, the instruction. There is no point in defining software operation at any lower level than this, unless dealing with a micro-programmable device. Ignoring that possibility, the processor takes care of everything below the instruction level. A language like C adds another layer of abstraction, in that each statement written in C will result in several instructions being created when passed through a compiler. The compiler understands both the input language and the instruction set and aspects of the processor architecture that will enable it to create efficient instruction sequences. These sequences should make use of any specific capabilities of the processor, such as pipelining, or multiple execution units.

In the hardware world, there are two primary means of abstraction. These are time and data. By looking at the model taxonomy in Section 1.9, you will see that these are the two orthogonal axes that generally represent abstraction.

You will also notice the other three axes that make up the entire taxonomy. These are communications, computation, and configurability. While configurability is in some ways tied to implementation decisions, computation and communication are characteristics of a model that directly enable several implementation choices to be made. For example, if communications has not yet been tied down, then it directly implies some large architectural choices that have yet to be made. It can thus be seen that the three C's directly represent the gross decisions that turn a specification into a fully instanced form of an implementation, and the abstraction axis – time and data abstractions take us further down into the fully defined implementation. This is not a strict distinction since it is possible to have a behavioral model – that implies timing – without having fully nailed down architecture. This is in fact a fairly common modeling solution that is used to investigate performance characteristics of a system. In other words, to nail down architecture, you must consider timing and this requires pushing down into the solution space in order to make those choices.

1.5.2 Structure

While a final chip is a flat structure, models can represent this in a hierarchy that enables us to group certain things together, organize by function, perform duplication, and divide a problem. These are pure artifacts of the descriptive process and do not really describe anything about the system itself. This is not to say that they are not necessary. They are very necessary, especially when considering their impact on tools. If you tried to describe a system in a fully flattened form, very few tools would be able to handle the size and complexity of the resulting system. Hierarchy is very necessary and helpful.

A model may tell you things about an implementation or may be implementation neutral. In general, models that do not talk about implementation are referred to as specifications. Models may define the architecture of a solution, which may enable multiple implementations. Implementations are themselves models as they do not describe the actual silicon, but represent an abstraction of it.

Most models are not precise. While this sounds strange it is absolutely necessary, as it is this imprecision that enables the higher performance of analysis and defers detailed decisions until the point where they must be made. A specification should contain no implied decisions. As we go through the design process, architectural decisions are made that refine the specification toward an implementation. Each time a choice is made, one or more possible implementations is eliminated. These choices exist at every level of the design process. In the early stages they will be large architectural decisions, such as how many processors, which processors, or how many and what kind of interconnect buses or other forms of interconnect. As we move through the design process, decisions such as which processor certain pieces of software will run on are made, or how a particular function is to be implemented, such as the communications of data between two processing elements. Will it use shared memory, a FIFO, or other dedicated hardware structure, or will there be a direct flow of information across wires? At the lower levels of design, decisions are still being made such as what type of adder to use, or how a state machine is going to be coded.

Some people distinguish between two main categories of models, namely computational and execution. The way they separate this is roughly related to whether they contain implementation information or not. We prefer to see this as a continuum rather than as a definite split. For example, is an architectural model a computational model, or is it an execution model? This separation also stops us from considering the possibility that a computational model is annotated or, in the case of an aspect-oriented language, extended to become an execution model. This is all part of the same modeling process and to split them in this way causes confusion.

However, it is difficult to create a pure computational model, as most modeling languages indirectly imply a form of execution. For example, the C language is designed to run on a sequential processor, and thus all software accepts this form of computation. Thus when the model is executed, it already knows the general semantics of an implementation. It is thus not possible to define a pure specification in C as it implies aspects of a solution. Thus a general characteristic that is looked for in a specification language is the ability to be written without implying an implementation. Most of the cases where this is possible forces us to look at theoretical models, such as Petri Nets.

Other theoretical models in reasonably wide use are actor-oriented models such as the various forms of dataflow. These have been used extensively for describing algorithms used in telecommunications, wireless and mobile communications, image and multimedia processing, and related industries. The solutions for these generally result in the creation of software for digital signal processors (DSPs), as well as dedicated hardware accelerators.

1.6 Models and Languages

A modeling language is any artificial language that can be used to express a set of concepts in a consistent way such that independent tools can be built upon that language and can in most cases lead to exactly the same interpretation of the model. This is not always the case and many badly constructed languages fail in this respect. Perhaps the most famous for this industry is the Verilog language that fails to define the semantics associated with zero time delays. In fact, for many years the definition of correct behavior in Verilog was to first look in the Language Reference Manual (LRM), and if not found there or ambiguous, to look at what Cadence's Verilog-XL simulator actually produced, since Gateway Design Automation, purchased by Cadence, defined Verilog and created the first simulator for it.

Languages are basically built from three sets of concepts, namely semantics, abstract syntax, and a concrete syntax. We have already discussed semantics, especially in the context of the models of computation. The abstract syntax is used to describe in a language neutral way how components are connected structurally, but says nothing about the semantics of those connections. And finally the language syntax itself defines the keywords, how are they formed, the detailed rules for expressions, etc.

In the next few sections some of the major categories of languages will be discussed. This is meant to be a brief overview only and many of these languages will be discussed in much greater detail in later chapters of the book.

1.6.1 Imperative Languages

Most programming languages in use today are imperative languages. This is a class of languages in which statements are executed in order and modify the state of the system. The dominance of imperative languages should come as no surprise because computers themselves are imperative machines. The processor at the heart of the computer takes instructions from memory and executes them in sequence until such time as it receives an instruction that tells it to start execution from a different place. The execution of those instructions changes the state of the processor or other locations in memory that hold the state of the program. Until recently the processor was only capable of executing one instruction at a time, and communications was so expensive, in a comparative sense, that it was avoided until absolutely necessary.

All of the early programming languages, such as Fortran, BASIC, COBOL, and most importantly C are all examples of general purpose languages that are imperative. The RTL languages that we use today, namely Verilog and VHDL, were inspired by the C and ADA languages, so it is not surprising that they too are based on the same programming principle.

But there are important differences between the HDLs and general purpose programming languages, the most important one being that statements are not executed in order, but all statements that are ready to be executed are executed as if they all happened at the same time. We all know that this is not physically possible on a processor that can only execute one instruction at a time, and thus the simulation kernel in effect modifies the underlying semantics of the execution environment.

By changing the semantics of the simulation kernel, we can completely modify the operation of the program and thus the HDLs are incomplete languages without being associated with a simulator. This was a problem for one of the common HDLs, namely Verilog as mentioned above, because the execution semantics were not properly defined. In a similar sense, the use of C-based languages for synthesis relies on it being able to detect any dependencies that exist in the code that requires that certain instructions be executed in a particular order. High-level synthesis thus attempts to modify the underlying execution semantics while maintaining the original functionality as defined by the semantics of a general purpose CPU.

1.6.2 Declarative Languages

While imperative languages describe a set of ordered statements that are to be executed, declarative languages state what needs to be accomplished, but not the way in which it will get done. Another name for these languages is functional languages. Few successful languages have ever been created in this space, but there are some notable examples, such as Haskell and Prolog.

One would think that these would be excellent languages for specifying a system, especially at the ES level where it is difficult to avoid embedding too many architectural choices in the description. We will see examples throughout this book where this unintentional architectural choice limits the solutions that can be explored.

One partially declarative HDL is seeing some usage. It is called Bluespec [9] and while not fully conforming to the notions of declarative languages, it does define when states can be entered but not how they will be entered. These are called guarded atomic actions. A system is defined by a set of concurrent rules that can transform the state of the system. Some examples of the types of rules that can be created are

- Pipeline stages:
  Rule: when (input FIFO has item) pop item, process and push on output FIFO
- Cache:
  Rule: when (LOAD request) initiate load-request-state-machine
- Image processing:
  Rule: when (pixel neighbors available) update pixel with f (neighborhood)

- DMA:
  Rule: when (bus available) burst data out to bus.

Each rule can fire when it's guard statement becomes true. The order in which statements are executed is left to the compiler to figure out. In fact Bluespec attempts to combine the attributes of both types of language with the basic functionality being defined imperatively, but the architecture with which it is to be implemented, defined declaratively.

There is another emerging use for declarative languages and this comes from the formal side of the verification process. Properties are the main input for formal verification tools. A property defines something that should or should not happen in a system and a tool will attempt to find a way to prove the property false. If it can, it either indicates a problem with the way in which the property was described or more importantly that the tool has managed to find a way to force the system into an unwanted state or behavior. This is indicative of a design error.

Recently, properties have entered into the more general verification space and are now a common feature of many simulation systems. Accellera has standardized two such languages, namely property specification language (PSL) [10] and part of the new SystemVerilog language called SystemVerilog assertions (SVA) [11]. We can thus expect the industry to gain more understanding of these types of languages over the next few years and perhaps continued development of languages that leverage the power that they have.

1.6.3 Functional

Functional languages, not to be confused with the functional languages defined in the previous section on declarative languages, are those languages that we use to define the function of a system. These are thus the most commonly used languages of our industry and the entirety of Chapter 3 will discuss them in detail.

Both imperative and declarative languages can be used to define functionality. However, most of the time imperative languages are used, or mixed graphical and textual notations (such as the Mathworks Simulink) that specify communication flows of data and control information using graphical notations, and specify computational transformations using textual imperative languages such as C, C++, or MATLAB m-code.

Although over the years there have been hundreds or thousands of functional languages proposed and used, there are only a few that have really seen extensive adoption in ESL design. These can be divided into dynamic languages and models (imperative), and formal languages and models (declarative). Among the formal languages, we can talk about the property languages PSL and SVA (SystemVerilog assertions), already discussed above.

The dynamic languages and models can be further subdivided into algorithmic approaches and architectural modeling approaches. The most prominent and widely used architectural modeling language is SystemC based on a set of C++

class libraries defining language constructs which extend those in basic C++ and a simulator kernel that provides execution semantics for ESL models. It is also possible to use SystemC subsets as inputs into other use models such as ESL synthesis, and because the SystemC source code is available under a community source model, it is possible to add extensions to SystemC, such as those proposed for Analog/Mixed-Signal (AMS) modeling, that may become a more formal part of the language itself. This is the case with the classes, methods, and approaches known as Transaction-Level Modeling (TLM), which has been adopted by OSCI, the SystemC organization, and will eventually be standardized by the IEEE.

Algorithmic modeling languages can be subdivided into mathematical modeling textual languages, the most famous being MATLAB from The Mathworks; and dataflow modeling languages, which are usually a graphics plus textual combination. The latter are also known as "actor-oriented" modeling approaches and include The Mathworks Simulink, CoWare's Signal Processing Designer; Synopsys's System Studio, and research efforts such as Ptolemy and Ptolemy II from UC Berkeley.

These languages and modeling styles are further discussed in Chapter 3.

1.6.4 Non-functional

Non-functional languages and models are intended to capture characteristics of a system such as constraints (e.g., on energy consumption, area, cost) that define boundaries on the allowable or feasible implementations of those systems. They may also define structural characteristics of such systems, such as relationships between conceptual subcomponents and their interactions. At times the boundary between non-functional and functional languages and models may be unclear.

For example, the UML (Unified Modeling Language) contains many different notations – some functional (such as activity diagrams and state diagrams); some non-functional (such as class diagrams, use-case diagrams, collaboration diagrams); and some lying on the boundary in between (such as sequence diagrams, themselves an evolution of SDL's message sequence charts). Sequence diagrams lie in-between because they carry enough semantics to be useful to generate test bench models for use in verification and can be considered to be an "outsider's view" of the functionality of a block – as witnessed by the messages it creates and consumes under various scenarios.

The UML [12–14] is an interesting case in point. It has long been used as a documentation language or set of notations for describing aspects of complex software systems. It has many tools available for it from vendors such as IBM (which ended up consolidating a whole host of independent tool vendors such as Rational and Telelogic). It has enough semantics in its various notations to allow simulation of functional aspects of systems to be carried out by those tools using the functional notations of the UML. It has spawned variants or profiles – intended to apply the UML in more specific contexts – such as SysML [15] and MARTE [16], UML profile for Modeling and Analysis of Real-Time and Embedded systems. It has interested a community for several years in UML for SoC design [17] including generating SystemC models from particular UML profiles.

However, actual uses of UML for ESL design of complex SoCs remain few and far-between, despite experiments by companies such as STMicroelectronics. Many of the applications of SysML, MARTE, and other profiles will be done in the mil/aerospace sector, or for large software systems, and will not become mainstream in ESL for some years to come – if it ever takes off. Thus this book will not try to cover a practical demonstration of UML in its later chapters.

Another non-functional language – and one of long extent – dates back over a decade to work sponsored by the EDA industry council, then Accellera, and finally IEEE DASC (Design Automation Standards Committee). This is the Rosetta language [18, 19]. The Rosetta web site contains a succinct statement of what the language is for:

> The goal of Rosetta project is to provide language support for true system-level design. The Rosetta view of system level design centers on the need to bring together information from multiple domains to predict the system-level impacts of local design decisions. In support of this, Rosetta provides mechanisms for defining system-level functional requirements and constraints, defining heterogeneous specifications and specification domains, and composing specifications to define complete systems.

This is truly a non-functional language. To "execute" a specification in Rosetta is to solve the constraints placed on the system and its components to derive predictions about the system in a variety of non-functional domains, such as power or cost. There has been substantial work on Rosetta over more than the last decade. It may well be standardized eventually by the IEEE. However, its usage has been extremely limited – mainly to academic research and consulting projects in the systems design area. There are a couple of companies with tools (Edaptive Computing and Cadstone) – but it is too early to say whether Rosetta will ever become a viable language for significant ESL specification and modeling, with a large community of users.

1.6.5 Meta

Meta-models are usually pragmatically defined aggregates of information about a system or its components that are "left out" of all the other models in use. For example, characteristics of IP blocks such as energy consumption, area, and speed could be captured in meta-models. In fact, most of the information in a data sheet could be formalized for capture in a meta-model – and much of it has been.

Meta-modeling information has been in particular identified as an adjunct to IP classification, selection, comparison, and integration into larger systems. The beginnings of the IP industry in the mid to late 1990s saw the rise of IP databases such as the one offered by Design and Reuse to aid in IP selection. Since the early 2000s, we have seen XML-based meta-model formats such as SPIRIT's IP-XACT used to support IP integration tools. More recent interest in register definition formats and languages is sparking a new round of tool interest. These will be discussed in more detail in Chapter 2 on meta-models.

Calling meta-modeling formats "languages" is a bit of a stretch, in that they are often semantically weak, lack abstract syntax, and often consist entirely of concrete syntax with at times ambiguous semantics. However, over time, the use of these formats and notations has led to some introduction of greater semantic rigor in their definition.

1.6.6 Testbench

Testbenches are not really a language but a set of models, potentially in several languages, that work together to accomplish the goal of verification that ensures that an implementation matches a specification. In some cases, the same language may be used for part of the testbench as was used in the design, but a number of specialized languages also exist that can be used to enhance those capabilities. The choice of languages will also be influenced by the methodologies that are to be adopted for verification. For example, if formal verification technologies are to be used, this will lead to very different choices than if simulation is to be the sole means of verification.

One of the testbench models that is still evolving is the verification plan. For most companies today it is still a document written in a natural language, but it is beginning to be formalized and used to collect results from the various verification activities and in the future will direct those activities as well.

For dynamic simulation a number of models are created, including a predictor, a set of checkers that may include assertions, stimulus constraints, and a coverage model. The major component that needs to be defined, independent of the verification methodology to be used, is a predictor model that can be used to check the results produced by the design. This is likely to be written in one of the functional languages already described. A set of stimulus is then required that can excite the design. This is normally accomplished by a set of constraints and scenarios that when fed into a constrained random test pattern generator, will produce a stream of random data that fits in with the notion of usefulness as defined by the constraints. Alternatively, stimulus sets can be created manually or by other specialized tools, which may require other types of models. A coverage model needs to be defined that will measure progress through the verification task. This model is often closely tied to the verification plan.

When formal verification is to be used, the predictor model is normally replaced by a set of declarative properties. These define attributes of the design that are to be tested for. Along with the properties, it is usually necessary to have a set of constraints. These limit the input sequences that the tool will use to try and find a way to make the design fail. Thus it can eliminate unrealistic input streams from consideration. While coverage models have not yet been devised for formal verification the results do tie back into the verification plan to measure progress against the specified goals.

1.7 The Desire for a New Language

The number of existing modeling languages is huge and there seemed a time when every academic project would start with the definition of a new one. Tools, such as lex and yacc, are distributed with the Unix operating system and these made it fairly quick and easy to put a language together. Yacc, an acronym for "Yet Another Compiler Compiler," generates a parser for a language description written in a language similar to BNF. Yacc needs to be fed a stream of tokens, and these are generated by lex, which performs a regular expression match on the language input stream. The result is a series of program fragments that are understandable by the parser. Since their introduction, many improvements have been made in them and today they are included in many operating systems, such as OpenSolaris.

Just because it is easy to create a language does not imply that it is easy to create a good language. Most languages are constructed on top of a model of computation (MoC). This defines the entities, relationships, and events available in the language and the relative cost of those operations. Thus it is possible to define a language in which certain kinds of tasks would be performed more efficiently than would be the case if a different MoC were used. In many cases, the costs are associated with the underlying architecture of the hardware system on which they will execute.

In Section 1.6 we discussed the difference between imperative and declarative languages. The imperative model mimics the operation of the computer on which it is executing. Instructions are executed in order by a processing unit which manipulates variables held in a randomly accessible memory. This, however, does not mimic the operation of the hardware components from which the computer is constructed.

For the gate and RT levels of abstraction, the predominant MoC on which all languages have been based is the discrete event model. The fundamental aspect of this MoC is that events communicate the value of a variable with a timestamp. All processing elements can potentially execute in parallel and there is a consistent notion of time among the elements, which can be considered to be the global time. This is the fundamental operation of an event-driven simulator, on top of which languages such as VHDL and Verilog can be executed. Again, the language mimics the underlying execution semantics of the "computer" it is executing on. In this case the computer is the simulator, and its job is to convert those semantics into the execution semantics of the computer that it is executing on.

With the migration to ESL, the discrete event semantic is too limiting. We require a language that is capable of handling untimed or partially ordered models as well as being able to handle data in more abstract forms. The complete definition for these can be found in the taxonomy definition in Section 1.9.

If we look at systems that were the precursors to today's ESL, we quickly identify the digital signal processor (DSP) automation tools that have existed since the 1980s. Examples include SPW and Cossap. These utilize dataflow process networks and in most cases a subset of that called synchronous dataflow (SDF). A dataflow program specifies a directed graph where the nodes, which are generally called actors, represent computation and the arcs represent FIFO channels. These communication queues contain objects called tokens. The computation is meant to be purely functional in that the output value is determined solely by its input values.

Each computation fires when it receives the right number of tokens on each of its inputs and produces a token on its outputs. A goal of many tools that implement these languages is to find a consistent ordering of the firing so that they can be statically scheduled at compile time as this leads to an efficient execution of the model. The SDF variant requires processes to consume and produce a fixed number of tokens for each firing and a static schedule can always be found that will return the system to a state identical to its original state, meaning that the same number of tokens exists on each of the arcs. These are good for describing data intensive systems, but lack the ability to model control.

Another set of precursors include higher level software tools and graphical functional tools, such as SDL and StateCharts, which are all variants of finite state machines (FSM) with enhanced communications capabilities. While these are similar to the state machines that we build in hardware they do not specify time as an attribute associated with state changes. In other words state changes happen when the inputs direct that it should happen. The various systems contain extension to handle things such as hierarchy and concurrency. These languages are good for describing control intensive systems, but have limited capabilities for modeling dataflow.

The limitation with most of these earlier systems was that they could either handle data intensive operations well, or control systems well, but not both. For ESL to become real and effective, an environment was needed that could handle both equally well, and in addition would be capable of co-existing with the existing MoC, namely the discrete event models.

ESL also spans two domains, namely the hardware and software domains. We have just discussed the MoCs for the hardware domain, but the software domain is heavily entrenched in the imperative execution model. While there is speculation that the migration to multiprocessing will cause this to change over time (others think that this will never happen), it is necessary to support the huge software legacy that has been built up in the C language, or it's more recent evolution, C++.

Is it possible to have a single language that can do all of this?

1.8 Big Shoes to Fill

It is probably not possible for a single language to perform all of these requirements in an efficient manner. Attempts have been made to extend languages by adding other features and capabilities, but at the end of the day they are really multiple languages that borrow some syntactic elements from each other. For example, the Verilog language was extended to include capabilities for performing sophisticated verification activities. This required the integration of a constraint language, a coverage language, a declarative property language as well as extensions to the underlying hardware description language. Is this really one language? Not really.

Another example of this is the Unified Modeling Language (UML) that is an amalgam of many existing languages. These models are even more loosely tied together than SystemVerilog in some ways, but they do have semantic interactions between them whereas in SystemVerilog there is none.

1.8.1 Ptolemy Simulator

So, a single language is probably not the right direction to go in. What are the other choices? The University of California at Berkeley has had a long history of working on the integration of multiple MoCs. Under the guidance of Edward Lee, they have assumed that the model of computation is part of the framework within which components are embedded rather than part of the components themselves as shown in Figure 1.5.

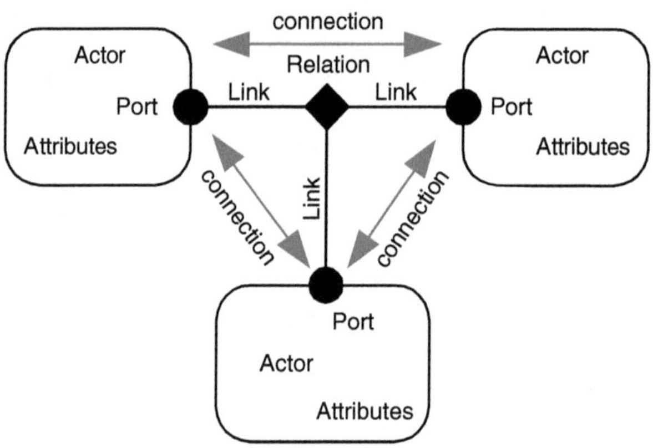

Fig. 1.5 MoC framework

With this kind of framework the components can be any MoC so long as they can be made to look like actors with ports. The interaction between them is governed by the framework MoC which handles the flow of control, the communications protocols, and the management of non-functional properties such as timing and resource management.

The Ptolemy simulator [20] is one manifestation of that work and was developed to simulate complex systems using multiple heterogeneous models of computation in a hierarchical design environment operating at multiple levels of abstraction. That is, the simulator enables the description, integration, and execution of diverse functionality – such as signal processing, feedback control, and sequential decision making, both digital and analog – with different, complex timing constraints and communication requirements. Consequently, the simulator supports multiple computation models, including continuous time, discrete event, FSM, synchronous dataflow, and synchronous/reactive and time-triggered synchronous execution [21].

The original Ptolemy simulator – later renamed Ptolemy Classic – was introduced in 1993, with the "stable production" version released in 1998. Ptolemy Classic was succeeded by Ptolemy II, some facets of which were demonstrated in 1999. The latest version of Ptolemy II was released in 2005. It has also been commercialized by Mirabilis.

1.8.2 SystemC

At first it may seem strange to bring up SystemC as a solution to this problem. Indeed SystemC is based on a single fixed model of computation, the same as the existing HDLs: namely, it follows the discrete event semantics. But because SystemC is built on an extensible framework, namely C++ it is much more flexible. In addition, given that it is based on C++, which itself is an extension of C, it fully supports the legacy software that exists.

The flexibility in SystemC comes from two mechanisms. The first is the use of the notify and wait primitives and the second is the sc_interface, sc_channel, and sc_prim_channel classes. Using these primitives, extension to the language can be built as libraries that support additional models of computation over the base that is shipped in the kernel. With these features, SystemC has been shown to support static and dynamic multi-rate dataflow, Kahn process networks as well as the discrete event model of computation [22], although others claim that it can only do this efficiently by the incorporation of kernel extensions [23].

Another capability in its favor is that it has been constructed to support multiple levels of abstraction within the discrete event domain ranging from RTL constructs up to transaction modeling.

An example of an extension is the TLM 2.0 library which creates additional communications capabilities that mimic bus-based semantics. While this still remains within the discrete event MoC, it illustrates how additional semantics can be built upon the base. To construct most of the untimed MoCs such as Kahn process networks, the computation would be mapped into SC_THREADs and while this MoC could use the sc_fifo channel, there are problems with the way in which tokens are placed and removed. Thus new FIFOs could be constructed which provide the correct token control [24].

For a more complete description of the SystemC language and the TLM 2.0 communications library you should refer to Chapter 3.

Thus while SystemC may not be the universal language, it is a language that directly supports the discrete event model of computation, supports imperative descriptions directly, and is capable of being a framework for several other models of computation. It is this adaptability that has made it the front-runner in becoming the first true ESL language.

Another requirement for a successful language is the rate at which it is adopted by the industry. A language may be perfect, but unless it is used by enough people it does not mean it was successful. SystemC is far from perfect and initially only received a tepid reception in the industry, although academia took to it more strongly.

With the introduction of TLM 2.0, another huge barrier was removed, which was model interoperability. SystemC does not define the semantics of communications between models as it only provides the essential primitives necessary for communications. Thus there was no agreement in the industry about how these interfaces should be constructed. Several EDA vendors, such as CoWare, attempted to create and proliferate communications libraries, but these saw no uptake because of the proprietary nature of them. Today we are seeing rapid adoption of TLM 2.0 by the industry with significant support coming from all of the major EDA players.

1.8.3 Function and Interface

The languages that were used to model at the RT level are somewhat homogenous in that they do not make any distinction between the internal functional logic of the model and any logic that is associated with communications between models. This is because the communications model for RTL is very simple. The interfaces are wires or groups of wires that usually have either two or four possible values each, and can either be sampled, or have events associated with them that correspond to transitions in values on those wires.

With the migration to higher levels of abstraction, communications become a lot more important, and as we shall see in several of the chapters in this book, there is a big advantage in being able to separate the notions of function and interface. This allows the two aspects of design to be separated and refined independently. It also allows a mix and match of model abstractions to be used such that an optimal set of performance and accuracy can be established for each component in a design during simulation.

1.9 Taxonomy

In the book *ESL Design and Verification* [1], the need for a model taxonomy for ESL was explained along with a description of earlier model taxonomies such as those developed by Rapid Prototyping of Application Specific Signal Processors (RASSP) and Virtual Socket Interface alliance (VSIA). From these earlier works, a new taxonomy was developed that extended upon these concepts, retaining two of the original axes, namely the temporal and data axes. While some of the labels on these axes are now showing their age, they still convey the right information and thus they have not been updated. As an example, the labels talk about tokens, whereas today we would probably substitute the term transaction. Together these two axes represent the abstraction of the model. Other axes were dropped as it was thought that the concepts that they attempted to convey were too vague and did not really add to the overall value of the classification. Some aspects of these axes were subsumed by the newly developed axes.

In the previous section, we discussed the desire to separate function and interface. That means that a model may contain different abstractions in each of those areas. For example, it is quite possible to have a very abstract computation model, with a more detailed model for its interface. This is often how processor models are put together. Similarly, we may have a detailed computation model communicating through an abstract interface. This is likely to happen when an implementation model is verified within the context of a system model.

1.9.1 Three New Axes

Three new axes were introduced that when added to the two that were retained from the previous taxonomy provide a complete representation of model functionality across all aspects of the ESL space. The three new axes were concurrency, communications, and configurability. While a full description of these axes, along with their justification can be found in that book, a summary is provided here.

1.9.1.1 Concurrency

Concurrency defines the amount of processing or execution of an application that can be performed simultaneously. An implementation does not have to utilize all of the available concurrency because this may produce a solution that does not fit the non-functional requirements of the design such as cost, size, or power consumption constraints. In addition, more concurrency increases the difficulty of verifying an implementation.

Along that axis some discrete points were defined. It should be noted that these points are representational only and do not cover every point that could exist. The defined points were

1. Sequential – No explicit concurrency.
2. Multi-application – Highly coarse grained such as would exist with two independent applications that share data.
3. Parallel – Functions that operate independently but occasionally need to cooperate for certain activities. An example would be a transaction-processing system.
4. Multi-thread – A single function that may have been defined with multiple threads of execution. The concurrency is explicitly built into the operation of the function and may have explicit points of synchronization.
5. Pipeline – The streaming of data from one operation to another in a controlled manner. Pipelining can exist at many different levels of a system. This is typical of an RTL model, where the combinatorial actions to be performed between registers are defined.
6. Signal level – Very fine level of concurrency that is typical in a gate level or asynchronous description.

1.9.1.2 Communications

When more than one processing element exists, there must be communications between them. That communication can take many forms and is highly dependent on the architecture of the final solution. Very fine-grained parallelism, such as operations within an instruction executed on a processor, or dedicated pieces of hardware are naturally handled by point-to-point communication or pipelines. At the other end of the spectrum, two software threads that need to communicate are likely to use some form of shared memory to communicate with one another, although they may also use point to point or buffered mechanisms such as hardware FIFOs connecting processors. In extreme cases, two software processes may communicate over the Internet, sharing data by transactions or as data within a file that is copied from one machine to another.

The discrete points that have been defined along the axis are

1. Point-to-point – This allows two concurrent functions to communicate with each other directly without any additional form of control between them.
2. Buffered – This is the most elemental form of communications that can be made between two concurrent tasks. If the buffer depth is greater than one, as is the case of a FIFO, then rate independence can also be achieved.
3. Coprocessor – Data is owned by one process but shared with another in order to perform a function. The coupling is usually very tight between them.
4. Memory – Memory is a very common way to transfer information between two functions. At this level, it is assumed to be multi-port such that each function can read or write to memory independently.
5. Bus based (high speed) – In software systems where the function is run on a processor, almost all communications are made though one or more buses, each of which may have different transfer rates or latencies. This category does not make a distinction about the performance of a bus.
6. Bus based (low speed) – In this context, a low-speed bus means that there is a more extensive protocol being used on the bus to ensure transport integrity, or that is capable of allowing things to be connected at much larger distances.
7. None – If the functions do not need to communicate, then no channel is necessary between them. It also implies that there is no concurrency within the system that has to be handled.

1.9.1.3 Configurability

Most people would consider software to be configurable, but the configurability of the platform on which the software runs, or in which it is embedded is just as important. The only difference between hardware and software in this regard is the delivery mechanism. Software, written in such languages as C or C++, is modified at design time and compiled to work on existing hardware solutions, but once it is shipped it cannot be trivially modified in the field, although mechanisms increasingly exist on many devices for downloading new or modified compiled executables from a network. Java code, on the other hand, can be modified or retargeted for different hardware because the compilation step is deferred until close to execution time and thus has more configurability.

Dedicated fixed-function hardware solutions are often very rigid and can only perform a single task. A processor with a fixed instruction set is also considered a fixed solution. However, a processor with an extensible instruction set has some degree of configurability. This configurability is handled by the tools in the design path. Communications processors can support many different protocols. A reprogrammable fabric, such as an field programmable gate array (FPGA) or programmable logic device (PLD), can be reprogrammed at start-up, or even possibly dynamically reprogrammed during operation of the system. Although run-time reconfiguration is not in common use today, there is no reason to think fully dynamically reprogrammable systems will not exist in the future that can change personalities to configure themselves for the task or computation loads that they face at any given time.

The discrete points that have been defined along the axis are

1. Fixed – The user of a fixed block or device has no ability to make any kind of changes to it. It is considered a black box.
2. Design – Configurability at the design level is possible for any design where the original source code is available.
3. Configurable – Blocks that can be personalized to take on many different roles. Once the options are set, they cannot be modified during operation.
4. Programmable – Programmable devices such as FPGAs and PLDs can take on any number of personalities by loading configuration information at start-up.
5. Dynamically reprogrammable – These devices may be reconfigured with new personalities by loading configuration data dynamically during operation.

1.9.2 Application to Models and Languages

When applying the taxonomy, it can either be applied to languages or to models. When applied to languages, it describes the range of possibilities that can be expressed in that language. So, for example, the C language can be used to describe any sequential arithmetic computation and when coupled to the execution semantics of a typical computer, can solve any Turing complete problem. However, a hardware component written in the C language adopts a different set of execution semantics (including limitations and assumptions) and even these change depending if we are considering simulation or synthesis.

We can map the capabilities of the C language as shown in Figure 1.6. Looking at the temporal axis, the C language is capable of defining order but not timing. We have shown it extending from partial order down to instruction cycle. This perhaps

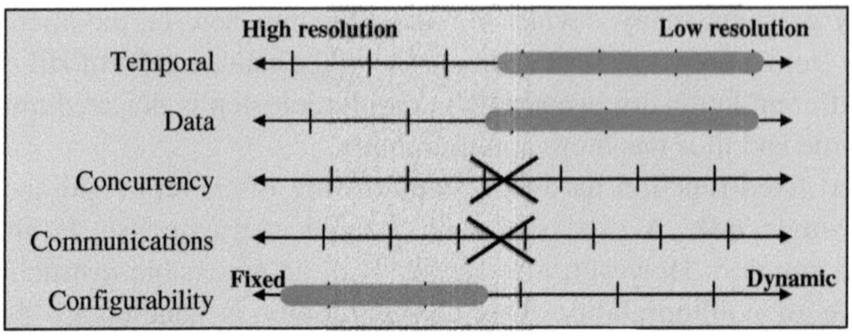

Fig. 1.6 Taxonomy diagram for "C"

slightly generous in defining its capabilities and does illustrate some of the problems associated with defining named points on a scale. For the data axis, the scale is more definite and C is capable of defining data from tokens down to values. It cannot define the way in which those values would be implemented in hardware, nor can it define things such as fixed point numbers without the use of additional libraries.

C has no capabilities for expressing concurrency or communications. While the language can be used in both of these environments, it relies on the existence of an operating system, libraries, and artifacts of the hardware that it is running on to provide these capabilities. They are fundamentally not present in the language itself. In the case of concurrency, this could have been represented as a small circle over the sequential marker, but the nomenclature of using the X across the axis was adopted to indicate that this is not a choice in the language, but a determination of the semantics of the language as implied by the hardware platform it was designed to execute on. It is possible, as many high-level synthesis tools do, to extract concurrency from the sequential language. The language does not say that the instructions must be sequential.

Finally for configurability, while it is possible to define almost anything in the C language, the language itself has no support or even concept of configurability of the hardware platform on which it is meant to operate. It is possible that a tool could infer these from a model written in C, but just like the previous two axes, the language itself does nothing to support it. It thus ranges from being configurable at design time, or being fixed if it is delivered in a compiled form.

When a model is written using C, it may well have a much more constrained representation in the taxonomy. This is because it is now being used to define the actual abstraction and capability as supplied by that model. A model cannot add capabilities that were not present in the original language unless it can assume the existence of certain capabilities coming from a library or other mechanism that becomes an integral part of the environment in which it is intended to be used. As an example, the C++ language does not specify arbitrary precision of variables. It is constrained to the built in types such as short or integer. However, as we will see in Chapter 9 Mentor provides a library containing a set of arbitrary precision datatypes that can be utilized to extend the capabilities of the language.

In fact all of the additional capabilities of SystemC come from the extension of the C++ language by adding certain class libraries.

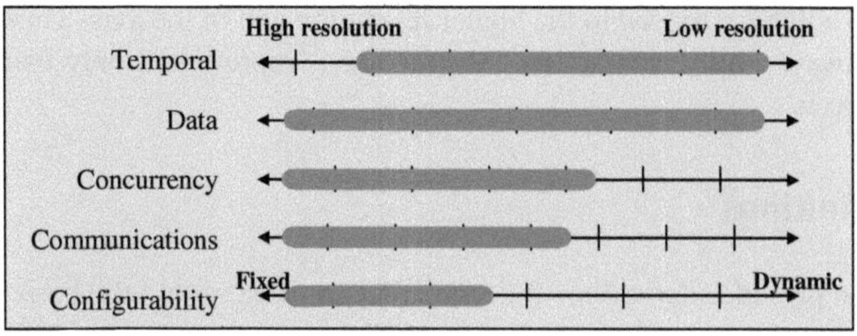

Fig. 1.7 Taxonomy diagram for SystemC

As can be seen from the taxonomy diagram for SystemC in Figure 1.7, the abstractions that can be represented by it are a pure superset of those represented by the C language. In addition, direct support for communications and concurrency is provided. All useful concurrency models for the support of hardware and most software are supported, but the communications is still deficient in that it cannot model some of the bus-based communications properly. A large part of this deficiency was overcome by the release of TLM 2.0 which specifically adds libraries of additional functionality to target this problem.

It is likely that a specific model instance will utilize a small range of attributes on each of the taxonomy axes. In addition, there are some models whose job is to convert between one level of abstraction and another. These converters, often called transactors, will represent two or more points or ranges for each model. This could either be diagrammed as several discrete points, or using separate diagrams for each of the interfaces that it exposes.

1.9.3 Transformation of Models

When tools operate on models, they can make transformations that impart additional knowledge or design decisions into the models. In most cases this will result in the resolution of the model increasing. For example, during a high-level synthesis process, variables that were expressed as integers will become registers or locations in memory. In this step, the format of the data storage is defined. This is the RT level of abstraction. When it is in turn synthesized, those registers or memories get transformed into flip-flops or other suitable structures at the bit level.

However, there are cases when a transformation can appear to move in the reverse direction. This is particularly true when considering communications. A high-level model may define point to point or buffered communications between two computation components. As the system is refined, it may be decided that many of these inter-block communications will share a common bus, and thus the resolution of the model may appear to have gone in the wrong direction. This is not the case because while we may have moved to the right on this axis, other markers that represent abstraction will have moved to the higher resolution end of the axis. Thus resolution does not always imply abstraction. Abstraction is represented only using the first two of the axes.

1.10 Definitions

This section provides definitions for terms that are used within the book and points out where there is industry divergence in terms.

Abstraction – Describing an object using a model where some of the low-level details are ignored.

Accuracy – The degree of confidence that an engineer can have in the results obtained from a model.

Architecture – The way in which aspects of a design relate to each other either structurally or functionally.

Assertion – The implementation of a property, evaluated or executed by a tool. See also Property.

Assertion coverage – Observing an assertion evaluated or executed, passing or failing, and recording the possible paths of evaluation through the assertion.

Behavioral – A model of a system at any level of abstraction that includes timing information.

Clear box (white box) – A term used to define the amount of visibility and/or control an engineer has into a block for the purposes of verification. In this case all internals are visible. The opposite of this is opaque or black box.

Compiled code model – The transformation of code, either hardware or software that was intended to run in a simulator, into code that will run directly on the host machine without an interpreter.

Corner case – A corner case is one or more data values or sequential events that, in combination, lead to a substantial change in design behavior. It is often an exceptional condition – hence rare – that is difficult to predict.

Coverage – A measure of how thoroughly a design has been exercised during verification to minimize the probability that latent errors remain in the design.

Coverage model – An abstract representation of device behavior composed of attributes and their relationships. The relationships may be either data or temporal in nature.

Declarative languages – A programming paradigm that describes what the program should accomplish, rather than describing how to go about accomplishing it.

Design space exploration – The exploration of multiple possible solutions while attempting to find the one that best fits the requirements.

Dynamic verification – Demonstrating that a design conforms to its functional specification through execution. This form of verification relies on the progression of time to allow the design's internal functionality to propagate through the design-specific values placed on the design's inputs at a given time. This algorithm requires some specific mechanism (e.g., simulation, emulation, or prototype) to run the design and a specific methodology to apply stimulus and check the output of the design (verification environment) to verify the correctness of the design.

Electronic system level – The use of appropriate abstractions to increase comprehension of a system and enhance the probability of successfully implementing its functionality in a cost-effective manner, while meeting necessary constraints.

ESL synthesis – A process that takes as input an executable software representation of a function or system in the form of SystemC, C, C++, or some other higher level language, and produces an implemention of the function or system. See also high-level synthesis.

Executable design – An executable version of a design that includes the required behavior that characterizes the targeted digital subsystem or component. It may imply an implementation architecture, but does not define it. Neither does it define hardware–software partitioning. The executable design provides a simulation model, whose defined external interfaces exhibit the required I/O behavior with selected accuracy.

Functional – A model of a system at any level of abstraction that does not include timing information.

Hard real-time constraint – A system has events that have a valid lifetime before their effects become invalid. A hard real-time constraint is defined to be one that if not processed before a certain deadline then their effect or result becomes dangerous for the system.

High-level synthesis – A form of ESL synthesis where the input description only describes the hardware aspects of a system or component.

Imperative languages – A programming paradigm that describes computation in terms of statements that change a program state. Each statement is executed in sequential order.

Intellectual property (IP) – A block of code that describes any aspect of a system, including its hardware, software, or the verification environment, which is reused between multiple designs or parts of a design.

Interpreted model – A model that uses the services of a program, such as a simulator, to execute the necessary behaviors.

Mapping – The relationship between an application and elements of a platform.

Meta-model – A pragmatically defined aggregates of information about a system or its components that are "left out" of all the other models in use.

Model – A way of capturing certain functional, behavioral, and/or other attribute of a system. It normally involves the application of some amount of abstraction such that adequate performance can be obtained at a desired level of accuracy.

Model of computation – A model of computation is an attempt to distil the fundamental aspects of a system's ability to solve problems.

Modeling language – Any artificial language that can be used to express a set of concepts in a consistent way such that independent tools can be built upon that language and can in most cases lead to exactly the same interpretation of the model.

Opaque box (black box) – A term used to define the amount of visibility and/or control an engineer has into a block for the purposes of verification. In this case NO internals are visible. The opposite of this is clear or white box.

Partitioning – Partitioning is the process of subdividing an initial specification, into a set of concurrent cooperating processes, and assigning them to a set of more or less abstract resources, representing processors for software, silicon area, or intellectual property blocks for hardware, communication channels, and storage resources.

Performance – Any quantity that can be defined to characterize a system, either physical (area, power consumption, etc.) or logical (quality of service, fault tolerance, etc.).

Platform-based design –

- A reuse-intensive design style for embedded systems where large portions of the design are based on pre-designed and pre-verified SoC components.
- An integration-oriented design approach emphasizing systematic reuse, for developing complex products based upon platforms and compatible hardware and software virtual components, intended to reduce development risks, costs, and time to market.
- A single abstract model that hides the details of a set of different possible implementations as clusters of lower level components.

Property

- A statement of an expected liveness or safety behavior. A liveness (including fairness) behavior says that something should eventually happen and is often called an eventuality. A safety behavior says that something should never happen and is often called an invariant. As such liveness and safety properties define valid or invalid paths through the state space of a design.
- A collection of logical and temporal relationships between and among subordinate Boolean expressions, sequential expressions, and other properties that in aggregate represent a set of behavior (that is, a path).

Requirement – A requirement is a

- condition or capability needed by a user to solve a problem or achieve an objective;
- condition or capability that must be met or possessed by a system or a system component to satisfy a contract, standard, specification, or other formally imposed document;
- documented representation of a condition or capability as above.

Scheduling – The synchronization and ordering of functional tasks.

Static verification – The process of demonstrating that a design conforms to its functional specification through comparative analysis and proof, as opposed to design execution. Sometimes referred to as formal verification.

Synthesis – The transformation from one model to another where the target model is usually of a lower level of abstraction than the source model.

Taxonomy – A systematic arrangement of objects or concepts showing the relations between them, especially one including a hierarchical arrangement of types in which categories of objects are classified as subtypes of more abstract categories, starting from one or a small number of top categories, and descending to more specific types through an arbitrary number of levels. (Webster Dictionary, http://www.webster-dictionary.net 2004).

Testbench – A set of models that work together to accomplish the goal of verifying that an implementation matches a specification.

Traceability (IEEE) – The degree to which a relationship can be established between two or more products of the development process, especially products having a predecessor–successor or master–subordinate relationship to one another.

Traffic generator – A particular kind of model that is able to generate stimulus sequences that would mimic an actual device, even though the details of that device are not modeled.

Transaction – An abstract model where both data and control are conveyed together.

Validation (IEEE) – Confirmation by examination and provisions of objective evidence that the particular requirements for a specific intended use are fulfilled.

Verification (IEEE) – Confirmation by examination and provisions of objective evidence that specified requirements have been fulfilled.

References

1. Brian Bailey, Grant Martin, Andrew Piziali. ESL Design and Verification: A Prescription for Electronic System Level Methodology. Morgan Kaufmann Elsevier 2007.
2. Bran Selic. “Models, Software Models and UML”, Chapter 1 of UML for Real: Design of Embedded Real-Time Systems, edited by Luciano Lavagno, Grant Martin and Bran Selic, Kluwer (now Springer), 2003.
3. Definition of model, consulted 22 May, 2009: http://en.wikipedia.org/wiki/Model
4. Definition of computer model, consulted 22 May, 2009: http://en.wikipedia.org/wiki/Computer_model
5. http://celebrating200years.noaa.gov/foundations/numerical_wx_pred/S1Chart06.html Consulted October 7th 2008.
6. http://www.sciencedaily.com/videos/2008/0507-perfect_weather_predictions.htm Consulted October 7th 2008
7. Stewart Robinson. Issues in Conceptual Modelling for Simulation: Setting A Research Agenda. Proceedings of the 2006 OR Society Simulation Workshop.
8. Axel Jantsch, Modeling Embedded Systems and SoC’s: Concurrency and Time in Models of Computation, Elsevier Morgan Kaufmann, 2004.

9. Abhinav Agarval, Man Cheuk Ng, Arvind. Comparison of High Level Design Methodologies for Algorithmic IPs: Bluespec and C-based Synthesis, MIT Computer Science and Artificial Intelligence Laboratory (CSAIL)
10. Property Specification Language Reference Manual http://www.eda.org/vfv/docs/PSL-v1.1.pdf
11. Srikanth Vijayaraghavan, Meyyappan Ramanathan. A Practical Guide for SystemVerilog Assertions, Springer 2005.
12. Grady Booch, James Rumbaugh, Ivar Jacobson. The Unified Modeling Language User Guide, 2nd Edition, Addison-Wesley, 2005.
13. James Rumbaugh, Ivar Jacobson, Grady Booch. The Unified Modeling Language Reference Manual, 2nd Edition, Addison-Wesley, 2005.
14. Martin Fowler. UML Distilled: A Brief Guide to the Standard Object Modeling Language, 3rd Edition, Addison-Wesley, 2004.
15. Sanford Friedenthal, Alan Moore, Rick Steiner. A Practical Guide to SysML: The Systems Modeling Language, Elsevier Morgan Kaufmann, 2008.
16. http://www.omgmarte.org/
17. Grant Martin, Wolfgang Müller (Ed.), UML for SOC Design, Springer, 2005.
18. Perry Alexander. System Level Design with Rosetta, Elsevier Morgan Kaufmann, 2006.
19. https://wiki.ittc.ku.edu/rosetta_wiki/index.php/Main_Page/
20. Ptolemy Group, Ptolemy II, Website: http://ptolemy.eecs.berkeley.edu/ptolemyII/
21. Louis Scheffer, Luciano Lavagno, Grant Martin, Electronic Design Automation for Integrated Circuits Handbook, vol. EDA for IC System Design and Testing, Taylor & Francis/CRC Press, Boca Raton, FL, 2006, pp. 3.9–3.10, 9.34–9.36.
22. Thorsten Grötker, Stan Liao, Grant Martin, Stuart Swan. System Design with SystemC. Kluwer 2002.
23. Hiren D. Patel, Sandeep K. Shukla. SystemC kernel extensions for Heterogeneous System Modeling. Kluwer 2004.
24. Fernando Herrera, Eugenio Villar. A Framework for Embedded System Specification under Different Models of Computation in SystemC. Design Automation Conference 2006.

Chapter 2
IP Meta-Models for SoC Assembly and HW/SW Interfaces

A functional model contains a lot of information, but it does not often contain the information about how it was meant to be used, what restrictions are placed on its usage or the way in which it is meant to be connected. This information is considered to be metadata about that block and was often the principal information that would have been found on a specification sheet for a device. Once that information is captured in a formalized manner it can be used by tools to help with things such as system construction, enable system consistency to be analyzed, or reduce the burden of things such as documentation.

2.1 Introduction

Metadata is especially relevant for the level of IP blocks and the associated models – whether hardware, software, or verification IP. As mentioned above, meta-models convert the unstructured data found on component datasheets or specifications into a more structured format that may be amenable to tool-based processing.

2.2 IP Databases

The IP revolution had barely begun in the mid-1990s, when people began looking at the conversion of unstructured datasheet information (or its web equivalents) into more formal databases for IP components or blocks. Indeed, by the year 2000, a collaborative paper in the DATE conference user's forum was able to cite several formats, or projects to create new formats for exchanging, storing, and searching data on IP blocks [1]. Some of these were based on the eXtensible Markup Language (XML) [2]. Participants in the collaboration included several of the more prominent companies or groups looking at IP metadata for selection and interchange

- Design and Reuse [3] – a French company based in Grenoble that emerged in the late 1990s (1997) with an IP database and offers both the IP database and collaboration tools and repositories for IP-based design

B. Bailey, G. Martin, *ESL Models and their Application*, Embedded Systems,
DOI 10.1007/978-1-4419-0965-7_2,

- RAPID – an early attempt at an IP industry trade association from the late 1990s, RAPID attempted to set up a database of IP information. It disappeared early in the 2000s
- Silicon Integration Initiative (Si2) [4] whose ECIX (Electronic Component Information eXchange) [5] project developed an XML-based IP metadata format. This no longer appears to be active although its documents are available for download from Si2
- Synchronicity – a commercial company selling design management tools, which at that time also offered tools for virtual component management. Synchronicity was later purchased by MatrixOne which was itself absorbed by Dassault Systemes
- VCX – the Virtual Component eXchange, set up in Scotland as part of Project Alba to foster commercial infrastructure and a trading floor for IP (Virtual Components). It developed some tools that were later picked up by Beach Solutions, but went defunct after several years
- The Virtual Socket Interface Alliance (VSIA) [6] Virtual Component Transfer Development Working Group (VCT DWG). The VSIA was an early industry-based standards group set up in 1996 to foster the use and interchange of virtual components by defining technical standards. Unusually for a standards group, it actually decided to fold up in 2008, while leaving its legacy documents on a web site so that future users would still have access to them. The Virtual Component Transfer defined a reasonably comprehensive set of metadata for IP transfer but not a specific format.

The collaboration discussed in the VSIA VCT paper used XML as a mechanism for defining IP metadata. XML has attracted significant attention over the years as a flexible way to define structured information exchange or languages in various domains. As one of many mark-up languages, it trades off brevity, readability, and human comprehension for ease of programmatic parsing and use in automated tools and systems. XML is a meta-format from which particular languages or notations can be constructed either using a DTD (Document Type Definition) or XML schema. XML in conjunction with IP metadata turns up several times.

The situation in 2009 with respect to IP metadata and IP databases has narrowed considerably. Right now, there are two commercial IP databases surviving – the Design and Reuse one and one from Chipestimate.com [7] (bought by Cadence in early 2008). Usage of both of them is unclear.

2.3 SPIRIT/IP-XACT

2.3.1 History of SPIRIT

SPIRIT [8] was set up as an industry consortium to standardize IP metadata for design and design tools. It has an interesting history. It was established in

2003 by an initial group of members including ARM, Beach Solutions, Cadence Design Systems, Mentor Graphics, Philips Semiconductor, STMicroelectronics, and Synopsys and announced at the 2003 Design Automation Conference. SPIRIT originally was an acronym, standing for Structure for Packaging, Integrating, and Reusing IP within Tool-flows. In July 2006, SPIRIT reorganized itself as a non-profit California corporation. At that time it dropped the use of the meaning of the SPIRIT acronym and decided to call its IP exchange format henceforth by the name IP-XACT – a name that unfortunately only a lawyer could love.

The initial goal of SPIRIT was to develop and demonstrate an IP metadata standard within a year, donate it to the IEEE for standardization, and then dissolve. However, rather than dissolving, it has continued its efforts in areas such as the recently announced SystemRDL in May 2009. IEEE standardization of the SPIRIT metadata format began in 2006 under the auspices of IEEE DASC (Design Automation Standards Committee) [9], as committee P1685, but appeared relatively moribund as of May 2009 when this was being written [10]. Archived email traffic on its web site seems to have stopped as of October 2007 (although the SPIRIT consortium also has a P1685 web site whose last email traffic appears to be February 2008, and there is a note on the web site saying "this group has been deactivated"). As of June 1, 2009, there was a note from SPIRIT indicating that the group was being reactivated from its hiatus, would restart weekly meetings beginning July 1, 2009, and was intended to "reactivate the group, evaluate any contributions, set up a balloting group, prepare for ballot, view ballot responses, prepare ballot response," which implies a newly awakened interest in the IEEE standardization process. As of June 2009, the SPIRIT consortium announced a merger with Accellera [11] which may accelerate this standardization process.

Another aspect of SPIRIT is the rationale for it coming into existence. Although the first six members included the three major EDA tool vendors, two semiconductor companies in Europe and one IP company, it was really motivated originally by Philips Semiconductor's wish to create a tool for SoC design via IP integration called NXPBuilder and Mentor Graphics' desire to sell their Platform Express technology to Philips as a basis for NXPBuilder. Platform Express was provided for free for individual use and they tried to sell the technology to companies who wanted to build IP integration flows. It relied on an XML format for IP metadata.

Philips Semiconductor did not want to build their IP integration flow on top of a proprietary Mentor Graphics IP metadata format and thus be trapped into single supplier support; it insisted that the Mentor format become the basis for an interoperable IP metadata format. Mentor signed up ARM as an important IP supplier to Philips Semiconductor, and to gain the credibility of the largest source of independent IP in the industry. Neither Cadence nor Synopsys had ESL IP integration tools at that time but thought that it might be an important new market. Given the common European interest in standards based design processes and tools, and given their own interest in IP integration flows, and desire to influence the evolution of commercial EDA tools, it was relatively easy to have STMicroelectronics come on board. Thus a veneer of industry non-partisanship was added to what was essentially a Mentor standard done for the benefit of Philips Semiconductor.

Platform Express and its XML-based metadata formats became the basis for all the early SPIRIT standardization effort. Other EDA vendors signed up for SPIRIT and some even developed SPIRIT-based flows. Synopsys wanted to sell its coreAssembler tool and incorporated SPIRIT-based IP import and SPIRIT subsystem export interfaces into the tool. STMicroelectronics was a user of this tool and talked about it at the Synopsys EDA Interoperability Developers Forum in 2006 [12]. CoWare supported SPIRIT-based IP model import into its Platform Creator tool – although not SPIRIT export (interoperability was a one-way street).

Given that Platform Express's IP metadata was essentially at the RTL signal/pin/port level, the early SPIRIT formats encompassed a signal level of abstraction and thus really did not represent much of a difference from what is possible with Verilog and VHDL. The initial SPIRIT 1.0 and IP-XACT 1.2 levels were essentially signal-level netlists and pin-interface lists dressed up with some marginal abstractions at the system level such as "bus definitions" (which really correspond to named clusters of signals). They did, however, allow initial formalization of the HW/SW interface by describing registers and bitfields albeit in a very verbose format.

IP-XACT 1.4, which came out in March 2008, supports a notion of ESL transaction-level modeling, although this is implemented in a weak fashion that relies on other groups such as OSCI defining the true semantics of various levels of TLM.

In May 2009, SPIRIT released a Register Definition Language standard that will be discussed in more detail later.

While the SPIRIT consortium defined not just the metadata model but the EDA environment within which it was used (the IP-XACT Design Environment or DE), the adoption of IP-XACT by vendors has led to two different approaches. One approach was to implement the design environment directly on top of the metadata model and become a so-called IP-XACT editor. This approach exposed many low-level eccentricities of IP-XACT data to users and workflows have not been very successful. The other approach was to have tools that have been categorized on the SPIRIT web site as "Tools with IP-XACT Interface." There were tools that could read or write IP-XACT but that would focus on specific features like IP fabric generation, register management, or chip assembly and use IP-XACT like a physical/communications layer.

SPIRIT adoption has been relatively weak. The only widely accepted and adopted level has been the RTL level IP-XACT 1.2. In addition to the examples cited earlier, other vendors such as Synplicity (now part of Synopsys) supported IP-XACT 1.2 in its IP-based System Designer tool, part of Synplify Premier, for IP-based system definition and assembly of FPGA. Duolog provides tools with IP-XACT interfaces through their Socrates integration platform. Beach Solutions supported it but was absorbed into Duolog. Magillem Design Services and MataiTech both offered IP-XACT design environments, although neither have been particularly successful. Synopsys allows platforms to be assembled using IP-XACT in their Innovator system-level prototyping tool. Some interest in SPIRIT was shown in Japan when STARC became a member of the SPIRIT consortium. Although Mentor Graphics announced support for IP-XACT 1.4 in its Platform Express tool in 2008

[13], currently Platform Express is not listed as a product on Mentor Graphics web site (consulted May 2009), so it does not appear to be actively supported. However support is being added along with other features of Platform Express into their Vista product line.

Beyond NXP (the renamed Philips Semiconductors), STMicroelectronics and Texas Instruments, industrial users of SPIRIT/IP-XACT who have named themselves publicly are few and far between. However, the industry needs standardization in this area and IP-XACT does seem to be gaining momentum as an accepted and maturing IP interchange format. The merger with Accellera in June 2009 will very likely attract new members, increase adoption, and give it a new lease of life.

2.3.2 RTL Assembly Level

The following is an example of a SPIRIT IP-XACT 1.2 description of a Tensilica Diamond DC_ B_106micro processor. This was created to use with the Synopsys Synplicity Synplify Premier System Designer tool. Major portions of this description have been removed. A complete version of this is available on the web site http://ElectronicSystemLevel.com.

```
<?xml version="1.0" encoding="UTF-8"?>
<spirit:component
xmlns:spirit="http://www.spiritconsortium.org/XMLSchema/SPIRIT/1.2"
xmlns:xsi="http://www.w3.org/2001/XMLSchema-instance"
xsi:schemaLocation="http://www.spiritconsortium.org/XMLSchema/SPIRIT/1.2
http://www.spiritconsortium.org/XMLSchema/SPIRIT/1.2/index.xsd">
    <spirit:vendor>tensilica.com</spirit:vendor>
    <spirit:library>DC_B_106micro_system</spirit:library>
    <spirit:name>Xm_DC_B_106micro_noCGXttop</spirit:name>
    <spirit:version>RB2007.2</spirit:version>
    <spirit:busInterfaces>
        <!-- Group: AHB-Lite Master -->
        <spirit:busInterface>
            <spirit:name>AHBLitemaster</spirit:name>
            <spirit:busType spirit:vendor="amba.com" spirit:library="AMBA3"
spirit:name="AHBLite" spirit:version="r1p0_6"/>
            <spirit:master>
                <spirit:addressSpaceRef spirit:addressSpaceRef="main"/>
            </spirit:master>
            <spirit:signalMap>
                <spirit:signalName>

    <spirit:componentSignalName>CLK</spirit:componentSignalName>
                  <spirit:busSignalName>HCLK</spirit:busSignalName>
                </spirit:signalName>
                <spirit:signalName>

    <spirit:componentSignalName>HADDR</spirit:componentSignalName>
                   <spirit:busSignalName>HADDR</spirit:busSignalName>
                </spirit:signalName>
                <spirit:signalName>

    <spirit:componentSignalName>HRDATA</spirit:componentSignalName>
                   <!-- MORE AMBA AHB BUS follows -->
            </spirit:signalMap>
        </spirit:busInterface>

        <!--  Group: DRam0 -->
```

```
        <spirit:busInterface>

            **** Removed for conciseness

        </spirit:busInterface>

            **** Other bus interfaces removed for conciseness

        <!--  Group: Interrupt -->
        <spirit:busInterface>
            <spirit:name>BInterrupt0</spirit:name>
            <spirit:busType spirit:vendor="spiritconsortium.org"
spirit:library="busdef.interrupt" spirit:name="interrupt"
spirit:version="1.0"/>
            <spirit:slave>
            </spirit:slave>
            <spirit:signalMap>
                <spirit:signalName>

    <spirit:componentSignalName>BInterrupt</spirit:componentSignalName>
                    <spirit:busSignalName>IRQ</spirit:busSignalName>
                    <spirit:left>0</spirit:left>
                    <spirit:right>0</spirit:right>
                </spirit:signalName>
            </spirit:signalMap>
        </spirit:busInterface>

        <!-- MORE INTERRUPT DEFINITION FOLLOWS →

        <!-- Group: OCDHaltOnReset -->
        <spirit:busInterface>
            <spirit:name>OCDHaltOnReset</spirit:name>
            <spirit:busType spirit:vendor="tensilica.com"
spirit:library="busdef.tensilica.OCDHaltOnReset"spirit:name="OCDHaltOnReset"
spirit:version="RB2007.2"/>
            <spirit:slave>
            </spirit:slave>
            <spirit:signalMap>
                <spirit:signalName>

    <spirit:componentSignalName>OCDHaltOnReset</spirit:componentSignalName>
                    <spirit:busSignalName>OCDHaltOnReset</spirit:busSignalName>
                </spirit:signalName>
            </spirit:signalMap>
        </spirit:busInterface>

        <!-- Group: TAP -->
        <spirit:busInterface>
            <spirit:name>TAP</spirit:name>
            <spirit:busType spirit:vendor="spiritconsortium.org"
spirit:library="busdef.jtag" spirit:name="jtag" spirit:version="1.0"/>
            <spirit:slave>
            </spirit:slave>
            <spirit:signalMap>
                <spirit:signalName>

    <spirit:componentSignalName>JTCK</spirit:componentSignalName>
                    <spirit:busSignalName>TCK</spirit:busSignalName>
                </spirit:signalName>
                <spirit:signalName>

   <spirit:componentSignalName>JTDI</spirit:componentSignalName>
                    <spirit:busSignalName>TDI</spirit:busSignalName>
                </spirit:signalName>
                <spirit:signalName>

    <spirit:componentSignalName>JTDO</spirit:componentSignalName>
                    <spirit:busSignalName>TDO</spirit:busSignalName>
                </spirit:signalName>
                <spirit:signalName>
```

```
<spirit:componentSignalName>JTMS</spirit:componentSignalName>
                <spirit:busSignalName>TMS</spirit:busSignalName>
            </spirit:signalName>
            <spirit:signalName>

<spirit:componentSignalName>JTRST</spirit:componentSignalName>
                <spirit:busSignalName>NTRST</spirit:busSignalName>
            </spirit:signalName>
        </spirit:signalMap>
    </spirit:busInterface>
</spirit:busInterfaces>
<spirit:addressSpaces>
  <spirit:addressSpace>
      <spirit:name>main</spirit:name>
      <spirit:range>4G</spirit:range>
      <spirit:width>32</spirit:width>
  </spirit:addressSpace>
  <spirit:addressSpace>
      <spirit:name>iram</spirit:name>
      <spirit:range>128K</spirit:range>
      <spirit:width>16</spirit:width>
  </spirit:addressSpace>
  <spirit:addressSpace>
      <spirit:name>dram</spirit:name>
      <spirit:range>128K</spirit:range>
      <spirit:width>16</spirit:width>
    </spirit:addressSpace>
</spirit:addressSpaces>
<spirit:memoryMaps>
</spirit:memoryMaps>
<spirit:model>
  <spirit:views>
      <spirit:view>
          <spirit:name>verilogSource</spirit:name>

<spirit:envIdentifier>:modelsim.mentor.com:</spirit:envIdentifier>

   <spirit:envIdentifier>:ncsim.cadence.com:</spirit:envIdentifier>
          <spirit:envIdentifier>:vcs.synopsys.com:</spirit:envIdentifier>

<spirit:envIdentifier>:designcompiler.synopsys.com:</spirit:envIdentifier>

<spirit:envIdentifier>verilogSource:synplify.synplicity.com:device.xilinx
    </spirit:envIdentifier>

<spirit:envIdentifier>verilogSource:synplify.synplicity.com:device.altera
    </spirit:envIdentifier>

<spirit:envIdentifier>verilogSource:synplify.synplicity.com:device.actel
    </spirit:envIdentifier>
          <spirit:language spirit:strict="true">verilog</spirit:language>
          <spirit:modelName>Xm_DC_B_106micro_noCGXttop</spirit:modelName>
          <spirit:fileSetRef>fs-verilogSource</spirit:fileSetRef>
      </spirit:view>
  </spirit:views>
  <spirit:signals>
      <!--  Group: BreakIn -->
      <spirit:signal>
          <spirit:name>TDebugInterrupt</spirit:name>
          <spirit:direction>in</spirit:direction>
      </spirit:signal>

      **** Many Signals Removed for conciseness

    <spirit:modelParameters>
    </spirit:modelParameters>
</spirit:model>
<spirit:fileSets>
   <spirit:fileSet spirit:fileSetId="fs-verilogSource">
```

```
            <spirit:file>
                <spirit:name>../../../../rtl_enc/bytebreaker.v</spirit:name>
                <spirit:fileType>verilogSource</spirit:fileType>
            </spirit:file>
            <spirit:file>
                <spirit:name>../../../../rtl_enc/ClkGate.v</spirit:name>
                <spirit:fileType>verilogSource</spirit:fileType>
            </spirit:file>
            <spirit:file>
                <!--MORE VERILOG FILE REFERENCES FOLLOW →
        </spirit:fileSet>
    </spirit:fileSets>
    <spirit:cpus>
        <spirit:cpu>
            <spirit:name>DC_B_106micro</spirit:name>
            <spirit:addressSpaceRef spirit:addressSpaceRef="main"/>
        </spirit:cpu>
    </spirit:cpus>
</spirit:component>
```

The component example above relies on bus definitions, some of which are standard examples from ARM or the SPIRIT consortium, and some of which are defined by Tensilica. For example, a DRAM0 bus definition is as follows:

```
<?xml version="1.0" encoding="UTF-8"?>
<!--  Group: DRam0 -->
<spirit:busDefinition
xmlns:spirit="http://www.spiritconsortium.org/XMLSchema/SPIRIT/1.2"
xmlns:xsi="http://www.w3.org/2001/XMLSchema-instance"
xsi:schemaLocation="http://www.spiritconsortium.org/XMLSchema/SPIRIT/1.2
http://www.spiritconsortium.org/XMLSchema/SPIRIT/1.2/index.xsd">
    <spirit:vendor>tensilica.com</spirit:vendor>
    <spirit:library>busdef.tensilica.DRam0</spirit:library>
    <spirit:name>DRam0</spirit:name>
    <spirit:version>RB2007.2</spirit:version>
    <spirit:directConnection>true</spirit:directConnection>
    <spirit:maxMasters>1</spirit:maxMasters>
    <spirit:signals>
        <spirit:signal>
            <spirit:logicalName>DRam0Addr0</spirit:logicalName>
            <spirit:isAddress>true</spirit:isAddress>
            <spirit:onMaster>
                <spirit:bitWidth>15</spirit:bitWidth>
                <spirit:direction>out</spirit:direction>
            </spirit:onMaster>
            <spirit:onSlave>
                <spirit:bitWidth>15</spirit:bitWidth>
                <spirit:direction>in</spirit:direction>
            </spirit:onSlave>
        </spirit:signal>
        <spirit:signal>
            <spirit:logicalName>DRam0Busy0</spirit:logicalName>
            <spirit:onMaster>
                <spirit:bitWidth>1</spirit:bitWidth>
                <spirit:direction>in</spirit:direction>
            </spirit:onMaster>
            <spirit:onSlave>
                <spirit:bitWidth>1</spirit:bitWidth>
                <spirit:direction>out</spirit:direction>
            </spirit:onSlave>
            <spirit:defaultValue>
                <spirit:value>0</spirit:value>
            </spirit:defaultValue>
```

```
        </spirit:signal>
        <*****  Other signalexamples removed for conciseness ****>
    </spirit:signals>
</spirit:busDefinition>
```

2.3.3 System Modeling Level

The system modeling level in IP-XACT 1.4 provides a small enhancement to the SPIRIT format. As excerpted from the 1.4 specification [14], Section 5.11.3, p. 42

"The following example shows a custom transactional port within an abstraction definition. Constraints are defined for transactional port used in master or slave interfaces.

```
<spirit:port>
    <spirit:logicalName>custom_tlm_port</spirit:logicalName>
    <spirit:transactional>
        <spirit:onMaster>
            <spirit:service>
                <spirit:initiative>provides</spirit:initiative>
                <spirit:typeName implicit="true">TLM</spirit:typeName>
            </spirit:service>
        </spirit:onMaster>
        <spirit:onSlave>
            <spirit:service>
                <spirit:initiative>requires</spirit:initiative>
                <spirit:typeName implicit="true">TLM</spirit:typeName>
            </spirit:service>
        </spirit:onSlave>
    </spirit:transactional>
</spirit:port>"
```

The rest of SPIRIT 1.4 follows similar principles in adding an "ESL" or transaction modeling layer. It defines a primitive add-on syntax for TLM or ESL ports, bus definitions, etc., but leaves the semantics of these undefined. The syntax ensures that ports of a new type such as TLM are the same and interconnected appropriately, and assumes the semantics of the interface to be defined by a group such as OSCI with TLM 1 or TLM 2. Because this level of definition does not add much value, it is not surprising that adoption of SPIRIT IP-XACT 1.4 is minimal. With the merger of SPIRIT with Accellera, IP-XACT may readdress its links back to implementation including focus on verification, power, and analog/mixed-signal packaging, and with that solid foundation, it may then provide proper abstractions to TLM2 and beyond.

2.4 Register Definition Languages

Since the dawn of complex systems, there have been attempts made to build formats and tools for defining the HW/SW interfaces of such systems and to generate useful artifacts from these definitions that will be used in downstream flows. It is interesting to note that some of the earliest tools that have been offered for sale came from large systems companies. In 2001, Lucent Technologies offered its HSI designer tool that was based on internal technology used since 1995 [15]. This was a commercial failure. SystemRDL originated in Cisco Systems, then used by Denali, and then donated to SPIRIT for standardization in 2007 [16].

Large companies building large complex systems often developed internal register definition technologies and languages because their system could have hundreds, thousands, or tens of thousands of HW/SW register interfaces. Managing this complexity and modifying all the relevant HW and SW interface header and documentation files (such as C header files and Verilog interface signal files) would be tedious and error-prone if reliant on entirely manual methods. After developing internal tools of this type, a natural assumption by the creators was that they could become commercially viable tools used by many companies willing to pay for them. This is the often erroneous assumption that sparked many attempts to commercialize tools developed inside individual companies or university research projects – attempts that usually failed to be commercially viable.

Other register definition formats and associated tools came from Beach Solutions in England (now part of Duolog), PDTi of Vancouver with SpectaReg, and Agnisys of India who announced IDesignSpec in the spring of 2009.

The rest of this section will discuss register definition motivations, HW/SW interfaces, and solutions in more detail, illustrated with technology from Duolog.

2.4.1 Motivation: Modeling the HW/SW Interface

Register definitions provide a high-level view of the HW/SW interface. Throughout system development, this interface has multiple perspectives including design and verification views from the hardware and software domains, and covers different scopes including hardware register elements and processor views of the system. We discuss the HW/SW interface at a high level and many of the common formats are presented within the context of a chip design flow. We also discuss the problems associated with maintaining multiple formats, and outline some emerging solutions including HW/SW interface management tools. A case study of one of the solutions illustrates many of the points raised.

2.4.1.1 What Is the HW/SW Interface?

In its simplest form, software accesses hardware by converting software instructions into hardware transactions as depicted in Figure 2.1. The software instructions run on a dedicated processor which may be discretely packaged apart from the rest of the hardware, or can also run on a processor embedded within the hardware system itself. In the former case, the software can run within a PC on, for example, an Intel processor and the hardware could be a wireless card with a PCI interface.

In the latter case, the software could run on an embedded processor within a SoC as shown in Figure 2.2.

In this section, we will focus on the HW/SW interface within the context of an SoC.

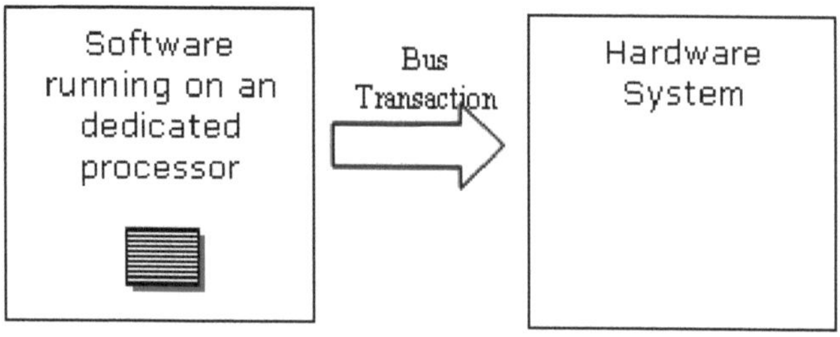

Fig. 2.1 Software running on a dedicated processor

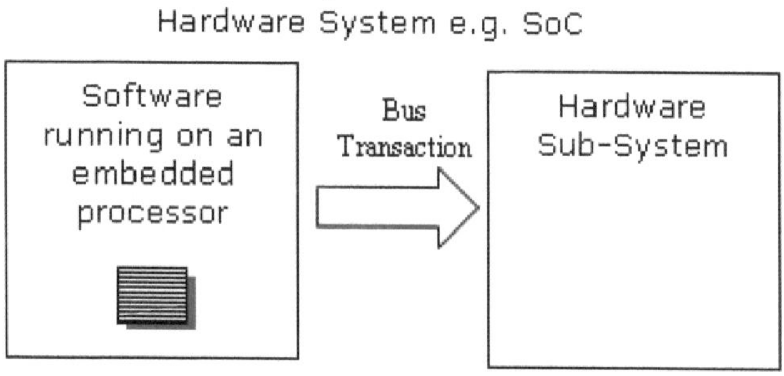

Fig. 2.2 Software running on an embedded processor

2.4.1.2 Hardware Configuration and Control Using Software

Software can interact with hardware to provide functions such as

- Resetting the hardware to a known state
- Configuration of the hardware to perform a specific function
- Getting status of the hardware
- Reading/writing data to/from the hardware.

Fundamentally, software can access hardware and either transfer data to the hardware (a write access) or get information from the hardware (a read access). This is typically done by transforming software instructions run on a processor to memory-mapped transactions that target a specific piece of hardware with either read or write transactions. These transactions are either coded directly as processor assembler instructions or compiled from a high-level programming language such as C. Of course, processors may also support other kinds of interfaces such as HW FIFOs, possibly using specialized instructions accessed as macros or pragmas.

For a write access, this information is typically either data that will be processed by the hardware or hardware configuration and control settings. In these cases, the hardware elements could be a memory, a hardware register that holds a specific configuration parameter or some type of input to a hardware process. For memories and hardware registers, the values written may be non-volatile (they will keep their value) or they could be volatile where the hardware may dynamically change the value written. For a read access, the information returned to the processor is typically either data or status information for the hardware block.

The data that is written or read during a software access may have little or no meaning to the corresponding hardware: for example, the data may be part of a message or may need to be stored in memory. It could also have a very specific meaning for the hardware block – for example, a hardware configuration parameter; and it may be defined using enumerations. To illustrate, a power control parameter for an IP block that is 2 bits wide may have the following enumerations (Table 2.1):

Table 2.1 Sample power control hardware element

Enumeration Value	Description
0x0:	Turn Power on
0x1:	Turn Power off
0x2:	Set Power to standby
0x3:	Set Power to idle

Those generic storage mechanisms, configuration parameters, or status values that can be accessed by the processor, via a memory-mapped transaction, can be generically described as Software Accessible Hardware Elements (SAHEs). This may be implemented in hardware with constructs such as memories, registers, or bitfields. While the representation of SAHEs can be straightforward for both hardware and software, best-in-class implementation of the HW/SW interface means effectively handling SAHEs while dealing with constraints from both the software and hardware perspectives.

2.4.1.3 Software Perspective

The software perspective of the hardware interface can be defined very simply. From the processor point of view the system can be viewed as an accessible address space. This address space is not defined randomly (although it can often appear as such) but is generally an ordered layout or map of the hardware. A hardware block such as a UART module which is part of a peripheral subsystem would appear within the peripheral subsystem memory map as indicated in Figure 2.3.

As mentioned previously the processor can perform write transactions which involves writing data to a particular address within the processor address space.

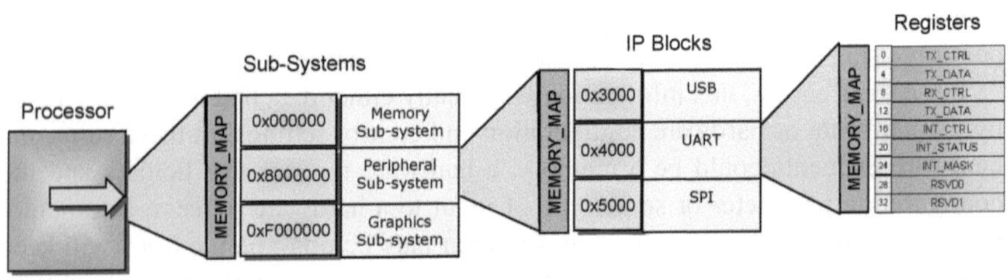

Fig. 2.3 Software perspective of the hardware

The processor can also perform read transactions which involve reading from a particular address within the processor address space and will typically return some value from the hardware. Usually the best practice in embedded software design is to ensure optimal processor performance by reducing the number of transactions between hardware and software.

The processor typically works with a certain data bus size, e.g., 32 bit/64 bit. with some limited control over the granularity of the access. Some processors allow a 32-bit data transaction with 4 byte enables meaning that the minimum data access for the system is 8 bits. This is also known as the least addressable unit (LAU) and is the smallest size of data the processor can handle on a transaction [17]. Of course, hardware IP blocks such as the UART may not implement a byte enable mechanism so the LAU may be a physical constraint of the IP itself.

The LAU imposes a constraint on the HW/SW interface. If we consider the previous power control parameter which can be represented by a 2-bit-wide SAHE that is contained in a hardware block with a LAU of 32 bits – how do we write this value? There are two options to consider in implementing a write transaction to that specific SAHE.

The first option is to assign this SAHE a dedicated address space, meaning that the other 30 bits will not be used, or probably made "reserved." However if each SAHE is given its own address space, this will, although allowing ease of access to the SAHEs, carry some drawbacks. First, the available address space can fill up very quickly if there is one address per SAHE. This also has implications on the hardware design – an increase in the amount of address decoding is required. From a performance point of view a significant increase in the number of transactions is required. For example, if there are 32 single bit SAHEs, accessing each of them will require 32 separate transactions. If they shared the same address space, they could be accessed in a single write transaction, thereby optimizing the number of reads and writes. Our second option therefore could be to "pack" as many SAHEs as possible into a shared address.

While there are advantages to letting SAHEs share the same address space, what happens when just one of the SAHEs needs to be written? If a write transaction is performed across a number of SAHEs then they will all get written this data, so either the software retains a copy of the hardware SAHEs so that it can construct the correct write data or the software must perform a read–modify–write type of transaction. This is a type of transaction where the processor reads the data from a specific address, updates the relevant bits specific to that SAHE, and then writes the modified data back, preserving the other SAHEs at that address. This type of transaction requires some latency between the initial read and the final updating write; this introduces an opportunity for non-deterministic errors. For example, other processors or other tasks on the processor may try to alter other SAHEs in the same address space during the interval between the read and the write. Good software development practices, robust transaction protocols, and good hardware implementation typically minimize the risk of such errors. Other mechanisms include specific coherency approaches or bus locking during read–modify–write transactions to minimize the risk of occurrence.

In the case where the SAHE is bigger than the maximum addressable unit for example, a processor that handles 32-bit data may need to set a 256-bit DSP coefficient, then accessing this HW element may involve several accesses to write or read the data. The software can utilize certain transaction optimizations such as burst data or block reads/writes to obtain better performance. Because of these SAHE mapping constraints, there are many other optimizations that can exist in the HW/SW interface. These will be discussed in the next section when we look at the interface from the hardware perspective.

2.4.1.4 Interrupts

Another important part of the HW/SW interface is the ability of the hardware to notify the software when something has occurred. This is typically done using a hardware interrupt. From the software perspective, interrupts are asynchronous breaks in program flow that occur as a result of events outside the running program [18]. Examples of interrupts include a button press, infrared detect, battery power exhausted, timer expiration, or completion of a data transfer. Software typically handles interrupts through an interrupt service routine (ISR) which can become extremely cumbersome for a large system. For instance, on an interrupt, the ISR may need to read information from a hardware block called an interrupt controller to determine the source block of the interrupt. Once the interrupt source is known the software may have to query the hardware block to get more granularity on the exact cause of the interrupt. The ARM UART Primecell [19] has 11 interrupt triggers that can be combined to produce a single interrupt. If this single interrupt is used within the interrupt controller the ISR will have to read the UART interrupt status register in order to know the true source of the interrupt and to therefore respond accordingly.

In a system, interrupts can have different priorities and thus there is the potential for one interrupt to "interrupt" the processing of another interrupt (this is called a nested interrupt). The ISR has the possibility of disabling interrupts in the hardware at many different levels to help manage priorities. As interrupts can be considered random occurrences from the software point of view, complex systems with multiple levels of interrupts, the actual interaction between the hardware and software can be non-deterministic. As always, SW/HW performance needs to be considered and in this case the metric that may be critical in interrupt processing is the response time between an interrupt being asserted in the hardware and the corresponding ISR being executed. This is known as the *interrupt latency*.

2.4.1.5 Software API

Software–hardware interaction needs efficient access to the SAHEs. The hardware domain presents these SAHEs as memory, registers and bitfields within a processor address space thus allowing software interaction. In order for embedded software developers to write applications they require a means to interact, control, and monitor hardware elements, as well as potentially accessing and processing data streams. Software developers must understand and use these elements to bridge the gap between the hardware and software domains. This is achieved using a software API (application programmers interface). The lowest level of API will typically contain functions for accessing the SAHEs in an efficient manner. Thus if a single bit SAHE requires independent access, it may be captured in a function that implements the read–modify–write scenario as described previously. Similarly higher level functions can be created from these low-level functions to produce a higher level API. If we consider that a specific hardware IP can be sourced from different IP providers it may be possible to create a level of abstraction that would be relatively hardware-independent. With a programming language like C, there are several choices about how to implement the low-level API. It could be done via standard C constructs such as functions or macros, or using C++ structures such as classes and class methods [20].

2.4.1.6 Hardware Perspective

When software accesses the hardware it does so by means of a processor transaction; a read or write access to a specific address location. This transaction is implemented by a signaling protocol that involves the toggling of wires (representing address, data, and control elements) according to a well-defined specification. The address, write data, and control may be distributed throughout the system to different subsystem and IP blocks while the read data typically needs to be routed efficiently back to the processor by means of data multiplexers. Figure 2.4 shows a sample system where the processor can access a UART, contained in a peripheral subsystem, through a Network-on-Chip (NoC) and an AHB2APB bridge.

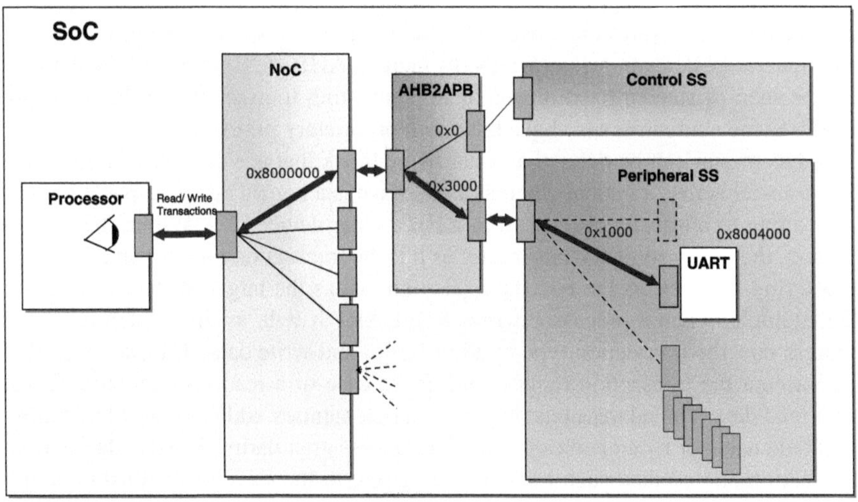

Fig. 2.4 Processor view through an NoC (Network-on-Chip)

While the diagram may appear quite simple there are many implementation-specific hardware scenarios that make this more complex.

2.4.1.7 Transaction Bus Protocol

In the hardware domain, this bus transaction is implemented by a signaling protocol involving the toggling of wires. Figure 2.5 shows an example of the signaling protocol for an ARM AHB bus.

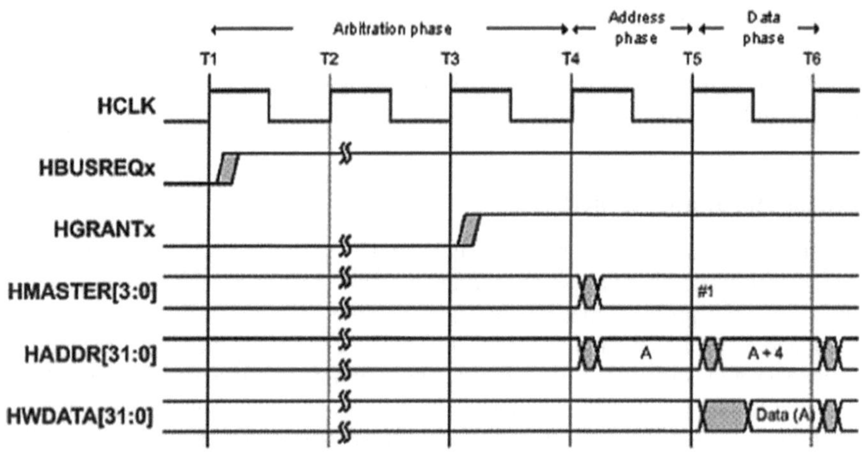

Fig. 2.5 AHB write transaction [21, pp. 3–30]

The protocol is a specification of how the transaction is realized in hardware and involves addressing, data, handshaking and control, clocking, and timing; in essence the logical rules that govern the communication between the processor and a peripheral [22]. Examples of protocols include AHB, AXI, PCI, and OCP which can be seen as standardized means of implementing transactions in the hardware world. Some companies also have their own proprietary protocols.

A bus master can be defined as a hardware block that is a capable of initiating a data transfer across a system whereas a bus slave is a logical device capable only of responding to a transfer request from a bus master device [22]. Within the current context, the processor is the bus master as it is the transaction initiator and the block it is trying to access is the bus slave (also known as the target of the transaction). For example, when a processor writes to a UART it will, as the master, drive the address bus, the transaction type (read or write) and write data. The slave (UART) may accept the transaction request and in the case of a read transaction, respond with read data. The address bus may have supplementary address control including a global enable or byte enables to refine the access granularity. The data bus carries transaction-related data such as write data going to the slave or read data coming from the slave. Typically in on-chip buses, the read and write data are independent signals.

The bus protocol can have simple read/write transactions which involve implementing a single access to an address. More complex bus protocols include the ability to perform burst transactions or do interleaved or out-of-order transactions. Different protocols have different control options, for example,

- The clocking for the protocol
- Bus arbitration control if there are multiple masters
- Byte control for accessing specific bytes within the read or write data
- Control needed for master–slave handshaking such as extending transactions by inserting wait states
- Indicating when a write burst has been completed
- Error control.

The specific detail of each bus protocol is not in the scope of this chapter but it is important to note that in the hardware system there may be a mix of protocols used between the bus master and a specific slave leading to increased complexity through protocol translation.

2.4.1.8 Protocol Translation

It is likely in a SoC with IP blocks coming from a range of different sources that a variety of different protocols may be used. Figure 2.6 shows one such system

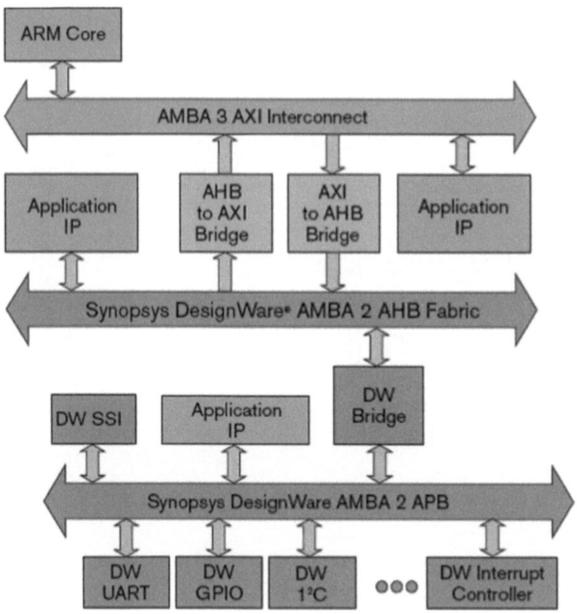

Fig. 2.6 Bridging from the AMBA 3 AXI protocol to AMBA 2 protocols [23]

where the ARM core (with an AMBA3 AXI protocol interface) at the top of the diagram needs to access the UART (with an AMBA2 APB interface) at the bottom of the diagram. The AXI protocol is first translated to the AHB protocol, through an AXI2AHB bridge, which is then translated to an APB protocol via a bridge, which is then compatible with the UART.

As a hardware transaction progresses through the system, it may require several transformations to other protocols. In addition other transformations may be required, such as

- It may need to be re-synchronized if the transaction crosses clock domains
- If the protocol crosses power or voltage domains it may require isolation cells or level shifters to be inserted
- It may need registers inserted to combat logic delays or race conditions (also known as pipelining.

All of these transformations and bridging will result in a delay or latency between the master initiating a transaction request and the slave's response. This latency is an important consideration from a software perspective as it might make more sense to do a read burst across a slave device rather than individual and "slow" read writes.

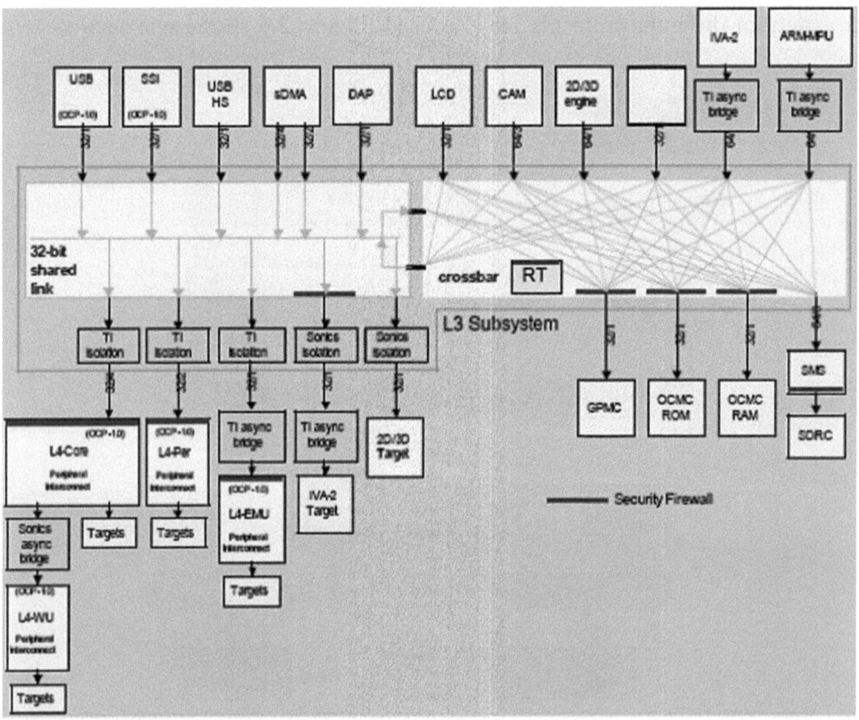

Fig. 2.7 OMAP multimedia processor [24]

A general rule of thumb in HW/SW interface design is to minimize the number of transactions required to implement a specific function.

Another example of protocol transformation is shown in Figure 2.7, an OMAP multimedia processor [24].

In Figure 2.7, the ARM Cortex processor is connected to an L3 high-speed NoC. Any AMBA transaction needs to go through crossbars, shared links, bridges, isolation blocks, firewalls, through subsystems until finally reaching targets such as peripheral IP blocks.

As the address is distributed throughout the hierarchy it goes through various levels of decode where address paths are enabled or disabled depending on the decoding of the address. This is coded in RTL and may be an implicit part of the NoC type of block or it can be coded explicitly in RTL. An address decoder decodes an address range and usually generates a set of enables for the next level of targets in the hierarchy until the address is resolved down to the specific leaf components of the system. These leaves can range from simple components with a few registers to complex subsystems with thousands of registers. It can be assumed for simplification purposes that leaf components are IP blocks that fully terminate the address path and contain registers.

2.4.1.9 Registers and Bitfields

Memory-mapped registers are the lowest direct level of HW/SW interface and are usually the last addressable elements in the HW/SW address decoding to have direct read/write communications. The difference between a SAHE and a register is that the SAHE can be any size, whereas memory-mapped registers are constrained by the LAU as described earlier.

Registers can have different access constraints. For example, if a particular memory-mapped register is used to store IP status information, the register may have read-only access. Similarly a register that controls an IP configuration may have read–write access. Because of the volatile nature of hardware it is typical for most writable registers to have a read back capability.

While registers are the common unit of granularity in the HW/SW interface, they can be composed of bitfields which are considered subsections of the register and can range from a single bit wide to the width of the register. Most bitfields have a one-to-one correspondence with SAHEs, in that a 1-bit SAHE can be represented by a 1-bit bitfield. Bitfields, like SAHEs, can have the following characteristics

- The width of the bitfield
- The offset of the bitfield within the register
- The access type of the bitfield (e.g., read/write)
- The value of the bitfield at reset.

Another characteristic of a bitfield that can be defined is the formal encoding of the bitfield values using enumerations. For example, a single-bit reset bitfield may have two values called reset_off=0 and reset_on=1.

Like a register, a bitfield may have a specific type of access characteristic, depending on the SAHE. In fact a register can contain bitfields that have different access types. In Figure 2.8, there is space in the register that is unused and typically marked reserved. This bitfield has no write capability but it can be read (and would usually return 0). Therefore, while the access type of that particular register is read–write, the access type of the reserved bitfield may be considered read-only.

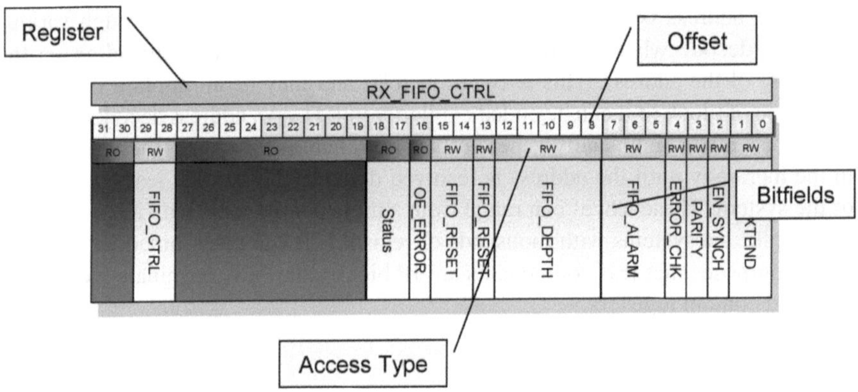

Fig. 2.8 Hardware/software register

In order to write to a register, the processor must execute an instruction that references a particular memory-mapped address within the processor address space. This instruction initiates a write transaction and the address, control, and write data are routed through the system until they arrive at the target IP Block.

Some of the same dynamics of accessing a memory map at the chip level occur within the blocks, albeit at a microlevel. In an IP block that contains registers the processor bus may interface to some type of hardware state machine to ensure the correct handshaking mechanism is implemented. The write data from the processor bus interface is broadcast to each register. However, for a standard read/write register, the register contents will only be updated (written) if its address is on the bus and a write transaction is enabled. The address is typically decoded to register select signals as shown in Figure 2.9. Read data coming from all of the registers or internally from the logic block needs to be multiplexed, based on the address at the bus interface to ensure that the read data from a selected register is presented back to the processor. The state machine decode, enabling and multiplexing logic, is the responsibility of the IP design hardware engineer and is typically captured in RTL code. Note that while the diagram shows the HW/SW interface logic in an explicit block this may not be the case and sometimes decoders and registers are mixed with other logic. From a HW/SW verification perspective, it is useful to keep this as a self-contained block with a standard notation for registers and bitfields so that it is easier to verify accesses to the registers.

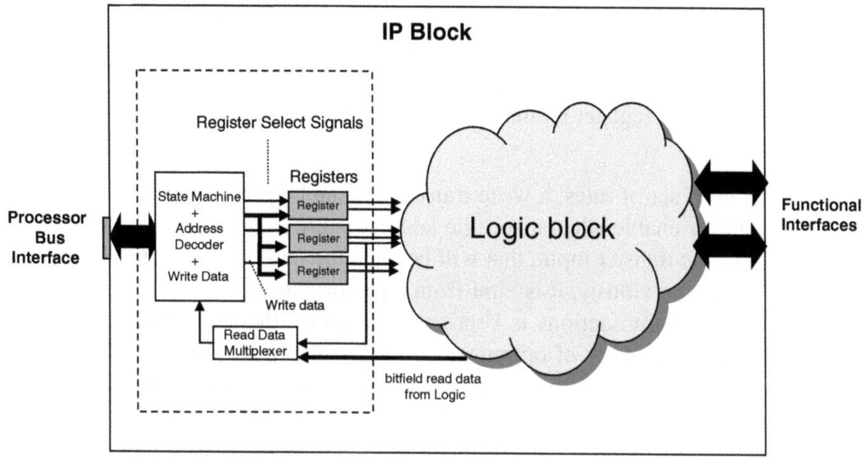

Fig. 2.9 Registers in a hardware IP block

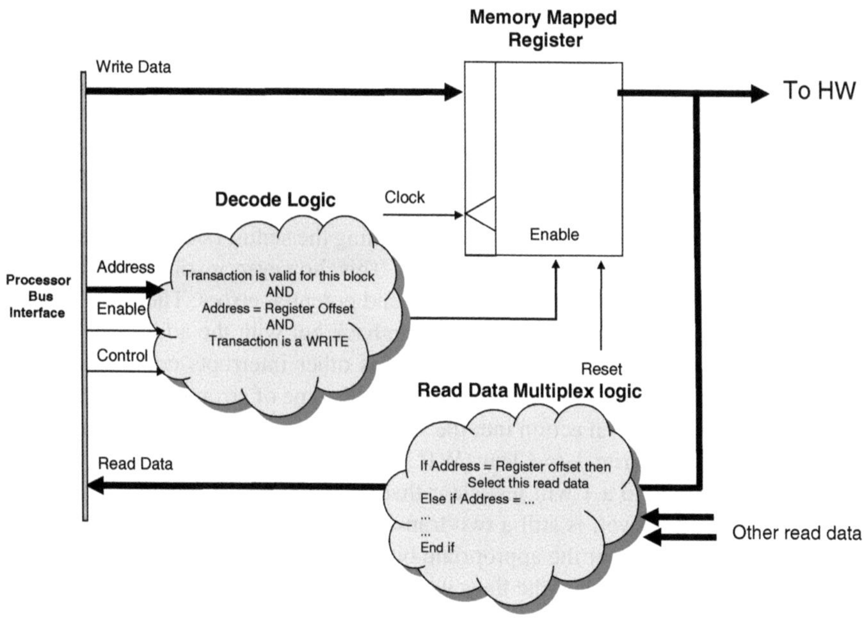

Fig. 2.10 Example RTL view of a simple memory-mapped register

The RTL implementation of a simple memory mapped register is shown in Figure 2.10.

This logic typically involves a clocked logic process with the following characteristics

- The output of a register can be set to a particular value when reset is asserted
- If reset is not asserted then the output of the register will be assigned the input of the register on a specific clock edge, if enabled
- If not enabled the register maintains its current value.

With this brief set of rules, a write transaction can be processed and decoded to assert the register enable when a specific address is presented, and if the write data is presented as the register input, that will become the new register data.

As mentioned previously, it is vital from a performance point of view to ensure that the number of transactions is kept to a minimum. Because of this goal, it is possible to make a number of optimizations in the HW/SW interface to minimize the number of transactions. It may be possible to provide a read or write burst across the registers and the effects of the transaction latency can be reduced (averaged). Another method for reducing transactions is to superimpose additional behavior with the transaction so that a read or write does more than simply read or write. This additional behavior is also known as side effects [20].

Hardware interrupt status registers are good examples of side effects. If hardware wants to interrupt the software it typically sets an interrupt status SAHE to "1" which can be used to trigger an interrupt line resulting in the software becoming aware that a hardware block needs to be serviced. In order to find out what caused this particular interrupt, the software will run an ISR to determine the cause of the interrupt. However, once the processor reads the data and knows what the hardware is indicating, the SAHE needs to be set to 0 so that it can be used to interrupt again at a later stage.

With no optimizations the software, after reading the status, could execute a write transaction to clear the specific interrupt to 0. This, however, involves two transactions which would lead to additional latency and potential errors. There would be a calculation required to write the correct value back but with the additional latency the value to write back might not be correct as other interrupts could have been asserted between the read and the write. To stop this type of error, the first optimization would be to encode an action into the write data to explicitly clear the interrupt bit. This is called a Write-1-to-Clear (W1C) type of side effect and setting a particular bit in the data to a 1 will just clear that bit in the register, without affecting any others. This, however, is still a two-transaction solution, one to read the interrupt status and one to clear the appropriate bits. This can be further optimized by a mechanism which would clear the flags when they are read in the first place. This is a specific type of access behavior that can be seen as an extension of a standard read access and is typically called Read-to-Clear or R2C. This type of behavior is called an access side effect. Figure 2.11 shows the type of logic required to implement this in RTL.

In this example an interrupt from the hardware will "Set" the register to a "1" and a read from the processor will clear it to "0." These types of optimizations can challenge verification engineers to cover HW/SW corner cases, such as ensuring that if a second interrupt assertion happens in the same clock cycle as the access clear, that the interrupt is not lost. Hardware design must ensure that the correct data is presented back to the processor while ensuring any pending interrupts are not lost.

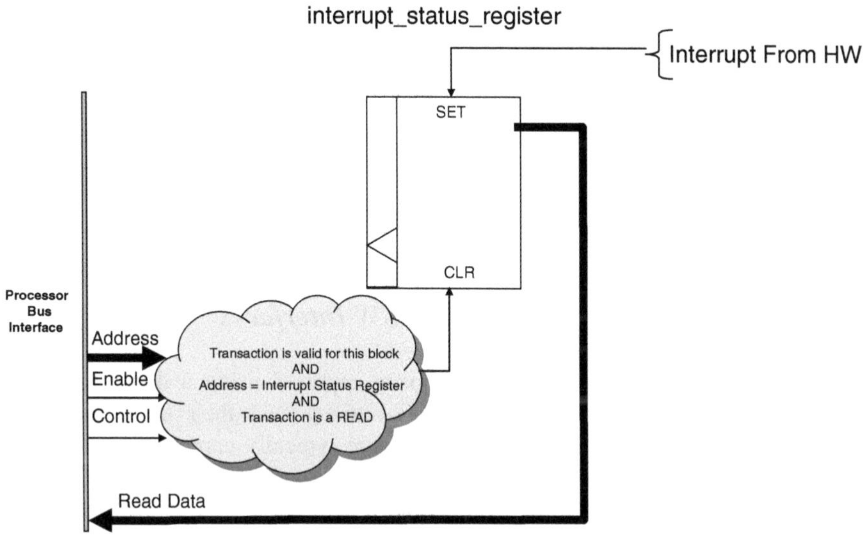

Fig. 2.11 Example RTL of Read-to-Clear on an interrupt status register

These are examples of different read and write behaviors that allow optimized and granular interaction with the hardware. There are a wide range of nuances that exist in the real HW/SW interface including

- W1C – Write 1 to clear
- W0C – Write 0 to clear
- R2C – Read to clear
- R2S – Read to set
- RRR – Read returns remote.

Some of these nuances can be considered best practices because the result may have better performance. However, they make the HW/SW interface implementation more complex for both domains. Also for hardware IP in systems, there are unlimited possibilities to what can actually be implemented and corner cases exist in both domains. For example,

- Registers that are mapped at multiple addresses
- Registers with one address for read and one for write
- Read-only registers that mimic a FIFO: every time they are read, a different value is returned
- Registers that change characteristics depending on the state of an internal signal or other register. This is also known as auto-shadow or modal
- Indirect memory access where a set of registers is used to mimic a transaction protocol that is used to access a bigger address space.

These are just a few examples of the huge variety of nuances at the HW/SW interface. From the software perspective, the developer needs to know everything about the register structure, how the bitfields are laid out, reset values, volatility, the access types, and any other nuances. With this they can develop a low-level hardware API. They will also need to know about the overall architecture of the system and what the main bottlenecks are to ensure that they get the best performance from their code.

2.4.2 HW/SW Design Flow for HW/SW Interfaces

The HW/SW interface has traditionally been captured using a document/paper specification. An IP block's registers and bitfields and their associated data, including offsets, resets, and access types, were typically captured in document tables. This document was then the source HW/SW interface reference for the design flow, including IP design/validation, system design/validation, software design/validation. Figure 2.12 shows a sequential HW/SW design flow, a waterfall type model, where software development is the last stage of product development, often only starting when the hardware is ready. A more parallelized flow will be presented in a later section.

In the following section, a simple design example is detailed and a few examples of some common HW/SW interface formats are presented. This is neither an exhaustive set of representative examples, nor is it reflective of the various styles of

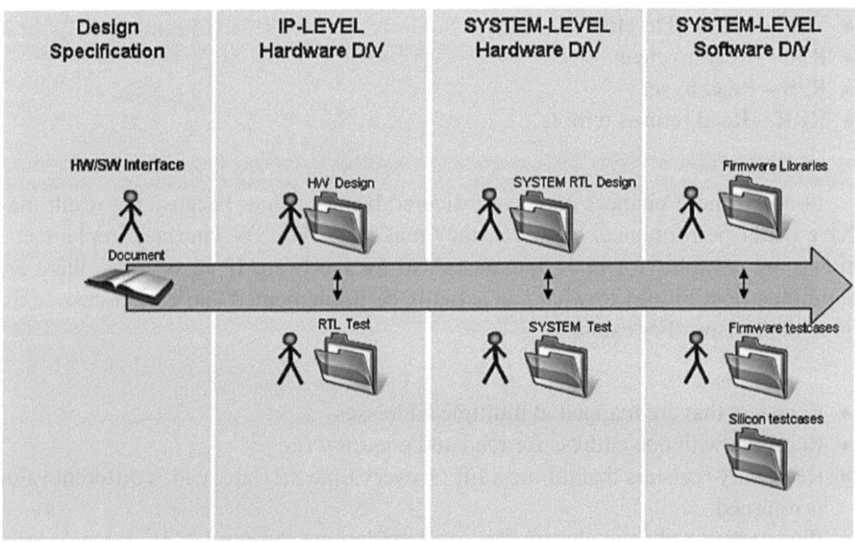

Fig. 2.12 Traditional HW/SW interface design flow

the format. Also the example is quite simple so not all register or bitfield characteristics are shown. The objective is to show a representative set of many common formats found in a system design flow.

2.4.2.1 Example IP Design – The Requirements

The following block is an example of a generic hardware block with a simple HW/SW interface (Figure 2.13).

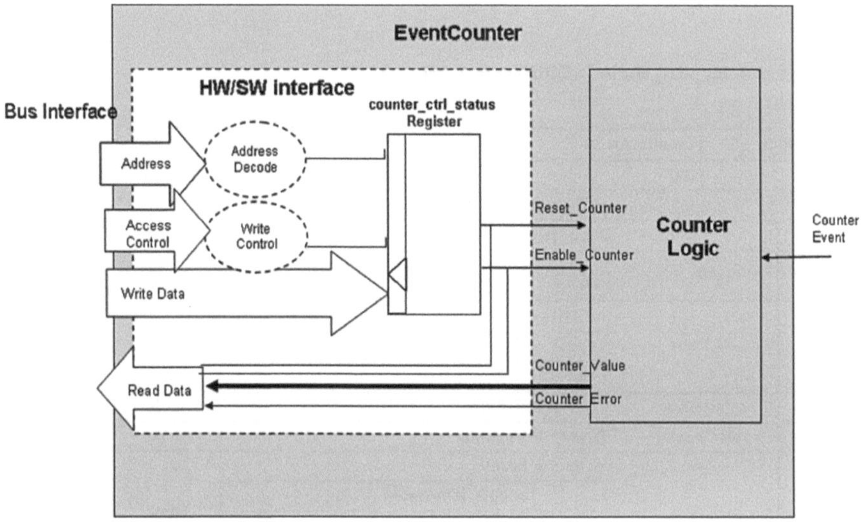

Fig. 2.13 Software running on a dedicated processor

This generic block counts events and can be used for applications such as counting the number of errors in a packet transmission or the number of interrupts that have occurred. The counter value is accessible via software and is 8 bits wide. The counter may need to be reset by the software at specific times and it can also be enabled and disabled. Also, if the counter does overflow, an error will be asserted by the counter logic via the Counter_Error flag which will be de-asserted when the counter is reset.

In total there are four SAHEs in this block

- SAHE 1 : Reset_Counter : Read/Write, single-bit
- SAHE 2 : Enable_Counter : Read/Write, single-bit
- SAHE 3 : Counter_Value : Read only : 8-bit
- SAHE 4 : Counter_Error : Read only : single-bit.

An IP designer will start to map these SAHEs into registers. From an IP design perspective, the sum total of these bits is 11 and if the data bus is 32 bits wide, they will easily fit into one register. The IP designer may capture these in a design specification document.

2.4.2.2 Specification – Documentation

The IP owner, or even the system architect, may capture the HW/SW requirements in a documentation format. Typical formats used for documentation are Microsoft Word, Microsoft Excel, Adobe Framemaker, Docbook [25], or XML. The following is an example of a register table that may be described in some form of IP specification (Figure 2.14).[1]

Register counter_ctrl_status1: COUNTER_CTRL_STATUS

Register Information			
Description	Counterstatus		
Offset	0x00	Type:	RW

Bitfield Details				
Field	Name	Description	Access	Default
31:11	Reserved	Not used	RO R returns 0s	0x000000
10	CounterError	Indicates if the counter has an error. Read 0: No error Read 1: FIFO Overflow	RO	0
9:2	CounterValue	Value of the counter	RO	0x00
1	EnableCounter	Enables the counter	RW	0
0	ResetCounter	resets the counter 0: counter it not reset 1: Reset the counter. This value should be cleared again in order for counter to start	RW	0

Fig. 2.14 Example register table

2.4.2.3 IP-XACT (SPIRIT)

IP-XACT from the Spirit consortium was introduced earlier in this chapter. As mentioned, it defines a schema and method for describing and handling IP components within a design environment. A subset of IP-XACT can be used to specify memory maps, register, and bitfield details. The following is a snippet of an IP-XACT description 1.2 for the EventCounter's counter_ctrl_status register.

```
...
...
<spirit:register>
    <spirit:name>counter_ctrl_status</spirit:name>
    <spirit:dim>1</spirit:dim>
```

[1] This is not the most optimal mapping of the register as the 8-bit counter value needs to be shifted every time it is read by the software. After finishing the chapter, consider what it would take to change this in a manual flow after it has been implemented in all the different views!

```
    <spirit:addressOffset>0</spirit:addressOffset>
    <spirit:size spirit:resolve="user">32</spirit:size>
    <spirit:access>read-write</spirit:access>
    <spirit:reset>
       <spirit:value>0</spirit:value>
       <spirit:mask>4294967295</spirit:mask>
    </spirit:reset>
    <spirit:field>
       <spirit:name>ResetCounter</spirit:name>
       <spirit:bitOffset>0</spirit:bitOffset>
       <spirit:bitWidth spirit:resolve="user">1</spirit:bitWidth>
       <spirit:access>read-write</spirit:access>
       <spirit:description>resets the counter</spirit:description>
       <spirit:values>
          <spirit:value>0</spirit:value>
          <spirit:description>counter it not reset</spirit:description>
          <spirit:name>No_effect</spirit:name>
       </spirit:values>
       <spirit:values>
        <spirit:value>1</spirit:value>
        <spirit:description>Reset the counter. This value...</spirit:description>
        <spirit:name>reset_counter</spirit:name>
       </spirit:values>
    </spirit:field>
    <spirit:field>
       <spirit:name>EnableCounter</spirit:name>
       <spirit:bitOffset>1</spirit:bitOffset>
       <spirit:bitWidth spirit:resolve="user">1</spirit:bitWidth>
       <spirit:access>read-write</spirit:access>
       <spirit:description>Enables the counter</spirit:description>
    </spirit:field>
    <spirit:field>
       <spirit:name>CounterValue</spirit:name>
       <spirit:bitOffset>2</spirit:bitOffset>
       <spirit:bitWidth spirit:resolve="user">8</spirit:bitWidth>
       <spirit:access>read-only</spirit:access>
       <spirit:description>Value of the counter</spirit:description>
    </spirit:field>
    <spirit:field>
       <spirit:name>CounterError</spirit:name>
       <spirit:bitOffset>10</spirit:bitOffset>
       <spirit:bitWidth spirit:resolve="user">1</spirit:bitWidth>
       <spirit:access>read-only</spirit:access>
       <spirit:description>Indicates if the counter has an error.
       </spirit:description>
       <spirit:values>
          <spirit:value>0</spirit:value>
          <spirit:description>No error</spirit:description>
          <spirit:name>NoError</spirit:name>
       </spirit:values>
       <spirit:values>
          <spirit:value>1</spirit:value>
          <spirit:description>FIFO Overflow </spirit:description>
          <spirit:name>Overflow</spirit:name>
       </spirit:values>
    </spirit:field>
    <spirit:field>
       <spirit:name>Reserved</spirit:name>
       <spirit:bitOffset>11</spirit:bitOffset>
       <spirit:bitWidth spirit:resolve="user">21</spirit:bitWidth>
       <spirit:access>read-only</spirit:access>
       <spirit:description>Not used</spirit:description>
    </spirit:field>
    <spirit:description>Counter status</spirit:description>
</spirit:register>
...
...
```

2.4.2.4 SystemRDL

In 2005 Denali released a tool called Blueprint that used a format called Register Description Language (RDL) for describing registers. At that time, it was stated that Denali did not create Blueprint, but that it came from an unnamed design house [26]. It became clear who it was when Cisco released an open-source version of RDL in March 2006 [27]. SystemRDL was then adopted by the SPIRIT and released as SystemRDL 1.0 in May 2009 [28]. The following is a snippet of a SystemRDL File.

```
reg { // REGISTER
      // ADDRESSOFFSET 0x0
      // ACCESS read-only
        name = "counter_ctrl_status";
        desc = "Counter status";
        regwidth = 32;
        default hw = rw; default sw = rw;

        field {
            name = "ResetCounter";
            desc = "resets the counter";
            hw = rw; sw = rw;
            fieldwidth = 1;
        } ResetCounter [0:0];

        field {
            name = "EnableCounter";
            desc = "Enables the counter";
            hw = rw; sw = rw;
            fieldwidth = 1;
        } EnableCounter [1:1];

        field {
            name = "CounterValue";
            desc = "Value of the counter";
            hw = rw; sw = r;
            fieldwidth = 8;
        } CounterValue [9:2];

        field {
            name = "CounterError";
            desc = "Indicates if the counter has an error";
            hw = rw; sw = rw;
            fieldwidth = 1;
        } counterError [10:10];

        field {
            name = "Reserved";
            desc = "Value of the counter";
            hw = r; sw = r;
            fieldwidth = 8;
        } Reserved [31:11];
    } counter_ctrl_status @0x0; // END REGISTER
```

2.4.2.5 IP Hardware Design

As described previously, an RTL model of the HW/SW interface requires logic that translates the bus protocol into read and write access to specific registers or memories. This logic typically includes

- Address decoding to enable write and read accesses
- A model of the register.

There are numerous ways and styles to implement register behavior in RTL. Once such way is D–Q modeling which stores the input and output of registers as separate signals but on a specific clock edge the output of the register (Q) is set to the input value (D).

The following is a sample of the VHDL code:

```
--
-- Registers process
-- Registers clocked on rising edge of PCLK
--
pclk_rise : process (pclk, rst_n)
begin
  if (rst_n /= '1') then
    q_counter_ctrl_status_countererror  <= '0';
    q_counter_ctrl_status_countervalue  <= conv_std_logic_vector(16#0#,8);
    q_counter_ctrl_status_enablecounter <= '0';
    q_counter_ctrl_status_resetcounter  <= '0';
  elsif pclk'event and (pclk = '1') then
    q_counter_ctrl_status_countererror  <= d_counter_ctrl_status_countererror;
    q_counter_ctrl_status_countervalue  <= d_counter_ctrl_status_countervalue;
    q_counter_ctrl_status_enablecounter <= d_counter_ctrl_status_enablecounter;
    q_counter_ctrl_status_resetcounter  <= d_counter_ctrl_status_resetcounter;
  end if;
end process pclk_rise;
```

Note in this example that the output of the register bits are set to 0 during reset. Also we see here that the register output is always assigned to the input value. Below is a sample of the code in Verilog:

```
//
// Registers process
// Registers clocked on rising edge of PCLK
//
always @ (posedge pclk or negedge rst_n)
begin
  if (rst_n == 0) begin
//
// REGISTER: counter_ctrl_status
//
    q_counter_ctrl_status_CounterError  = 'b0 ;
    q_counter_ctrl_status_CounterValue  = 'h0 ;
    q_counter_ctrl_status_EnableCounter = 'b0 ;
    q_counter_ctrl_status_ResetCounter  = 'b0 ;
  end
  else begin
//
// REGISTER: counter_ctrl_status
//
    q_counter_ctrl_status_CounterError  = d_counter_ctrl_status_CounterError;
    q_counter_ctrl_status_CounterValue  = d_counter_ctrl_status_CounterValue;
    q_counter_ctrl_status_EnableCounter = d_counter_ctrl_status_EnableCounter;
    q_counter_ctrl_status_ResetCounter  = d_counter_ctrl_status_ResetCounter;
  end
end
```

The address decode is used to set up the correct conditions for a read/write access.

```
--
-- Decode APB register accesses
--
apb_decode : process( penable, psel, pwdata, pwrite, paddr
     q_counter_ctrl_status_enablecounter, q_counter_ctrl_status_resetcounter)
```

```
begin
  -- These are the default assignments if the decode doesn't
  enable a write.
  d_counter_ctrl_status_enablecounter <= q_counter_ctrl_status_enablecounter;
  d_counter_ctrl_status_resetcounter  <= q_counter_ctrl_status_resetcounter;

  prdata                              <= (others => '0');
  --
  -- Access decode
  if (psel = '1') and (penable = '1') then
    case paddr is
      when counter_ctrl_status_REG =>
        if (pwrite = '0') then
          prdata(10)                  <= counter_ctrl_status_countererror;
          prdata(9 downto 2)          <= counter_ctrl_status_countervalue;
          prdata(1)                   <= q_counter_ctrl_status_enablecounter;
          prdata(0)                   <= q_counter_ctrl_status_resetcounter;
        else
          d_counter_ctrl_status_enablecounter <= pwdata(1);
          d_counter_ctrl_status_resetcounter  <= pwdata(0);
        end if;
      when others =>
    end case;
  end if;
end process apb_decode;
```

In this code the default input to the enablecounter and resetcounter bitfields are the values stored in the registered values ("q_"). When the address (paddr) is 0 and the IP is selected (psel="1") and enabled (penable="1"), then if it is a read access (pwrite=0), the read data (prdata) is set as defined. Otherwise for a write access (pwrite=1), the inputs to the counter are set to the write data (pwdata). Again, the same behavior in Verilog is as follows:

```
//
// Decode APB register accesses
//
always @( penable or psel or pwdata or pwrite or paddr
          or counter_ctrl_status_CounterError
          or counter_ctrl_status_CounterValue
          or counter_ctrl_status_EnableCounter
          or counter_ctrl_status_ResetCounter
)

begin
  //
  // Default assignments
  //
  d_counter_ctrl_status_CounterError  = counter_ctrl_status_CounterError;
  d_counter_ctrl_status_CounterValue  = counter_ctrl_status_CounterValue;
  d_counter_ctrl_status_EnableCounter = q_counter_ctrl_status_EnableCounter;
  d_counter_ctrl_status_ResetCounter  = q_counter_ctrl_status_ResetCounter;

  w_prdata <= 0;

  //
  // Access decode
  //
  if ((psel == 1) & (penable == 1)) begin
    case (paddr)

      //
      // REGISTER: counter_ctrl_status
      //
```

```
            counter_ctrl_status_REG:

            if (pwrite == 0) begin
              w_prdata[10] = q_counter_ctrl_status_CounterError;
              w_prdata[9:2] = q_counter_ctrl_status_CounterValue;
              w_prdata[1] = q_counter_ctrl_status_EnableCounter;
              w_prdata[0] = q_counter_ctrl_status_ResetCounter;
            end
            else begin
              d_counter_ctrl_status_EnableCounter = pwdata[1];
              d_counter_ctrl_status_ResetCounter = pwdata[0];
            end
        endcase
      end
    end
```

This RTL behavior can also be modeled in SystemC.

2.4.2.6 IP Verification

From an IP (block level) point of view, HW/SW interface verification is typically done using simulation-based approaches, where transaction accesses are modeled using low-level signaling protocols. At this level, one of the main verification objectives is to exhaustively test the write and read accesses to the IP block. This may typically involve

- Testing the address decode to ensure that not only are all the registers at the correct address but that accessing invalid addresses has no effect
- Testing the reset value of the registers
- Testing that all valid bits of a register can be accessed. This may include writing patterns to every write-accessible register bit to ensure that the bit is writable in an independent fashion
- Testing that all side effects behave as specified.

While the key IP verification objective is to ensure that the hardware interface is fully accessible and behaves in a correct manner when accessed by a specific type of transaction, verification from the software perspective may not even be considered. The IP verification owner may not know (or care) if there is a host processor or embedded processor, what OS is running or the type of bus protocol used in the system. The objective is to ensure that the given protocol at the IP interface can access the registers and bitfields in the correct manner and the overall behavior of the IP is correct.

This level of verification will typically require a high degree of control over the testbench and DUT including emulating the full bus transaction protocol (e.g., AHB). This means being able to force certain signaling behavior, e.g., read/write accesses and other conditions such as interrupts, clocks, and reset. There is also a corresponding need to be able to directly observe the effects of the register access by monitoring signals within the RTL code. This is commonly known as backdoor access [29]. The verification environments that permit these include standard HDLs like VHDL or Verilog or hardware verification languages (HVLs). HVLs typically allow a better level of control and observability of the verification environment than pure HDL methods.

2.4.2.7 HDL Verification Environments

HDL-based verification environments were the traditional IP testbenches. They typically had a bus functional model (BFM) to emulate the processor transactions at the IP interface that through procedures could be used to perform read and write accesses. The testbench could be scripted to do reset, read/write tests etc. The main drawback with these types of testbenches is that they are not verification focused and the users are restricted to extremely low-level capabilities.

2.4.2.8 HVL Environments

HVLs on the other hand have powerful verification capabilities at block and chip levels. They typically have characteristics of high-level programming object-oriented languages such as C++ or Java while maintaining the HDL-like features at the bit level. SystemVerilog, OpenVera, *e*, and SystemC are the most commonly used HVLs. The OO characteristics allow great programming flexibility prompting the EDA industry to promote and leverage best-practice methodologies.

A high-level verification methodology consists of an HVL running in a simulation environment which can interact with the RTL. The HVL's object-oriented infrastructure can be organized and extended to deliver high levels of productivity for specific user applications. The key mechanisms are stimulus generation, monitoring, and coverage [30, p. 319] (Figure 2.15).

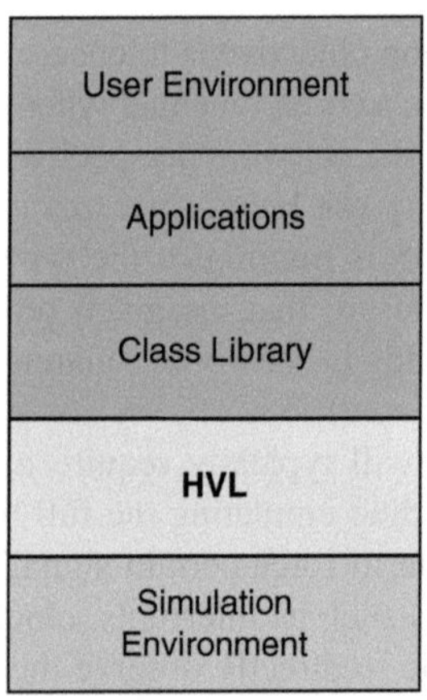

Fig. 2.15 High-level verification environment

In 2002 Verisity (since acquired by Cadence) was the front runner of HVL methodologies when it released the *e* Reuse Methodology (eRM) [31]. Synopsys followed later with the Verification Methodology Manual (VMM). Mentor released the Advanced Verification Methodology (AVM) in 2006 while Cadence released the Universal Reuse Methodology (URM) in 2006 and their Incisive® Plan-to-Closure Methodology (IPCM) in 2007. In 2007, Cadence and Mentor consolidated their verification methodologies into the Open Verification Methodology (OVM).

Currently the main runners in HVL methodology are VMM and OVM based with SystemVerilog implementations and *e*RM, with its' *e* implementation. Cadence also has a version of OVM with e support what is interesting from the perspective of the HW/SW topic is that all of these methodologies have produced applications and solutions that address the HW/SW interface and as such they have infrastructures that model registers.

2.4.2.9 VMM – Verification Methodology Manual

The VMM from Synopsys was released in 2005 and is currently based on the IEEE 1800 SystemVerilog standard. It offers a methodology to develop powerful transaction-level, constrained-random verification environments and includes a library of classes and applications to provide higher level functions for improved productivity. One such application is the Register Abstraction Layer (RAL) which defines blocks, registers, bitfields, addresses, etc., in a format called RALF [32]. The RALF can be used to create a VMM infrastructure.

```
block EventMonitor {
  bytes 4;
    register counter_ctrl_status {
      bytes 4;
        field ResetCounter @0 {
        bits 1;
        access rw;
        reset 1'b0;
      }
        field EnableCounter @1 {
        bits 1;
        access rw;
        reset 1'b0;
      }
        field CounterValue @2 {
        bits 8;

        access ro;
        reset 8'b00000000;
      }
        field CounterError @10 {
        bits 1;

        access ro;
        reset 1'b0;
      }
      field reserved @11 {
        bits 21;
      }
    }
}
```

2.4.2.10 OVM – Open Verification Methodology

OVM is also based on the IEEE 1800 SystemVerilog standard and offers a methodology to create block and system-level verification environments. Released initially in January 2008, OVM is jointly developed by Cadence and Mentor Graphics and architected for multiple languages to promote reuse and interoperability using standard libraries based on a proven methodology. OVM is aiming to support HVLs other than SystemVerilog, including SystemC and e.

In OVM, registers are mapped in a register definition file. These models can then be used by macros in the verification environment, which can be used by some common HW/SW interface tests – for example, a hardware reset test. The user has the capability to add additional test cases in a highly productive environment.

```
package EventMonitor_pkg;

  import ovm_pkg::*;
  import ovm_register_pkg::*;
//
// Register type definitions
//
  typedef struct packed {
       bit              ResetCounter;
       bit              EnableCounter;
       bit [7:0]        CounterValue;
       bit              CounterError;
       bit [20:0]       Reserved;
  } counter_ctrl_status_t;

//
// Register object declarations
//

 class EventMonitor_counter_ctrl_status extends ovm_register #(counter_ctrl_status_t);
    function new (string name, ovm_component p, counter_ctrl_status_t val);
      super.new (name, p, val);
      WMASK = 'b00000000000000000000000000000011;
    endfunction
 endclass

//
// Registerfile declaration
//
class EventMonitor_register_file extends ovm_register_file;

  // Register class declarations

    EventMonitor_counter_ctrl_status    EventMonitor_counter_ctrl_status_reg;

    function new (string name, ovm_component p);
      super.new (name, p);
    endfunction

    function void build ();
      ovm_report_message ("EventMonitor_register_file", "build ()");
      // Register construction
      EventMonitor_counter_ctrl_status    = new ("counter_ctrl_status", this,
                                             {1'b0, 1'b0, 8'b00000000, 1'b0,
                                             21'b000000000000000000000} );
      // Add registers to the register file
      add_register (
        EventMonitor_counter_ctrl_status_reg.get_fullname (),
        'h0, EventMonitor_counter_ctrl_status_reg);
    endfunction
  endclass
endpackage
```

2.4.2.11 eRM – Specman "e" Reuse Methodology

As part of its eRM, Verisity promoted a methodology called VR_AD based on the Specman e language which is used to define register and memory-mapping information.

```
//===========================================================
// counter_ctrl_status register//
//-==========================================================

reg_def EventMonitor__counter_ctrl_status EventMonitor 0x0 {
     reg_fld Reserved       : uint(bits : 21)   : R    : 0x000000    : cov;
     reg_fld CounterError   : uint(bits : 1)    : R    : 0x0         : cov;
     reg_fld CounterValue   : uint(bits : 8)    : R    : 0x00        : cov;
     reg_fld EnableCounter  : uint(bits : 1)    : RW   : 0x0         : cov;
     reg_fld ResetCounter   : uint(bits : 1)    : RW   : 0x0         : cov;
};
```

2.4.2.12 OVM vs. VMM Interoperability

Each methodology claims that they are fully interoperable with any SystemVerilog simulator but it still remains to be seen how truly interoperable OVM and VMM will be. Will we see a standardization of verification middleware in SystemVerilog? So far interoperability between the vendors and methodologies is only rumored, but hopefully the gaps between the various verification methodologies and concepts will be closed. Accellera has recently release a document that shows how these two methodologies can be used together.

Also, the evolution of these methodologies promises to be interesting as the support for other languages such as SystemC or e comes on board as well as the maturing of the various applications. For now though, the focus on the HW/SW interface verification could not have come at a better time.

2.4.2.13 Chip-Level Verification

Within a full chip-level view, the IP now has a system-level context and sits somewhere in a memory map (or maps for multi-processor systems) and is directly accessible by a processor.

The following are some typical errors in the HW/SW interface [23]:

1. Inconsistencies between a register specification and its corresponding implementation including bitfield layout, register accesses, and reset values
2. Address mirroring of registers because internal address buses are defined to be too small
3. Incorrect address decoding
4. Incorrect memory map specification causing overlaps.

As described earlier in the hardware domain there can be a wide range of address, data, and control transformations. This leaves the possibility of implementation errors through the incorrect hook-up of address, data, or control signals or even incorrect logic or logic connectivity across the transaction path. An example of this could be a power isolation block not working properly leading to a whole subsystem being inaccessible.

Errors such as those in item 1 above show that there may be mismatches between a specification and an implementation. Even a correct match, at block-level, between a register specification and its implementation does not guarantee the same match at chip level. The block-level triad of information (specification, design, and verification) needs to be cross-referenced with the chip-level triad and if not automatically generated from a single source can be the source of mismatches. The block level may be correct and coherent but which correct and coherent version is being used at the chip level?

Because of the potential for errors and mismatches, the low-level HW/SW interface therefore still needs a strong focus, even at chip level. While the testing required may not be as extensive, exhaustive, or even feasible as block-level validation, there still needs to be a substantial effort in ensuring chip-level HW/SW interface coherency. Some sample questions to answer could be

- Can I access all of the IP blocks in the memory map?
- Is the address decoding correct – Can I access all of the registers?
- Are the register reset values as defined?
- Is the data coming back in the correct format?

Both IP-level and system-level verification should ensure that the HW/SW interface for each IP block is fully tested and that the processor has proven access to the specified address space.

At chip-level the HW/SW access can be tested between the origin of the transaction (the processor interface) and the transaction target, e.g., a particular register. Chip-level testbenches, modeling environments, and test-case format can be very different. For instance,

1. Test cases can be written using a HVL (described in the previous section) to emulate the processor transactions by focusing on read/write transactions
2. The test cases may also be written in C and compiled as binaries which are then run on a model of the embedded processor. The processor model may be implemented at different levels of abstraction, e.g., an instruction-accurate instruction set simulator (ISS) model, cycle-accurate ISS model, or RTL model. The modeling level is dependent on the simulation goals – simulation speed vs. simulation accuracy.

In order to achieve an acceptable level of coverage in the HW/SW interface, it is important that registers are modeled from the perspective of the processor. The examples given in the HVL section can be reused except that the address of the registers from the chip-level memory map (also known as the physical address) would need to be used. In the case of item 2 above, the registers would need to be modeled in C. This C code does not need to be considered firmware as the objective is focused on HW/SW interface testing rather than developing optimized run-time code. An example of a C structure that can be used by a generic system register test is as follows:

```
const registerType event_monitor_counter_ctrl_status = {
             0x0,                  /* Address */
             0xfffffffu,           /* ResetDefined */
             0x0,                  /* ResetValue */
             0x7ffu,               /* ReadableBits */
             0x3ffu,               /* WriteableBits */
             0x0,                  /* ReservedBits */
             32,                   /* Size */
             0x2u,                 /* SupportedAccessWidths */
             0xfffffffu,           /* resetTestBits */
             0x3ffu                /* RWTestBits */
};
```

Structures like these can be used as the input to generic access test sequences – for example, checking of reset values and checking of writeable and read-only bits. This level of exhaustive checking can be costly in terms of simulation time. One of the main problems with this type of verification is that this type of simulation-based methodology will yield poor simulation times in an RTL simulator environment. Simulation time can go from days to weeks and needs to be repeated for every change [33]. There are various emerging solutions to this problem that range from running the chip on a prototyping platform to providing a high-speed virtual prototype. These will be detailed in a later section.

Another interesting approach is with formal verification which aims to prove functional correctness through formal methods, requiring no simulation. [33] shows how the HW/SW interface of a chip with 3,217 registers and 12,600 bitfields can be formally verified using SystemVerilog and requires 70% less effort. This methodology defines SystemVerilog property checkers (write and read) which can be referenced for each bitfield register as follows:

```
property eventMonitor_test;
eventMonitor_reset_counter : assert property (
   write('h0,"i_EventMonitor/hw_sw_interface/reset_counter",0,0)
);
eventMonitor_enable_counter : assert property (
   write('h0,"i_EventMonitor/hw_sw_interface/enable_counter",1,1)
);
eventMonitor_reset_counter : assert property (
   read('h0,"i_EventMonitor/hw_sw_interface/reset_counter",0,0)
);
eventMonitor_enable_counter : assert property (
   read('h0,"i_EventMonitor/hw_sw_interface/enable_counter",1,1)
);
endproperty
```

The formal methods have the benefit that they can prove correctness whereas simulation is non-exhaustive.

2.4.2.14 Software Development – Firmware

As stated previously, the lowest level SW/HW interaction needs efficient access to the SAHEs and can do so by means of an API. This code layer contains the base-level accesses and essentially they are about reading and writing to a memory map. These accesses, however, can be coded an infinite number of ways [34] with varying degrees of portability, readability, and usability as well as performance and memory characteristics. For instance, if the EventMonitor block appears at 0xfffe0000 in a memory map the COUNTER_CTRL_STATUS register can be written as follows:

```
*((volatile uint32_t *)0xffee0000) = 8;
```

This is considered poor coding style because any incremental changes in the system would require a compete rewrite. The use of macros can decouple the static (but potentially changing) constants from the implementation code. For instance, the same code using macros could be

```
#define EVENTMONITOR_COUNTER_CTRL_STATUS 0xffee0000
*(volatile uint32_t *) EVENTMONITOR_COUNTER_CTRL_STATUS = 8;
```

Even including the cast inside the macro itself leaves the code more readable.

```
#define EVENTMONITOR_COUNTER_CTRL_STATUS ((volatile uint32_t*) 0xffee0000)
*EVENTMONITOR_COUNTER_CTRL_STATUS = 8;
```

While macros have no overhead in terms of code speed they are an extremely low-level mechanism and can provide a non-intuitive programming interface. Creating pointer variables to point at each register, however, will add performance and additional memory impacts. Using structs can provide a more logical view of a device's registers as follows:

```
/************************************************************
 *  TYPE            EventMonitor Registers
 *  DESCRIPTION     Structure describing registers in a component
 ************************************************************/
struct eventmonitor_reg_t
{
    volatile uint32_t COUNTER_CTRL_STATUS;
    ... etc...
} *eventmonitor_reg_ptr;
eventmonitor_reg_ptr->COUNTER_CTRL_STATUS = 8 ;
```

There is a drawback as this code may not be portable; neither C nor C++ standards specify how the contents of a struct are laid out in memory and this can be compiler dependent [34]. While the struct is guaranteed to be laid out in memory with the fields in the order of the declaration, it is possible for compilers to insert padding between the structure subelements and therefore the exact address used for a specific register may be incorrect. The Register model could also be described in a variety of different C++ formats [34, 35] and while these have the benefits of OO programming, they may have a lot of overhead.

Functions are often used to encapsulate the low-level accesses and when used with macros can be a good general method. Consider a write to a register bitfield. Programmers may need to use "read–modify–write operations" or store local copies of register values if implementation speed is critical and registers are non-volatile. The following is a typical function to write to a specific bitfield (ENABLE_COUNTER) within a register (COUNTER_CTRL_STATUS) of a specific component (EVENTMONITOR). Depending on the granularity of control required, there may be a function like this for every bitfield in the component.

```
#define EVENTMONITOR_COUNTER_CTRL_STATUS_ENABLECOUNTER_MASK    (U32T)0x00000002u
#define EVENTMONITOR_COUNTER_CTRL_STATUS_ENABLECOUNTER_OFFSET 1
extern void EventMonitor_WriteCounter_ctrl_status_EnableCounter  (const U32T baseAddress,
const U32T value)
{
    const U32T offset = baseAddress + EVENTMONITOR_COUNTER_CTRL_STATUS_OFFSET;
    register U32T data = RD_MEM_32_VOLATILE(offset);
    register U32T newValue = value;
    data &= ~(EVENTMONITOR_COUNTER_CTRL_STATUS_ENABLECOUNTER_MASK);
    newValue <<= EVENTMONITOR_COUNTER_CTRL_STATUS_ENABLECOUNTER_OFFSET;
    newValue &= EVENTMONITOR_COUNTER_CTRL_STATUS_ENABLECOUNTER_MASK;
    newValue |= data;
    WR_MEM_32_VOLATILE(offset, newValue);
}
```

The bitfield write function performs a classic read–modify–write operation in order to change the values of a specific bitfield within the complete register. It expects two values to be supplied when it is called: the base address of the component instance and the value to be written. Macros for register offset addresses (e.g., EVENTMONITOR_COUNTER_CTRL_STATUS_OFFSET) and bitfield positions are typically stored in a header file allowing a degree of stability of the code. The include file may also contain bitfield mask values indicating which bits of the register are associated with the bitfield that the function is accessing. This bitfield read–modify–write function then operates as follows:

- READ
  - The absolute offset address (offset) for the register is constructed from the local register offset address and the component instance base address
  - The complete register value is read via the offset and its contents are stored in data so that it can be manipulated. A copy of the value to be written is also made (newValue) so that it can be later shifted into the correct bit position.
- MODIFY
  - The appropriate bits of data value are cleared by ANDing with the inverted bitfield mask value

  - newValue is shifted to align it with the position of the bitfield in the register. Other bits within the new value which are not associated with the bitfield that the function is addressing are cleared by ANDing with the bitfield mask
  - With the appropriate bits in the existing register now cleared and the new value shifted into the correct bit position and all other bits cleared – the existing data can be ORed with the new value.

- WRITE

  - Finally the new register value can be written back to the component via the absolute offset address, thus finishing the read–modify–write transaction.

Parts of the hardware address space may be dynamically changed by the hardware and considered volatile from the software perspective, while other parts may be considered non-volatile – that is, relatively "unchangeable" by the hardware. While some guidelines recommend that all registers are declared volatile [36], for performance reasons it may be useful to store non-volatile information in memory to reduce the dependency on hardware transaction access.

Given the low-level of granularity required in writing to bitfields, it is essential that the software API and structures align perfectly with the hardware view otherwise a read–modify–write to a misaligned bitfield may seem to be working correctly from the software access but corrupt the entire register on the hardware side.

2.4.2.15 Firmware Verification

It is cheaper and easier to prevent a bug from creeping into code than it is to find and kill it after it has entered [37]. Therefore a strongly cast low-level of firmware will allow the compiler to stop certain bugs getting into the code. However, a substantial part of firmware development involves writing test cases to ensure that the SW/HW behaves according to its specification. For instance, a test may involve ensuring that a bitfield value can be set to all of its valid values, or that a pattern written into the register map and read back can be validated. These test cases can be developed in parallel to the hardware-oriented test cases to give the developer confidence that the hardware model is aligned to the specification and robust enough to run software against. The models themselves can be very similar to the coding for firmware design but because the target is test code there are fewer constraints than for real firmware code and thus many different approaches can be used.

Software can be developed and tested against a model of the hardware. The hardware model can be of different formats. It can range from an RTL format which is a very accurate low-level hardware model to a software model of the hardware, also called virtual models. Depending on the software development criteria there are a range of different models with different strengths and weaknesses.

2.4.2.16 RTL Models

The RTL, or in some cases a synthesizable ESL model, can be executed on a number of different environments including simulators, emulators or be synthesized into an FPGA. While RTL is very accurate, it has two main disadvantages:

- If the verification mechanism is simulation based, it is slow to simulate significant amounts of software against
- It can take a significant amount of schedule time to develop hardware to the point where the software can be integrated in an efficient manner and thus it may appear too late in the design flow to achieve a fully parallelized HW/SW codesign flow.

Synthesizing the hardware to an FPGA-based environment allows the model to be simulated very quickly and provide a rapid prototyping environment for software development. While the speed factor has been solved, the FPGA-prototyping environment needs robust hardware models. In addition, one of its main drawbacks is that it does not have the observability associated with a traditional simulation environment.

2.4.2.17 Virtual Models

Virtual models are high-level software models of the device hardware, which include models of the hardware system as well as the environment within which it exists. They are designed to allow software developers to develop real software without any dependency on the hardware design flow. The levels of abstraction as defined by OSCI includes the algorithmic view (AV), the programmers view (PV), the programmers view with timing (PVT) all of which are at the transaction level (TL) [38]. Within these views, memory-mapped registers may be modeled so that the virtual models remain "register accurate" and thus again we have another potential register modeling activity. While standards have emerged to capture the interface and transactional communication, e.g., TLM 2.0, the register view has yet to be standardized. Greensocs is promoting a SystemC infrastructure and format for defining registers called GreenReg [39].

2.4.2.18 Earlier Software Development

The mandate of developing and delivering software earlier in the chip design cycle has led to different types of emerging solutions. As shown, the main value metrics of these solutions are speed of execution, availability of hardware model, accuracy of hardware model, and observability of the hardware model.

Virtual models are register-accurate software models of the device hardware, and may include models of the hardware system as well as the environment within which it exists. This is also known as a virtual platform or a Virtual System Prototype (VSP) [30]. These systems allow users to develop real software without any tight dependency on the hardware implementation flow. Depending on the maturity of the virtual platform it might be possible to begin software development before the hardware design is even started. As the models are high-level software executable abstractions of the hardware, they can act like the hardware in real time. It is also possible to go beyond the hardware boundaries and integrate components such as cameras, LCDs, keyboards, or any peripheral devices to allow intuitive application development. The high-level of abstraction, however, has a cost – accuracy. The virtual model of the hardware may mismatch with the real model and the timing inaccuracy of the virtual prototype may have unforeseen side effects on the software running on the real system. Companies that produce environments on which to run virtual platforms are Synopsys, CoWare, Mentor Graphics, VaST, Imperas, and Virtutech. Several of these will be discussed in detail in later chapters of this book.

Hardware prototyping is another solution that is used to address the issues of early software development. The chip-level RTL or synthesizable ESL can be synthesized for an FPGA-based system and can be run at a significant fraction of the speed of the real device. This allows for early software integration and testing. It may not be as early to integrate as the virtual platform; however, the accuracy (functional and timing) is its key value. Confirma from Synopsys/Synplicity is an example of a hardware-prototyping platform [40] as well as Zebu-XXL system from EVE [41] which provides an FPGA-based prototyping system with debugging features. Of course custom FPGA-prototyping systems are used by companies who have their own specific requirements.

The emergence of FPGA prototyping has an interesting effect on front-end hardware design flows. The demand of early software requires a concept of a two-phase SoC integration flow. The first phase of this integration flow is a release that allows software integration. Functionality that is not associated with the processor may be left to a later design iteration. Activities such as test insertion or power insertion or IP not associated with the software may be omitted. The full hierarchical infrastructure of a SoC, however, needs to be put in place in order for back-end design flows to execute through the design flow. The second phase of integration can be delayed as the hierarchy now becomes a bottleneck to integration closure.

2.4.3 Emerging HW/SW Interface Tools and Design Flows

As described previously there are new models of the HW/SW interface emerging continuously. Verification methodologies such as OVM and VMM allow users to leverage the benefits of an object-oriented verification environment. Virtual modeling allows earlier SW development, even though there are different formats of models from different vendors. As design flows evolve, newer methodologies promise better efficiencies while legacy data and methodologies threaten to stagnate.

The HW/SW interface has traditionally been captured using some form of a document specification. Figure 2.16 shows how, in a document-based system, all HW/SW interface-related teams need to interpret the specification and translate it to their native format. It may be possible for them to give feedback

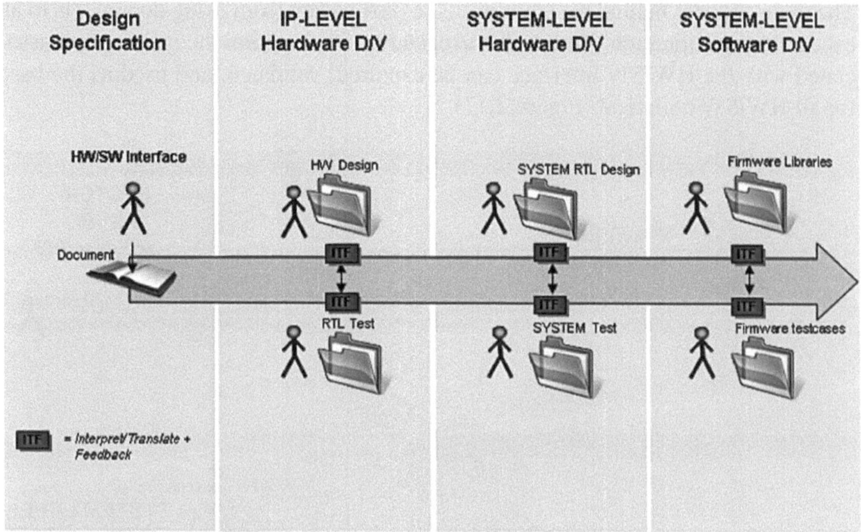

Fig. 2.16 Interpret/Translate/Feedback in a document-based system

where a change is required in the document. This is shown in the figure as an Interpret–Translate–Feedback (ITF) step.

If the document is open to interpretation the wrong information may get translated and cause a mismatch somewhere in the design flow. A more disturbing scenario, however, is the fact that the central specification, the document, requires translation into the different formats, making the effect of a design change non deterministic across the different views. For example, assume that the IP block designer needs to increase the width of an existing bitfield – change the width of the event counter from 8 to 9. They can do this easily in their RTL design, making the edits to a few lines of code, updating the design, and sending an email to the specification owner. It may be a few days before the document is updated. The change may have been broadcast to several teams via email but some people might miss it and some people might change it. Meanwhile RTL code is being validated and software written against a specification that may be out of date. Another disadvantage to this process is that the translation may be done incorrectly – for example, making the offset 9 instead of the width. A document-based process will introduce interpretation bugs, synchronization bugs, and translation bugs at various stages in the design flow. Some may be caught by verification activities early on in the flow, some will be found in the lab when the device comes back and some may even appear in the field.

In a highly parallelized design flow synchronization bugs become more widespread as all teams may be referencing the same register concurrently.

2.4.3.1 Register Management Tools

There are new tools emerging in the market that address this problem through automated, correct-by-construction methodologies dedicated to the HW/SW interface.

The main concept behind the solutions is to move away from using documents to an executable specification. Within this executable specification the information associated with the HW/SW interface can be captured, validated, and used as the base for all HW/SW collateral (Figure 2.17).

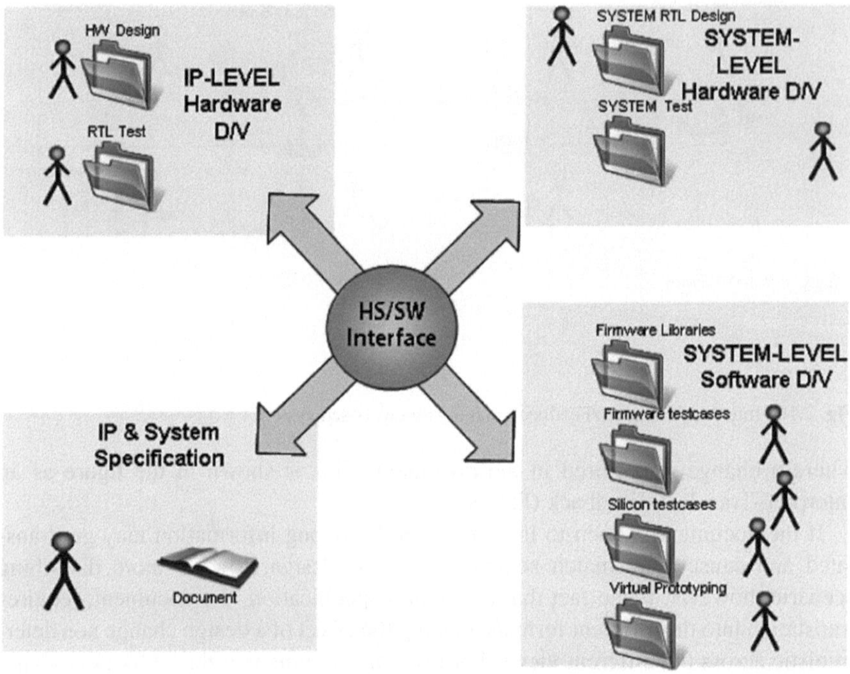

Fig. 2.17 Centralized HW/SW interface management

The main features of these tools are

- A user interface enabling efficient capture
- Usable from all perspectives – HW/SW/design/verification/documentation
- An extensive underlying model of the HW/SW interface
- Extensive coherency checking at all levels
- IP/sub-system and system capture and manipulation
- Interfaces to standards like IP-XACT
- Full regression capability to allow overnight builds of the HW/SW interface
- Flexible generators to generate a wide range of HW/SW design verification collateral including.
  - Documentation – programming reference guide
  - RTL code for modeling registers
  - HVL code for IP and chip-level testbenches.
  - SystemC for virtual prototyping
  - Firmware header files.

The main benefits of these solutions are

- Reduced front-end effort since all of the HW/SW collateral can be automated from a single executable specification and thus incremental changes do not require the effort that manual methods do
- Reduced verification effort because the HW/SW views are kept consistent, coherent, and synchronized. This correct-by-construction methodology increases quality and removes the need to find, identify, fix, update, and close HW/SW bugs
- Turn-around time between an update to a register bitfield of an IP that is used in a large system to fully updated design views is a matter of minutes and with the corresponding reduction in effort the overall result is that the hardware and software can be integrated earlier and easier
- With the advent of virtual platforms it is also possible to ensure alignment between a virtual hardware model, the hardware RTL model, and the real chip, therefore enabling streamlined software integration from the earliest software models to the final product binaries.

Before presenting EDA industry solutions it should be noted that the solution of formalizing and automating register descriptions is not a new concept. Many solutions exist within semiconductor and system companies as they often realized the problem a long time ago. Some solutions use machine readable documentation, e.g., tagged tables in documentation, spreadsheets, databases, or XML to formalize data and some form of scripting to both check the data coherency and generate specific outputs. The internal solutions, however, can become unwieldy and require a lot of investment to move to the next size devices and the next generation technologies. IP-XACT, from the SPIRIT consortium has emerged as the main standard for specifying registers and memory descriptions.

A short list of dedicated register management solutions includes

- Bitwise from Duolog Technologies [42]
- CSRCompiler from Semifore [43]
- Spectrareg from PDTI [44].

Beach Solutions were the initial creators of dedicated register management tools called the EASI tool suite. They were acquired by Duolog in January of 2009. There are other non-dedicated register management solutions in the industry that rely on IP-XACT editing – for example, as discussed, Platform Express from Mentor Graphics [45].

2.4.3.2 Case Study of a Register Management Solution: Bitwise

Duolog technologies provides a solution called Bitwise which manages the HW/SW interface from IP level to chip level across a range of disciplines including HW

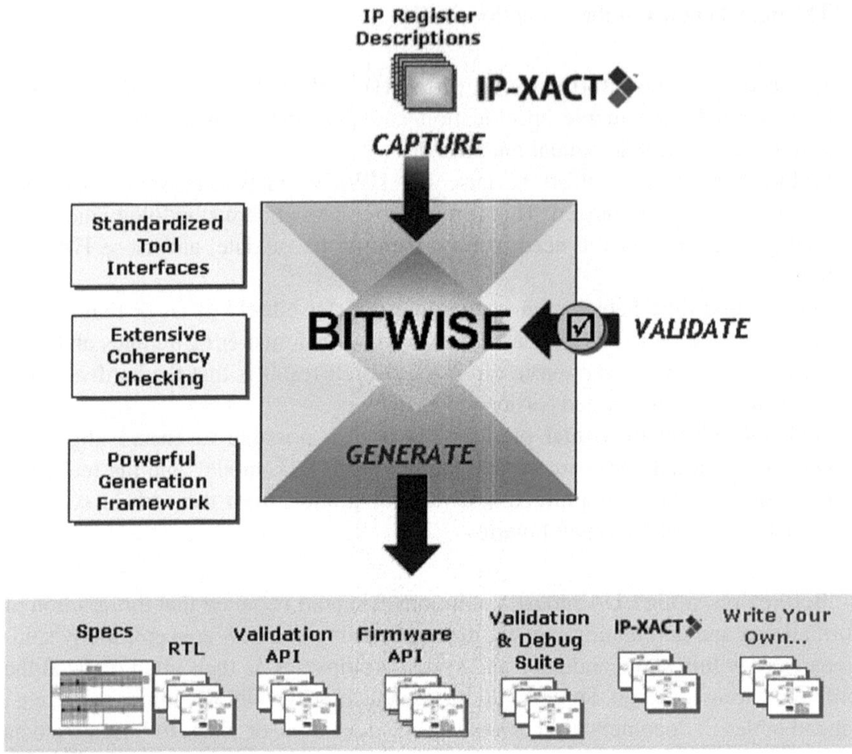

Fig. 2.18 Bitwise design flow

design and verification and software design and verification. The flow detailed in Figure 2.18 shows three distinctive phases of capture, validate, and generate.

The capture flow allows users to specify registers and memory maps manually via a GUI or also to import from IP-XACT or legacy formats. Once registers are captured the IP can be used and mapped within a hierarchical system as part of a system-level memory map. The validate flow ensures that the register and mapping information is coherent and checks for things like unused bits, bitfield, or memory map overlaps.

The generation flow allows a range of different information to be generated with a template-based framework and includes documentation (RTF, EXCEL, HTML, Framemaker, Docbook, or IP-XACT), RTL HDL, HVLs such as Specman, OVM, or VMM, C for system verification, firmware, and virtual platforms

Figure 2.19 shows an example of the Bitwise GUI showing registers and memory map of GUI capture.

Bitiwise is built on the Eclipse environment and has many Eclipse GUI features and concepts. The main GUI components are

- *Library/Design explorer*, this presents all of the components and is used to open and view different components

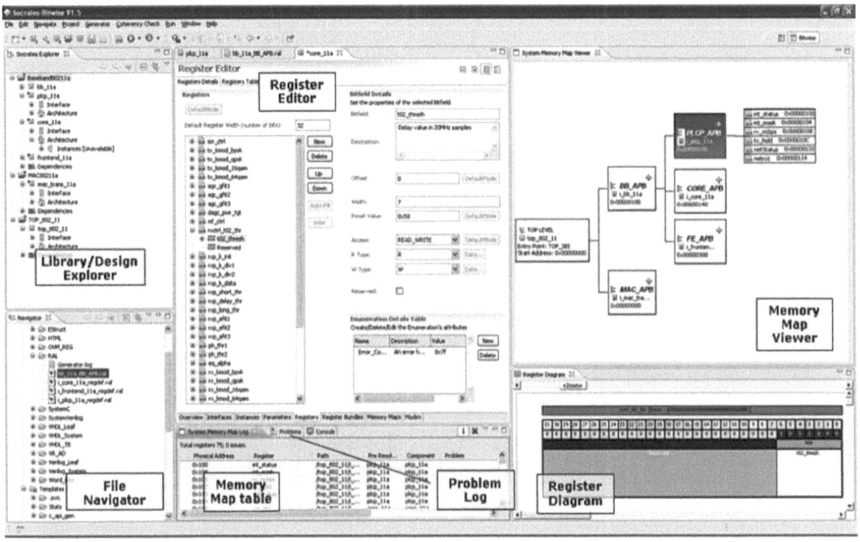

Fig. 2.19 Bitwise screenshot

- A *Register editor*, used to capture register details. This is part of a component editor within which it is possible to define interfaces, instances, and memory maps
- A *Memory map viewer* to give a quick view of the memory map hierarchy
- A *Problem viewer* to allow users to view problems and navigate quickly to the source
- A *System memory map log* to give users a table listing of all addressable elements
- The menu system can give access to all of the generators
- The file *Navigator* allows users to view the generated output or edit the generator templates.

Bitwise has been used in a case study of an OMAP- style multimedia application processor. The SoC contained five processors in the system including ARM cortex, DSP, and IVA, with each processor having shared and private address spaces and different data bus widths. The system has 253 object instances with 169 unique IP blocks. From the software perspective the HW/SW interface consists of 7,991 registers (32 and 64 bit register widths) with 36,105 bitfields over eight levels of memory map hierarchy [46] (Figure 2.20).

All of the IP register descriptions were formalized by translating legacy information, IP-XACT, and also manual capture. The system memory mapping system with multiple levels of hierarchy was captured using Bitwise. With a host system consisting of a 2.4 GHz, 2 Gbyte desktop machine with T3400 Intel dual core, running Windows XP professional, the time to do a full check of all registers, bitfields, and memory maps was less than 10 min.

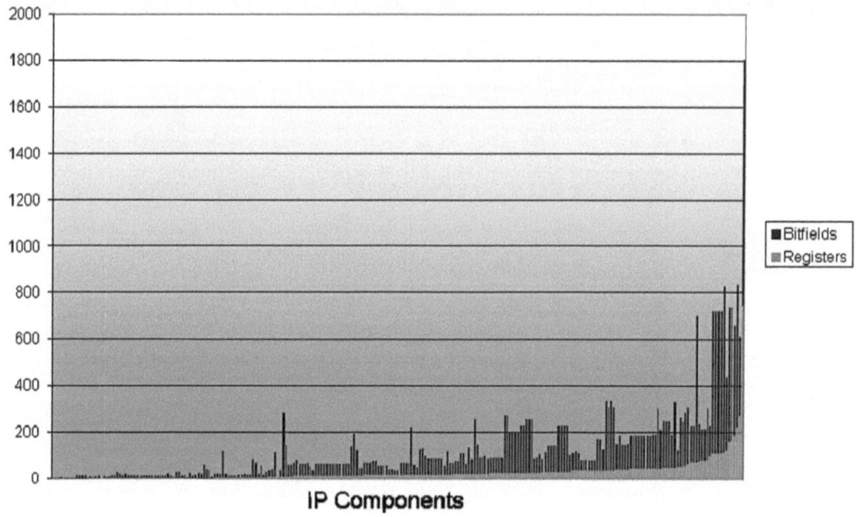

Fig. 2.20 Number of registers and bitfields per IP in large SoC

The automatically generated output consisted of

- 94 MB of documentation which included over 4,000 pages of a word document programmers reference as well as HTML
- 7.5 million lines of C code which included
  - C structures for exhaustive embedded validation
  - C API functions and macros for system validation
  - C header files for embedded SW development
  - Structures for virtual modeling.
- 1.26 million lines of Specman E code.

The generated collateral included both IP-level outputs for block-level validation and system-level outputs for top-level validation. The average time for running a generator was 5 minutes.

This level of complexity demonstrates the need for centralized dedicated register management solutions. With register management tools, a coherent specification can be maintained for the HW/SW view of the entire system and all collateral including documentation can be automatically generated ensuring a correct-by-construction methodology.

2.5 Conclusions

In this chapter we have seen how information about various aspects of IP blocks used in a system and HW/SW interfaces can be captured in an implementation neutral manner through the use of IP metadata models. Two areas have attracted the most attention so far, those being the interconnect of models or IP blocks into a system and the definition of registers that define the boundary between the hardware and software systems.

While the need for these models has existed for a long time, with examples going back more than a decade, there has yet been little standardization in those models. This is particularly true for system interconnect, where the only standard that has been defined (IP-XACT) has not attracted much attention or industry consolidation, although this may change with the merger of SPIRIT and Accellera as discussed. This is because its value, as defined today, is marginal and relies on the existence of other standards to provide the semantics. Given that the existing HDLs are fairly good at defining the syntactic nature of interconnect and are more succinct, and that many ES level virtual-prototyping systems rely on graphical interconnect, the implementation as generated by these tools, such as SystemC, provides an acceptable interchange format.

There has been more industry success with the standardization of register definition languages, primarily due to the higher levels of productivity that can be associated with their usage. They also aid in a reduction of verification effort and significantly reduce the documentation efforts.

References

1. Anssi Haverinen, Katherine Hsu, Gordon Price, Rano Rafidinitrimo, Kumar Venkatramani and Melanie Yunk, "A standard Internet ready VC Exchange System", Proceedings of DATE 2000 User Forum, Paris France pp. 85–91.
2. http://en.wikipedia.org/wiki/XML
3. http://www.design-reuse.com/
4. http://www.si2.org
5. http://www.si2.org/?page=435
6. http://www.vsi.org/
7. http://www.chipestimate.com/
8. http://www.spiritconsortium.org/home
9. http://www.dasc.org/
10. http://www.eda-stds.org/spirit-p1685/
11. http://www.edn.com/article/CA6664639.html
12. http://www.spiritconsortium.org/press/presentations/65nm_SoC_Design_by_ST_Micro.pdf
13. http://www.mentor.com/company/news/platformexpressipxactspiritconsortium
14. IP-XACT v1.4: A specification for XML meta-data and tool interfaces.
15. Richard Goering. "Lucent Enters EDA Arena with Interface Design Tool", EE Times, June 21, 2000 http://www.eetimes.com/news/design/showArticle.jhtml;jsessionid=BQ3MLEXGEHEW0QSNDLOSKH0CJUNN2JVN?articleID=17406057
16. Richard Goering. "Register Description Format Gets 'Spirit' of Standardization", EE Times, May 21, 2007, http://www.eetimes.com/news/design/showArticle.jhtml?articleID=199700202
17. Bertrand Blanc, Bob Maaraoui. Endianness or Where is Byte 0?, Distributed Network Applications Lab; http://www.dnal.gatech.edu/lane/misc/endianness05.pdf
18. Russell Massey, "Introduction to Interrupts", Embedded System Design, June 1, 2001. http://www.embedded.com/story/OEG20010518S0075
19. ARM Primecell UART PL011 Technical Reference Manual.

20. Wolfgang Ecker, Wolfgang Müller, Rainer Dömer (editors), Hardware-dependent Software – Principles and Practice. 2009. Springer, 2009. ISBN: 978-1-4020-9435-4
21. ARM Limited. AMBA specification (Rev 2.0), ARM IHI 0011A, 1999.
22. Andrew Sloss, Dominic Symes, Chris Wright. ARM System Developer's Guide: Designing and Optimizing System Software, Elsevier Morgan Kaufmann, 2004, ISBN- 1-55860-874-5.
23. Mick Posner, Darrin Mossor. "Designing Using the AMBA 3 AXI Protocol: Easing the Design Challenges and Putting the Verification Task on a Fast Track to Success", Synopsys Compiler Magazine, September, 2005. http://www.synopsys.com/news/pubs/compiler/art3_amba3axi-sep05.html?cmp=NLC-insight&Link=Sep05_Issue_Art3
24. David Witt, TI. "ASICs to ASSPs for Working Engineers: Building the OMAP 3430", HOT Chips, 2007 Tutorial. http://www.hotchips.org/archives/hc19/1_Sun/HC19.tutorial1.01.pdf
25. http://www.docbook.org/
26. Richard Goering. "Denali spreads new word in ESL mart", EETimes, August 8, 2005. http://www.eetimes.com/news/design/products/showArticle.jhtml?articleID=167600522
27. Richard Goering. "Register Language Available Through Open Source, EETimes, March 13, 2006. http://www.eetimes.com/news/design/showArticle.jhtml?articleID=181503390
28. http://www.spiritconsortium.org/press/SystemRDL_01Apr09_Final.pdf
29. Kevork V. Dikramanjian. "An Introduction to the VMM Register Abstraction Layer" http://www.soccentral.com/results.asp?CatID=488&EntryID=23383
30. Brian Bailey, Grant Martin, Andrew Piziali. ESL Design and Verification, Elsevier, 2007. ISBN-13:978-0-12-373551-5.
31. IP-Design, Reuse : http://www.design-reuse.com/news/3894/verisity-e-reuse-methodology-erm-reusable-quality-verification-component-development-erm-boosts-productivity-interoperability-all-evcs.html
32. Using the Register Abstraction Layer. http://www.vmmcentral.org/pdfs/using_register_abstract_layer.pdf
33. Joachim Knäblein, Hans Sahm. Automated Formal Method Verifies Highly-Configurable HW/SW Interface, SCD Source, April 30, 2009. http://www.scdsource.com/article.php?id=332
34. Pete Goodliffe. "Register Access in C++", *Dr. Dobb's Journal*, May 1, 2005. http://www.ddj.com/cpp/184401954
35. C. Ford, "Improving Hardware Access with C++", July 2007.
36. Timothy Stapko, "10 Tips for Writing More Maintainable Embedded Software Code", embedded.com, December 3, 2008 http://www.embedded.com/design/opensource/212201562?_requestid=174425
37. Michael Barr. "Bug-Killing Standards for Firmware Coding", embedded.com, May 1, 2009 http://www.embedded.com/design/testissue/217201405
38. http://www.systemc.org/
39. http://www.greensocs.com/en/Projects/GreenReg
40. http://www.synopsys.com/Tools/Verification/HardwareAssistedVerification/Pages/default.aspx
41. http://www.eve-team.com/products/zebu-xxl.php
42. http://www.duolog.com/products/bitwise-register-management
43. http://www.semifore.com/products/datasheet/
44. http://www.productive-eda.com/node/17
45. http://www.mentor.com/products/esl/news/platform_express_supports_spirit.cfm
46. David Murray, Zoltan Sugar. Register Management of complex SoCs http://duolog.com/files/pub/Duolog%20Register%20Management%20of%20Complex%20SoCs.pdf

Chapter 3
Functional Models

Functional models are the mainstay of many design and verification flows and there have been many languages defined to help improve the efficiency and/or effectiveness of these models. This chapter will explore the most common languages that are in use today and the ways in which they can be used. Small examples will be given in this chapter to show the fundamental aspects of these languages and will compare and contrast some of the languages so that informed decisions can be made regarding the best ones to choose for particular situations. In some of the later chapters of this book, we will also see examples of those languages in use.

We can divide the models into two general categories. First and most important are dynamic models – otherwise known as executable (or "simulatable") models. The second category is formal models. These tend to be declarative models as opposed to the imperative models used in the first category. We will illustrate declarative models with property and constraint languages.

The dynamic models can themselves be divided into two broad categories: algorithmic and architectural models. Associated with each of these are two categories of languages – the algorithmic ones feature, most prominently, mathematical modeling languages such as The Mathworks MATLAB; and dataflow-modeling languages such as The Mathworks Simulink, Signal Processing Designer (formerly SPW) from CoWare, System Studio from Synopsys, and research languages such as Ptolemy/Ptolemy II from UC Berkeley. For architectural modeling, which manifests itself on several levels ranging from IP modeling all the way up to full system modeling, dialects of C/C++, with SystemC being most prominent, are traditionally used.

C/C++/SystemC are also used for algorithmic modeling, with C (or C++) being used to model standards in signal and image processing. Standards groups very often develop and release reference implementations of the standard as a set of C or C++ models, with associated test stimuli and expected response files.

B. Bailey, G. Martin, *ESL Models and their Application*, Embedded Systems,
DOI 10.1007/978-1-4419-0965-7_3, © Springer Science+Business Media, LLC 2010

3.1 Dynamic Models and Languages

3.1.1 Algorithmic Languages

3.1.1.1 Mathematical Modeling Languages

The most prominent mathematical modeling language in the world is MATLAB's M language from The Mathworks. The Mathworks [1] was founded in 1984 in Natick Massachusetts by Jack Little, President, and Cleve Moler, Chief Scientist. It has over 2000 employees and unusually for a large and influential tools company has remained privately held. This enables the company to focus on the long-term, rather than short-term, quarter-by-quarter results as public companies too often do.

The Mathworks calls itself "the leading global provider of software for technical computing and Model-Based Design." It has two main products – MATLAB and Simulink, both of which are often confusingly called MATLAB by users (meaning that if someone says they use MATLAB, they may actually be using Simulink or both tools). There are over 1,000,000 users worldwide, with estimates of university use being about 90% of that, as The Mathworks has a very extensive university access program for both academic research and teaching, and inexpensive student versions widely available.

In the last couple of years The Mathworks has focused increasingly on building flows to ESL and EDA tools and has branched out into many additional application areas including finance and biology.

MATLAB is a mathematical toolbox "used for performing mathematical calculations, analyzing and visualizing data, and writing new software programs" and based on an interpreted language called M or m-code. This is a slightly "Fortran'ish" language (in that arrays start at an index of 1 rather than 0 as in C). Its default data type is double-precision 64-bit floating point, although it supports other normal types (ints, shorts, chars, etc.). It provides first-class support for vectors and matrices which makes it reminiscent of other interpretive languages such as APL, although without a lot of special characters and odd syntax. It is interpreted, not compiled, and therefore relatively slow to execute. Some companies (such as the now defunct Agility Design, formerly Catalytic) promoted MATLAB to C code generation partly to speed up execution of MATLAB m-code models.

Although basic MATLAB supports a good deal of analysis and algorithmic specification, The Mathworks offers up to 37 special "toolboxes" adding extended capabilities, libraries, models, and functions in many domains including fixed-point design, filter design, control system design, and RF design as well as the aforementioned areas such as finance, physics, and biology.

There are other competitive mathematics toolboxes, such as Maple, Mathematica, and MatrixX, as well as research capabilities offered by academics via the web in very specialized areas (such as cryptographic analysis). However, in this area, MATLAB is dominant. The Mathworks has had a very clever proliferation strategy, including its strong university access program, and its development of a sharing user community including a community download area with more than 10,000 entries in The Mathworks user community file exchange. In addition, it encourages and promotes book writing using The Mathworks tools as analysis toolboxes in many application areas. There are more than 1,000 MATLAB and Simulink-based books listed on The Mathworks web site, including those written in many languages other than English.

3.1.1.2 Example of MATLAB

The following example of MATLAB code was created by Abhijit Shah of Tensilica who has gracefully allowed it to be used in this book. It implements a half band filter, using inbuilt MATLAB filter functions.

```
Hhb = [ 1/8, -1/8, 1/2, 1, 1/2, -1/8, 1/8];
% Half Band Filter Coefficients; odd length, scaled such that
  center tap = 1
lenHhb = 11; % Length of Half Band Filter
lenHc = (lenHhb+1)/2; % length after removing zeros and center
  tap
Hc = Hhb(1:lenHc/2); % first half of the filter
Hc = [Hc, fliplr(Hc)]; % filter with no zeros and no center
  tap
%
%Setup for Cascaded Comb Integrator (CIC) filter
R = 2; % Upsampling rate
Mcic = 1; % CIC filter order
Ncic = 2; % Number of CIC stages

dsz= 32; % Data Size
bl = 16; % Block length
Nblk = dsz/bl; % number of blocks to be processed
zi = zeros(1,lenHc-1); % initial conditions set to zero

InHB = [1:dsz]; %ones(1, dsz); % define input data
OutE = []; % define output even phase
%______________________________________________________________
%Half Band filter recursion
for i = 0:Nblk-1
    Iblk = InHB(i*bl + (1:bl));  % select block of data
    [Oblk, zf] = filter(Hc, 1, Iblk, zi); % filter block of
      data
    %zi = Iblk(bl-lenHc+1:end);

    zi = zf; % set initial conditions for the next iteration
    OutE = [OutE, Oblk];
end
%
```

```
OutO = [zeros(1, (lenHc/2)-1), InHB]; % centertap output with
  filter delay
OutO = OutO(1:length(OutE)); % drop last few output samples to
  match length with oute

OutHB = [OutE; OutO];
OutHB = OutHB(:).'; % interlace outpus

InC = OutHB;
%_______________________________________________________________
bl = bl*2;
lenHc = Mcic+1;
OutC =[]; % reset out
% CIC Comb Filter: Comb Recursion_____________________________
for j = 1: Ncic
      zi = zeros(1,lenHc-1); % initial conditions set to zero
      OutC = [];
      for i = 0:Nblk-1
            Iblk = InC(i*bl + (1:bl)); % select block of data
            [Oblk, zf] = filter([1 -1], 1, Iblk, zi); % filter
                block of data
            zi = zf; % set initial conditions for the next
                iteration
            OutC = [OutC, Oblk];
      end
      InC = OutC;
      zi = zeros(1,lenHc-1); % initial conditions set to zero
end

InI = OutC;
InI = [InI;zeros(size(InI))]; % interlace with zeros
InI = InI(:).';
bl = 2*bl;   % data rate doubled due to interpolation
OutI = []; % reset out
% CIC Comb Filter: Integrator Recursion_______________________
for j = 1: Ncic
      zi = zeros(1,lenHc-1); % initial conditions set to zero
      OutI = []; % reset out
      for i = 0: Nblk-1
            Iblk = InI(i*bl + (1:bl)); % select block of data
            [Oblk, zf] = filter(1,[1 -1], Iblk, zi); % filter
                block of data
            zi = zf; % set initial conditions for the next
                iteration
            OutI = [OutI, Oblk];
      end
```

```
        InI = OutI;
        zi = zeros(1,lenHc-1); % initial conditions set to zero
end
OutI = OutI./2^(Ncic-1);
```

There are many use models with MATLAB and many possible flows. It is used extremely widely in ESL design teams, usually by algorithmic experts who may just toss the resulting algorithm over the wall to an implementation team, or may try to use or build flows to move artifacts from MATLAB into the ESL and EDA flows, at least for verification. The Mathworks has encouraged the use of MATLAB golden models to verify implementations on target processors, in HDL for hardware, or a mixture of both. Synopsys has recently released a tool called Synphony, that can automate the implementation flow, from M to RTL.

3.1.1.3 Example of C/C++ Reference Model

C/C++ or SystemC can also be used for algorithmic model specification. The following is a snippet of a reference piece of C code for JPEG decoding:

```
METHODDEF(void)
start_pass_huff_decoder (j_decompress_ptr cinfo)
{
 huff_entropy_ptr entropy = (huff_entropy_ptr) cinfo->entropy;
 int ci, blkn, dctbl, actbl;
 jpeg_component_info * compptr;

 /* Check that the scan parameters Ss, Se, Ah/Al are OK for
      sequential JPEG.
 * This ought to be an error condition, but we make it a
      warning because
 * there are some baseline files out there with all zeroes in
      these bytes.
 */
 if (cinfo->Ss != 0 || cinfo->Se != DCTSIZE2-1 ||
 cinfo->Ah != 0 || cinfo->Al != 0)
 WARNMS(cinfo, JWRN_NOT_SEQUENTIAL);

 for (ci = 0; ci < cinfo->comps_in_scan; ci++) {
 compptr = cinfo->cur_comp_info[ci];
 dctbl = compptr->dc_tbl_no;
 actbl = compptr->ac_tbl_no;
 /* Compute derived values for Huffman tables */
 /* We may do this more than once for a table, but it's not
      expensive */
 jpeg_make_d_derived_tbl(cinfo, TRUE, dctbl,
                         & entropy->dc_derived_tbls[dctbl]);
```

```
 jpeg_make_d_derived_tbl(cinfo, FALSE, actbl,
                    & entropy->ac_derived_tbls[actbl]);
 /* Initialize DC predictions to 0 */
 entropy->saved.last_dc_val[ci] = 0;
 }
 /* Precalculate decoding info for each block in an MCU of
this  scan */
 for (blkn = 0; blkn < cinfo->blocks_in_MCU; blkn++) {
 ci = cinfo->MCU_membership[blkn];
 compptr = cinfo->cur_comp_info[ci];
 /* Precalculate which table to use for each block */
 entropy->dc_cur_tbls[blkn] = entropy->dc_derived_tbls
      [compptr->dc_tbl_no];
 entropy->ac_cur_tbls[blkn] = entropy->ac_derived_tbls
      [compptr->ac_tbl_no];
 /* Decide whether we really care about the coefficient
        values */
 if (compptr->component_needed) {
 entropy->dc_needed[blkn] = TRUE;
 /* we don't need the ACs if producing a 1/8th-size image */
 entropy->ac_needed[blkn] = (compptr->DCT_scaled_size > 1);
 } else {
 entropy->dc_needed[blkn] = entropy->ac_needed[blkn] = FALSE;
 }
 }
........
```

3.1.1.4 Dataflow Modeling Languages

The other major class of algorithmic modeling languages are the actor-oriented dataflow modeling languages that usually implement one or several of the dataflow models of computation (such as synchronous dataflow (SDF) or dynamic dataflow (DDF) – or many other variants). Sometimes such languages and associated tools support a wider variety of computational models and may be a little informal with dataflow semantics.

Although the most prominent in this area is The Mathworks Simulink, it does not have the overwhelming dominance that MATLAB does over its rivals. Other commercial tools such as CoWare's Signal Processing Designer, Synopsys System Studio, and research tools such as UC Berkeley's Ptolemy and Ptolemy II have had, and continue to have, a fair user base over the years.

The Mathworks' Simulink is "used for modeling and simulating complex dynamic systems, such as a vehicle's automatic transmission system." Like most of the competitive tools, it is a mixed graphics–text based tool. Blocks or "actors" are interconnected in a graphics editor with signals or flows of tokens that pass from one block to another for data and control purposes. When control or data items arrive at a block in sufficient quantity, the block "fires," computes output results based on inputs and any retained internal state, and then produces output tokens as defined by the block semantics. Interestingly, Simulink was created initially to help model continuous time systems, not dataflow systems, but it has generalized block semantics that allow it to model various system semantics including dataflow. The block semantics are described in executable code – for example, M-code, C, or C++.

To support the various semantics in Simulink, there are a number of system simulation solvers for both continuous time and discrete time semantics. As with MATLAB, there are a number of additional "blocksets" to provide extra models and capabilities beyond the basic tool – for example, fixed-point models, state machines (stateflow), and real-time workshop (for code generation).

Simulink's competition includes CoWare's SPD – Signal Processing Designer. This started out as SPW (Signal Processing Workstation) from Comdisco in the late 1980s or early 1990s, was acquired by Cadence and formed part of the Alta Group, then was moved to CoWare in the early 2000s. Among its most notable early uses in the 1990s was to help design the basestation DSP hardware for the Iridium satellite-based communications system as described in [2]. During its years with Cadence, SPW significantly increased in sales as part of the Alta group, but after Cadence dissolved the Alta group, its poor management of the system level design tools including SPW almost destroyed SPW. When it was transferred to CoWare, it stabilized and CoWare continued to maintain and develop it, including the recent release of an LTE library. SPD, like Simulink, is a mixed graphics and text tool.

The same is also true of Synopsys's System Studio, which started out as a tool from CADIS in Europe called COSSAP; was bought by Synopsys in the 1990s and became Cocentric System Studio and in more recent years has just been called System Studio. Synopsys tried to generalize it for system modeling but in recent years has refocused on communications, including the recent release of an LTE modeling library developed by its partner, Steepest Ascent of Glasgow.

Ptolemy and Ptolemy II [3] are research projects at UC Berkeley that date back to the late 1980s. This has been an incredibly popular and influential research project and there have been several attempts to commercialize the tools. Two current tools based on Ptolemy are Agilent EESof's Advanced Design System (ADS) and Mirabilis. Copious information about Ptolemy and its uses both research and commercial is available from the web site reference.

3.1.1.5 Example of Simulink

Figure 3.1 shows a very simple example of a Simulink model, just for illustrative purposes.

In this model, random numbers (default 64-bit double-precision floating point numbers) are generated, divided by 3, saturated, added as floating point numbers, converted to fixed point numbers (sign plus 23 bits after the decimal point), and added as fixed point numbers. We subtract fixed from floating point result and take the absolute value. Scopes are added so we can see the error introduced by fixed point quantization. Of course, a real-system model in Simulink would be much more complex.

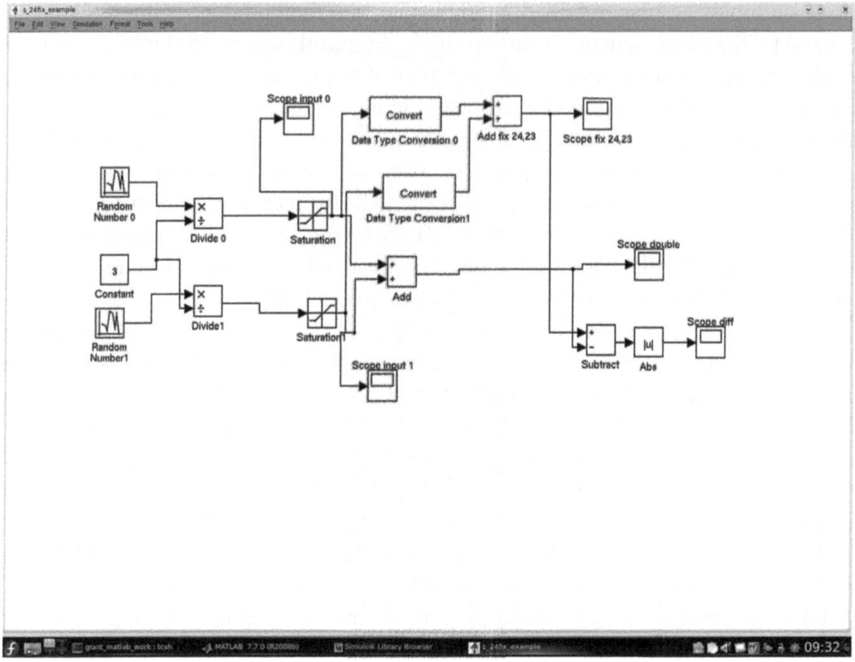

Fig. 3.1 Example of simulink model

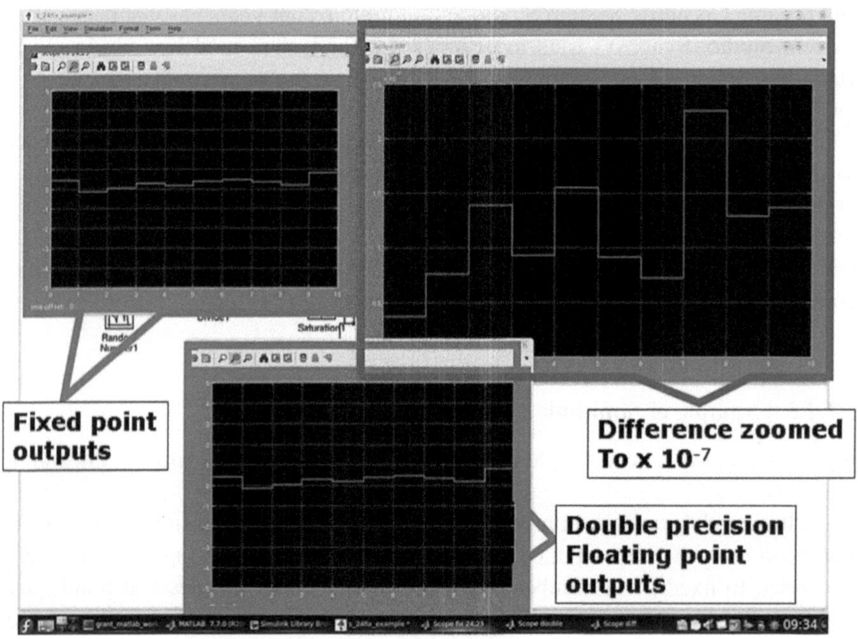

Fig. 3.2 Simulink results

In Figure 3.2 we see the results of running this on ten samples.

Here we show the fixed and floating point outputs Note that they appear identical in the window. To see the error imposed by quantization, we show the difference output zoomed to 10^{-7}. Plotting and an almost infinite amount of analysis are possible with Simulink and its inbuilt tools.

3.1.2 Architectural Modeling Languages: SystemC

In this section we will give an overview of the SystemC standard [4]. SystemC is a class library of constructs and methods that allow users to model hardware and abstract system-level behavior in C/C++. In the past, a variety of companies and academic research projects developed many C/C++ class libraries for system modeling. Many user companies and tool vendors invented proprietary methods to implement hardware concepts such as concurrency, communication, hardware data types, and timing. Standardization on approaches for modeling hardware and systems in C/C++ was necessary to allow easy integration of models and flows. Since product differentiation increasingly relied on software, C/C++ was the obvious choice, due to its wide use by software developers and system architects. The object-oriented features of C++ made it more suitable to define a modeling language than C or ANSI-C – for example, classes and objects map directly into C++ language concepts. In addition, C++ typing can permit more robust and reusable modeling code development.

Detailed information about SystemC can be found in [5, 6].

3.1.2.1 Scope of SystemC: Design Problems

SystemC addresses a number of design problems. They range from the functional specification of communicating blocks to validate algorithms and start partitioning the algorithms into a parallel structure, to virtual platform models for software development, to a more detailed architecture specification, which incorporates resource limitations for architectural analysis, and all the way down to reference models and testbenches for RTL verification and input models for high-level synthesis. Many of these usages will be demonstrated in the later chapters of this book.

The current sweet spot for SystemC modeling is applying it to multi-core systems. With multiple cores executing multiple applications, each with different bandwidth and processing power requirements, the complexity of system-on-chip reaches a level that can no longer be handled at the RT level. With the enormous rise in non-recurring engineering (NRE) costs, one multi-core platform must serve a much wider variety of products and markets, yet still be the right mix of features, cost and energy consumption so that there is a good return on investment (ROI).

With the growing number of multi-core systems, limitations in memory bandwidth, contention for shared resources, and inadequate processing performance are some of the critical issues that must be addressed. Choosing the right additional features during product development is critical, as the success of sales, product margins, and in many cases the survival of a group or even a company can depend on it. The main task in architectural exploration is to decide which component(s) need to be added to support these additional features, or if they can be handled by existing components. Even more critically, designers must analyze the existing memory subsystem to determine if it can support the additional data bandwidth and meet the targeted performance.

To limit project risk, one rule of thumb for today's designs says that designers should reuse 80% of the product from the previous design. Only 20% of the system is new development. For such platform-based design it is critical to identify what part of the system architecture will meet the new performance and power consumption goals, and what new functionality can be integrated without re-implementing major parts of the chip. SystemC has been developed to accurately model concurrent components accessing limited shared resources, thus identifying communication bottlenecks in the system. It also enables the quick and thorough exploration of architectural alternatives to meet system goals.

With shared buses and network-on-chip communication architectures, on-chip data communication is no longer a bunch of wires and buffers that transport information from one component to the next. Rather, on-chip interconnect is a complex system of controllers and shared resources, targeted to offer efficient and specialized data communication. In SystemC interconnect becomes an aggregate of specialized component models called channels. Channels specify interconnect behavior. The channel model also controls how components access shared resources such as communications and memory. Channel models arbitrate concurrent accesses and may provide additional intermediate storage for data and control information.

In most multi-core systems bandwidth requirements are highly dynamic. Not all applications are executing at all times. For example, some applications are fairly inactive. Others may have very high bandwidth requirements during initialization and low bandwidth requirements during normal operation. Since reducing energy consumption is increasingly important, especially for portable battery-powered devices, it is important to switch off unused hardware. The current device state must be stored when power is removed and restored when powered back up. This state might be stored in global shared memory – which further increases dynamic bandwidth requirements. For a full description of these low-power techniques, see [7].

In multi-core systems, software increasingly defines the behavior and determines the performance of the system. A growing number of operating system solutions are available to support multi-threading and resource virtualization. Each operating system is a bit different in how effectively it utilizes the hardware resources. Optimizations depend on the hardware architecture. If the hardware architecture is changed, previous operating system optimizations may not be effective. Analyzing impacts of hardware change on software performance, including operating systems, and vice versa, significantly reduces the risk of project failure.

Thus there are three critical challenges in system design:

1. Increasing number of multi-core systems
2. Advanced interconnect architectures
3. Increasing use of operating systems.

How is SystemC used to address these challenges? In the following sections we describe SystemC language concepts, and then show how those language concepts are used to build models that address the design problems described above.

3.1.2.2 SystemC 2.0

One goals of system-level modeling is to simulate the behavior of hardware–software systems two to three orders of magnitude faster than RTL, so that system-level performance analysis as well as software development and verification with the models is possible. After new advanced techniques were introduced to speed up instruction set simulation (ISS) performance, such as compiled code methods and binary translation approaches, the next challenge was to speed up the simulation of interconnect.

Implementing the communication between blocks with signals was not an acceptable solution. It is too slow and too cumbersome to connect individual signals to the ports of the components. To be able to explore various configurations and architectures it is critical to reduce the effort for rewiring the model interconnect. Combining signals, such as address, data, and controls into one single data structure is critical in the move to channel-based communications modeling. This turns interconnect communications into a more abstract transaction (as we also find in System-Verilog). However, SystemC channels went one step further than data abstraction by combining signals into transactions.

In SystemC channels for transactions are not just data structures, but also contain a set of methods, which are used by the components to initialize and configure the various features of the interconnect.

Before we go into more detail about communication and interconnect, let us look at other hardware-specific concepts in SystemC. These features include

1. Concurrent execution
2. Timing specifications
3. Hardware data types
4. Communication with channels.

3.1.2.3 SystemC Language Basics

Module Level Concurrency

Concurrency exists in SystemC at two levels. First, concurrency exists at the module level (SC_MODULE). Second, within the modules the user can specify processes (SC_THREAD, SC_CTHREAD, SC_METHOD). A module can have multiple processes, which are executed in parallel. Modules represent the structure of a system, and can be combined hierarchically. SystemC does not define any rules about the level of granularity of module definition – this typically depends on the design task at hand.

Using modules, the user documents the structure of the system under design. The level of granularity is defined by the design task the model is used for. For example, a CPU might be described as an SC_MODULE. The module might include the L1 cache as functionality inside the SC_MODULE or it might be described as a separate module. In the first case the interfaces between the CPU and the local memory are not exposed via SystemC to the user in a standard way. The information exchange and thus the transactions are internal. In this scenario, the user is not able to see the individual transactions between the CPU and the L1 cache, although the models still probably provides the user information about cache hit and miss rates. The level of information provided to the user thus depends on what vendors expose to the user in their tools and models. In a second scenario the information exchange, and thus the transaction's timing, is exposed to the user at the cost of slightly lower simulation speed.

Below we show a SystemC example for a hierarchical module of a multi-core system that illustrates concurrent execution at the module level. Here we model a multi-core system with a CPU and a DSP, which access two shared data memory banks through interconnect. Instruction memory is internal to the CPU and DSP. The interconnect uses a simple priority scheme to arbitrate concurrent accesses. So as well as specifying a name for the component, the user also specifies the bus priority during the creation of the CPU and DSP.

```
// SC_MODULE example
SC_MODULE(multi_core_system)
{
  // specify ports
  sc_in clock1;
  sc_in reset;

  // Declare all modules for the multi-core system
   MasterCPU                    *CPU;
  MasterDSP                     *DSP;
  MemoryBank                    *mem1;
  MemoryBank                    *mem2;
  Interconnect *interconnect;

  // The constructor create the multi-core system
  SC_CTOR(multi_core_system)
  : C("C")                              // C is placeholder for name at the
  //next level of hierarchy
  {
  // create all instances
  CPU = new MasterCPU ("CPU",1);                       // CPU(name,bus priority)
  DSP = new MasterDSP ("DSP",2);                       // DSP(name, bus priority)
  mem1 = new MemoryBank("Bank1",0x00,0x7f); // Memory1(name, startaddr, endaddr)
  mem2 = new MemoryBank("Bank2",0x80,0xff); // Memory2(name, startaddr, endaddr)
  bus = new Interconnect("interconnect");         // Interconnect has no parameter

  // connect CPU master
   MasterCPU->master_port(*interconnect);
  MasterCPU->clock(clock1);
  // connect DSP master
  MasterDSP->master_port(*interconnect);
```

```
MasterDSP->reset(reset);
// connect interconnect to 2 memory banks
interconnect->slave_port(*mem1);
interconnect->slave_port(*mem2);
}

// destructor, destroy all instances
~multi_core_system()
{
if (CPU) {delete CPU; CPU = 0;}
if (DSP) {delete DSP; DSP = 0;}
if (mem1) {delete mem1;mem1 = 0;}
if (interconnect) {delete interconnect; interconnect = 0;}
if (mem2) {delete mem2; mem2 = 0;}
}
}; // end multi_core_system
```

This platform model is used to identify whether the DSP, CPU, or any other master on the bus should get higher priority to meet real-time performance targets. The user runs various target application code on the CPU and DSP models, and measures the impact of the memory accesses by the target application running on the CPU and the performance of the DSP application. By extending the interconnect model with more advanced arbitration schemes, the user can explore the benefits of a more complex scheduler for the CPU and the DSP and the impact on application performance.

Process Level Concurrency in SystemC

The second level of concurrency defined in SystemC is that of reactive processes inside an SC_MODULE. For these processes, the user specifies activation conditions – i.e., events (sc_event). Changes in signal values are also treated as events, and the user specifies if the activation event is the positive or the negative edge of the signal.

To describe concurrent behavior the user specifies multiple processes. SystemC defines two different kinds of processes. The first one executes the full source code implemented inside the process sequentially to completion (run-to-completion semantics). This type of process is an SC_METHOD, and it simulates faster than the other type of process. In an SC_METHOD, no timing behavior is allowed inside the process – and waits are not allowed. Timing is based on the time when the activation event or events occur. The CPU example mentioned above uses a non-blocking interface and the method call returns immediately with no nominal simulation time elapsing within the model. Such a processor can execute further instructions before a data load from memory occurs, as in more advanced current processor architectures. Thus an SC_METHOD is an appropriate modeling style for the CPU.

```
// SC_METHOD example
SC_MODULE(MasterCPU)
{
// Specify ports
sc_in clock;                             // clock input port
```

```
sc_port<non_blocking_if> bus_port;     // details about ports
  declaration see (TODO: implement reference to section
  Communication in SystemC)

// Register all processes inside the Module MasterCPU with the
  SystemC kernel
SC_HAS_PROCESS(MasterCPU);

// constructor to create the system
MasterCPU (sc_module_name _name    //these are placeholders
  for the
//specification at the next level of hierarchy
    , unsigned int unique_priority)
: sc_module(_name) // this code connects the specification
  from the
// next level to this level
, m_unique_priority (unique_priority)
{
//Specify the SC_METHOD process, which is activated by the
  positive clock edge
SC_METHOD(execute_instruction)
dont_initialize();     // do not call during initialization
sensitive << clock.pos();
}

// process declaration
  void execute_instruction();
// declare internal variables
private:
  unsigned int m_unique_priority;
}; // end MasterCPU
```

The other kind of process has source code that can be interrupted and thus is re-entrant. These are SC_THREAD and SC_CTHREAD (for a clocked thread) processes. They can be suspended with one or more wait statements, and will resume when a condition specified in the wait statement becomes true. Such conditions may be events, signal changes, or specific time intervals.

SC_THREADs simulate somewhat slower than SC_METHODs. This is due to the overhead of context switches between the process and the SystemC simulation kernel. After activation the code inside the SC_(C)THREAD process is executed sequentially and with no imputed delays until the first wait-statement in the code. If the wait condition becomes true, the SystemC kernel re-activates the process and the next piece of code from the first instance of the wait until the next wait is executed. The master DSP example uses a blocking interface, waiting until the memory access returns, before it continues the execution of an instruction. In this case modeling using an SC_THREAD process is the right choice. The DSP model is activated by the positive edge of a reset signal.

```
// SC_THREAD example
SC_MODULE(MasterDSP)
{
  // Specify ports
  sc_in reset;     // declare reset port
  sc_port<blocking_if> bus_port; // details about ports
   declaration see

  // Register all processes inside the Module MASTERCPU
   with the SystemC kernel
  SC_HAS_PROCESS(MasterDSP);

// constructor to create the system
  MasterDSP (sc_module_name _name //these are placeholders
   for the
  //specification at the next level of hierarchy
      , unsigned int unique_priority)
  : sc_module(_name) // this code connects the specification
   from the
  // next level to this level
  , m_unique_priority(unique_priority)
  {
  // SC_THREAD process declaration with sensitivity
  SC_THREAD(execute_instruction);
  dont_initialize(); // do not call during initialization
  sensitive << reset.pos(); //reset is active high
  }

  // process declaration
     void execute_instruction();
// declare internal variables
private:
  unsigned int m_unique_priority;

}; // end MasterCPU
```

The activation of SC_METHODs and SC_THREAD is controlled by the SystemC simulation kernel. Processes are automatically registered with the SystemC kernel either by using the keyword SC_CTOR(<module name>) or SC_HAS_PROCESS(<module name>). SC_HAS_PROCESS is more flexible, because the module can have an arbitrary number of parameters. SC_CTOR can not accept any parameters, other than the name of the module. The SystemC kernel automatically calls all processes during initialization at simulation time 0. The user has to explicitly specify if a particular process should not run during initialization with dont_initialize(). After initialization the SystemC kernel maintains an event queue to guarantee the correct activation of the various processes in a simulation. The kernel also automatically ensures that the internal variables representing hardware states (the context) in an SC_THREAD or SC_CTHREAD are restored whenever the process continues execution.

Timing and Synchronization in SystemC

Another important aspect of modeling hardware behavior and system resource conflicts is timing. SystemC uses a 64-bit unsigned integer type (sc_time) to describe time values. The resolution of time can be set by the user with the function sc_set_time_resolution(t). The resolution values can be SC_FS, SC_PS, SC_NS, SC_US, SC_MS, and SC_SEC. The default resolution is nanoseconds (SC_NS).

The scheduler in the SystemC kernel controls the timing and the order of execution of the processes. The SystemC kernel maintains an event queue to identify when processes need to be scheduled and execution needs to be advanced. The user can advance time within a model explicitly using wait statements. The statement sc_wait(3, SC_NS) will advance the model time by 3 ns. The user can also specify implicit time advances with wait statements for events. The statement sc_wait(e1) waits for event e1 to happen. Events in SystemC carry timing information. They do not carry a value (sc_event e1). The user can create an event immediately by calling the function notify (e1.notify())

This immediate notification of events is useful to describe software, as software usually lacks the concept of delay or time unless explicitly specified. The statement e1.notify (3, SC_NS) will issue an event in 3 ns. To be able to simulate and synchronize concurrent hardware systems deterministically, the concept of delta cycles is required in SystemC, as in Verilog and VHDL simulators. The user can explicitly control the use of delta cycles and thus the order of execution by calling wait(SC_ZERO_TIME) or by creating events with a delayed notification (e1.notify_delayed()). The SystemC kernel uses a two-phase approach with an evaluation and update phase to accommodate the concept of delta cycles and achieve deterministic timing behavior. The following example illustrates the difference between notify() and notify_delayed(). The code `e1.notify(); e2.notify();` mimics two events, which are executed exactly in parallel. The order of execution of the events is undetermined. Any order is allowed and defined by the SystemC kernel. The code `e1.notify(); e2.notify_delayed();` mimics two events, which are executed in a deterministic order defined by the user. Event e2 has to be executed after event e1.

Hardware Data Types in SystemC

As well as the built-in C/C++ data types, such as int, uint, bool, SystemC standardized on a set of hardware-specific data types for bit-accurate behavior. This avoids tedious and error-prone translation from one proprietary data type to another. The data types specified in SystemC represent 2 value bits (sc_bit), 4 value bits (sc_logic), a 1–64 bit signed integer value (sc_int), a 1–64 bit unsigned integer value (sc_uint), an arbitrary length signed (sc_bigint) and unsigned (sc_biguint) value, as well as a bitvector (sc_bv) and logic bitvector (sc_lv) type. Beside these integer types SystemC also offers fixed point data types (sc_fixed, sc_ufixed, sc_fix, sc_ufix) to allow the exploration of various choices for bit width and resolution while translating floating point numbers into fixed point numbers, or when attempting to optimize a hardware implementation. The hardware data types are parameterized with the number of bits. So sc_uint<4>, specifies a 4 bit unsigned data type.

Communication in SystemC

To be able to analyze the performance of different kinds of interconnect architectures, SystemC uses ports, interfaces, and channels. The idea is to centralize the communication control functionality into a channel, which handles concurrent accesses from processing elements. With this concept the models of processing components, such as CPUs, DSPs, or DMA controllers, as well as those of target components, such as memory or peripherals, become much simpler. They may remain unchanged while exploring different interconnect architectures.

To mimic the resource constraints of communications, components communicate through ports (sc_port) with the rest of the system. Ports provide well-defined boundaries. The boundaries are defined in interfaces (sc_interface). An interface defines a set of function calls, which represent those boundaries. The interface function defines the agreed upon behavior between the processing elements and the channel. The processing element will use the interface functions to access a channel. Channels represent the behavior of the communication. Channels must implement all interface functions. How those functions are implemented is independent of the processing element. Different channels can implement those functions differently. The implementation depends on the specific features and architecture of the interconnect. The communication is separated from the behavior, so the processing element can remain unchanged. The code below shows an interface specification.

```
// interface example. Here the user defined how the function
// calls to the channels have to look like and what functions are available
class non_blocking_if
  : public virtual sc_interface      // all interfaces inherit from sc_interface
  // this makes sure that the interface is
  // connected to the port
{
public:
   // non-blocking single read interface call
   virtual void read(unsigned int unique_priority
   , int *data
   , unsigned int address) = 0;
   // non-blocking single write interface call
   virtual void write(unsigned int unique_priority
   , int *data
   , unsigned int address) = 0;
```

```
// non-blocking burst read interface call
  virtual void burst_read(unsigned int unique_priority
  , int *data
  , unsigned int start_address
  , unsigned int length) = 0;
  // non-blocking single write interface call
  virtual void burst_write(unsigned int unique_priority
  , int *data
  , unsigned int start_address
  , unsigned in length) = 0;

  // get actual status information from channel
  virtual simple_bus_status get_status(unsigned int unique_priority) = 0;

}; // end non_blocking_if
```

All interface functions have the C/C++ keyword "virtual" in front of them. Those functions are only virtual specifications. The keyword virtual ensures that the user must implement all functions specified here in the channel. If the user does not implement one or more of those function calls the compiler will issue an error message and will not finish compilation.

```
// SC_THREAD example
SC_MODULE(MasterCPU)
{
  // Specify ports
  sc_in reset;      // declare reset port
  sc_port<non_blocking_if> bus_port; // declare port to the
     interconnect
  ...
}
```

The user needs to specify the port inside the SC_MODULE with the respective interface it is using. Here for example this is done inside MasterCPU by using the keyword sc_port and the name of the interface inside the <>. The above example specifies a port of type "non_blocking_if." So the module can use all the function calls defined inside the interface "non_blocking_if" specified earlier. See below for an example of how to use the interface calls inside a process.

```
// Example of using the interface calls inside the process
     implement_instruction
// This is an SC_METHOD and is activated for each clock cycle
void MasterCPU:: implement_instruction()
{
  int data;
  bus_status status;

  // fetch next instruction word
  status = bus_port->read (m_unique_priority, data,
                                          m_address);
```

```
 // check for problems in the bus
          if (status ==INTERCONNECT_OK){
// Start decoding instruction in the next
          decode_instruction(mydata, m_address);
          }
          elseif (status == INTERCONNECT_ERROR){
             // Issue an error message with simulation time,
                    name of the module and address
       issue_error_message(sc_time_stamp(),name(), m_address);
  } // end if
  }
}
```

The channel will then implement the function calls, so that it represents the specific features of the interconnect. It deals with transaction scheduling and accessing the correct memory locations. If virtual address translation is used in the system, this is typically also implemented in the channels.

With the capability of SystemC to describe concurrency, timing, hardware data types, and communications in a standardized way using C/C++, a model developer is able to describe models at different levels of abstraction. Such models range from cycle accurate to cycle approximate to un-timed models. This allows a lot of flexibility. However, each user will use SystemC differently. In particular, the flexibility in the interface specification creates interoperability problems when a design team wants to use models from different sources. Each modeler defines their own set of interface functions and data formats, which they use for communication.

As a result, users were forced to write a lot of adapters, adapting each set of interface calls to other people's interface calls in order to integrate models into a system. A second issue is that each user may define and use different levels of abstraction. Again a lot of adapters need to be implemented to translate between more detailed or less-accurate models. In some cases adaptation of abstractions is not possible without losing accuracy in the overall system models.

These issues led to the transaction level modeling (TLM) initiative within OSCI. The TLM 2.0 specification standardizes on a set of interface calls and levels of abstraction to overcome interoperability issues. Later we will describe TLM 2.0, but before we go into the details of TLM 2.0 and discuss levels of abstraction, we will look at different use cases for SystemC models. We will discuss what kind of analysis is typically performed in these cases, and how much detail is necessary to describe them.

3.1.2.4 SystemC in Real Systems

We have already discussed design issues which may be resolved using SystemC models. We will now discuss the usage of SystemC in real systems. This allows a better understanding of the contents of these models, how real system characteristics are modeled in SystemC, and what the models can be used for. We first discuss the hardware portion of systems and then the software parts.

Hardware System Specification

For hardware system development, SystemC can be used for three design tasks:

1. High-level synthesis
2. Multi-application system-on-chip (SoC) development
3. Multi-chip development.

High-Level Synthesis

Today's technologies give us a wide design space within which to trade off area, timing, performance, and energy consumption when developing new hardware. It is possible to meet timing constraints in many applications while doing a significant amount of resource sharing. High-level synthesis enables the user to explore a much larger design space more efficiently than is possible with manual design methods. This subject will be explored in detail in Chapter 9.

High-level synthesis tools that support SystemC models, or any of the C/C++ language variants, allow users to explore various micro-architectures using constraints, rather than by modifying the specification. Today's synthesis tools take a functional specification of the system as the input. To explore the design space, the user sets design constraints and directives, or pragmas, to customize the design for a specific cost, performance, and power consumption target.

As well as modeling functionality, users of high-level synthesis tools are able to specify hardware concepts such as reset, hardware datatypes, concurrency, and various standard interfaces. The user typically describes functional blocks such as FIR filters, IDCTs, or motion estimation functions. Those blocks are often integrated into an existing system. SystemC allows the user to accurately specify the timing of the external interface behavior, so that the blocks can be integrated seamlessly into an existing system. For the internal communication, SystemC allows users to both specify and verify that the design meets the throughput requirements.

The following code example shows a synthesizable pixel processing component, showing how the synthesizable subset of SystemC may be used to create the input specification for high-level synthesis tools. It also shows an efficient way of separating the untimed functional behavior from the timing-accurate interface behavior. The example transforms RGB pixels data to luminance data, changes the brightness, and then converts it back to RGB pixels. This component is connected to an existing memory system. This system provides only a 16-bit memory component. So the 24-bit RGB information needs to be stored in two 16-bit memory locations.

```
// Pixel Processing example
SC_MODULE(saturation)
```

```
{
  // reset and clock are required ports
  sc_in<bool> rst;
  sc_in<bool> clk;

  //Input interface
  sc_in< illuminance::element_type > sat_in;

  // Separate out input and output controller
  // for data synchronization in "synchronization"
  synchronization::in input_sync;
  synchronization::outoutput_sync;

  // Group internal communication into a channel
  RV<pixel> chan1;     // Channel between input_synch and
                          saturation thread
  RV<pixel> chan2;          // Channel between output_synch and
                                saturation thread

  // Pixel Storage
  ram_type::port<PIN> in_ram;
  ram_type::port<PIN> out_ram;

  // result signal for saturated luminance value
  sc_signal< illuminance:: element_type > sat_val;

  // Constructor call with 3 parallel threads
  SC_CTOR(saturation)
  : rst("rst")
  , clk("clk")
  , input_sync("input_sync")
  , output_sync("output_sync")
  , chan1("chan1")
  , chan2("chan2")
  , in_ram("in_ram")
  , out_ram("out_ram")

  {
       // Timed input controller thread to in_ram
  SC_CTHREAD(in_thread, clk.pos());
  reset_signal_is(rst, false);

       // untimed functional thread for saturation
  SC_CTHREAD(saturation_thread, clk.pos());
  reset_signal_is(rst, false);

       // Timed output controller thread to out_ram
```

```
  SC_CTHREAD(out_thread, clk.pos());
  reset_signal_is(rst, false);

  } //SC_CTOR(saturation)

private:
  void in_thread();
  void saturation_thread();
  void out_thread();

  // extract high or low byte value for the 24 to 16-bit
       transformation
  sc_uint<8> highbyte( const sc_uint<16>& word );
  sc_uint<8> lowbyte( const sc_uint<16>& word );

}; //SC_MODULE(saturation)
```

The next piece of code shows the implementation of three concurrent threads: in_thread, saturation_thread and out_thread. Each thread starts by resetting the interfaces. Then the user describes the processing function in a while-loop. The thread "in_thread" reads values from the 16-bit RAM "in_ram," constructs the RGB representation of the pixel and passes it to the channel "chan1," which is used for internal communication. The channel code is shown later. By using this approach, the channel implementation can be varied without modifying the thread code.

```
// input thread implementation
void saturation::in_thread()
{
// reset the interfaces
input_sync.reset();
chan1.reset_out();
in_ram.reset();
wait();

 // processing functions
while( true )
{
// wait for sync
input_sync.wait_for_trig();
sat_val = sat_in.read();

 // 3-byte pixels are packed into a 2-byte wide memory.
// This thread reads words from memory and outputs
// pixels. It uses the extra_ variables to keep track of
// the second byte of the second word read.
sc_uint<8> extra_byte;
```

```
sc_uint<1> extra_count=0;
pixel pix;
const int numpix = 256*256;
const int numwords = numpix * 3 / 2;

 ram_type::address_type addr = 0;

 while( addr<numwords )
{
sc_uint<16> tmp;
switch( extra_count )
{
case 0:
tmp = in_ram[addr++];
pix.R = highbyte(tmp);
pix.G = lowbyte(tmp);
tmp = in_ram[addr++];
pix.B = highbyte(tmp);
extra_byte = lowbyte(tmp);
extra_count = 1;
break;
case 1:
pix.R = extra_byte;
tmp = in_ram[addr++];
pix.G = highbyte(tmp);
pix.B = lowbyte(tmp);
extra_count = 0;
break;
}
chan1.put( pix );
}
input_sync.done();
}
}
```

"saturation_thread" reads the RGB pixel values, transforms it to a luminance, saturates the luminance, transforms it back to RGB data, and puts the pixel in the channel connected to the output thread.

```
// this thread gets a pixel represented as RGB
void saturation::saturation_thread()
{
illuminance il;
pixel pix1, pix2;
```

```
// reset the interfaces
chan1.reset_in();
chan2.reset_out();
wait();

while (true)
{
pix1 = chan1.get(); // get a RGB pixel
il = pix1; // convert to luminance information
il.Y *= sat_val.read(); // saturate the luminance information
pix2 = il.to_pixel(); // convert back to RGB information
chan2.put( pix2 ); // put modified pixel into channel
}
}
```

"out_thread" writes the pixel information back to the 16-bit RAM "out_ram," transforming the 24-bit RGB pixel information to two 16-bit values.

```
// The out_thread reads pixels from the channel
// and writes them in packed form into the 16-bit input RAM,
void saturation::out_thread()
{
  // reset the interfaces
  output_sync.reset();
  chan2.reset_in();
  out_ram.reset();
  wait();

  while( true )
  {
  pixel pix;
  const int numpix = 256*256;
  const int numwords = numpix * 3 / 2;

  ram_type::address_type addr = 0;
  sc_uint<8> extra_byte;
  sc_uint<1> extra_count=0;
  while( addr<numwords )
  {
  if ( addr < numwords-1 )
  pix = chan2.get();
  switch( extra_count )
  {
  case 0:
  out_ram[addr++] = (pix.R,pix.G);
```

```
  extra_byte = pix.B;
  extra_count=1;
  break;
  case 1:
  out_ram[addr++] = (extra_byte,pix.R);
  out_ram[addr++] = (pix.G,pix.B);
  extra_count=0;
  break;
  }
  }
  // sync with next module
  output_sync.trig();
  output_sync.wait_for_done();
  }
}
```

The following code shows how to encapsulate communication protocols into a channel. In this case a simple ready signal (rdy) from the receiver indicates that data can be sent. A valid signal (vld) from the sender validates data.

```
// Channel example code. The channel uses a template to
   support any kind of data
// structure. In this case data is the RGB pixel information.
template <class T>
class RV
{
private:
  //
  // specify signals relevant for the protocol
  //
  sc_signal<bool> rdy;
  sc_signal<bool> vld;
  sc_signal<T> data;

public:

  RV( const char* name )
  : rdy( (sc_string(name) + "_rdy").c_str() ),
  vld( (sc_string(name) + "_vld").c_str() ),
  data( (sc_string(name) + "_data").c_str() )
  {}

  // reset input signals
  void reset_in()
  {
  rdy.write(0);
```

```
    }

    // reset output signals
    void reset_out()
    {
    T default_val;
    vld.write(0);
    data.write(default_val);
    }

    // Get data from the channel.
    T get()
    {
    wait();
    rdy.write(1);
        // wait until data is valid
    do { wait(); } while ( !vld.read() );
    rdy.write(0);
    return data.read();
    }

    // Push data into the channel
    void put( const T& val )
    {
        // wait until receiver is ready to receive data
    do { wait(); } while ( !rdy.read() );
    data.write( val );
    vld.write(1);
    wait();
    vld.write(0);
    }
};
```

The above example is synthesizable SystemC code for a real design, showing how to encapsulate interface control functionality for easy reuse using standard SystemC constructs. It also shows how to implement concurrent functionality as well as hardware resets. In this example, untimed functionality is combined with cycle-accurate interface behavior, so that the user has full control of the interface performance, which allows it to be integrated into existing hardware. The output of high-level synthesis is synthesizable RTL, which feeds into the standard RTL-to-GDSII design flow.

Multi-application SoCs

Today's products integrate various applications into one product and often in a single chip. Cell phones are cameras, video players, and music players; cameras capture pictures and video and directly connect to printers without a PC. This convergence of applications into a single product significantly increases the complexity of the hardware design tasks. Cost pressure requires more and more integration of the functionality into one SoC. The benefit of one chip is also that it removes the partitioning issues and pin limits associated with using multiple devices. This permits much higher on-chip communications bandwidth than is possible using off-chip communications. The downside of a higher level of integration is that it is impossible to repair modern products by replacing a single component that implements one specific application. Instead the full chip implementing all applications needs to be replaced. This leads to an additional requirement of more flexible software-based solutions, where defective chips or functionality can be replaced by software methods.

The reliable and correct functioning of such SoCs is critical. Since highly integrated SoCs have very high non-recurring engineering (NRE) costs to design, verify, and implement, it is critical to find performance bottlenecks early and allow analysis of customizations to meet cost targets. SystemC models of a multi-application SoC allow this and help to keep NRE cost and the risk of failure low.

Often one SoC must serve a variety of markets to get enough production volume to justify the design expense. This is offered by flexible solutions such as configurability and software programmability. As products mature, cost pressures often require further customization of the design. Good examples of multi-application SoCs are wireless applications processor platform chips such as TI OMAP, STMicroelectronics Nomadik, or Qualcomm's Snapdragon. These platforms have various versions customized to a specific segment of the market. The differentiation comes from variations in the communications infrastructure, interconnect and memory, the application processors and the accelerators added to the SoC. The combination of the applications processor and accelerators customizes each platform for a specific market or end customer.

RTL simulations are too slow to identify data communication or processing bottlenecks in a multi-application platform. Thus faster SystemC simulations are used to identify those bottlenecks early in the design cycle, before committing to a specific architecture. The development of a multi-application platform needs to be hierarchical. One can distinguish three different system analysis tasks for multi-applications platforms:

1. IP component development
2. Single application subsystems integration
3. Full multi-application platform integration.

Each of the three design tasks has specific requirements and use models for SystemC.

IP development needs to generate RTL quickly to accurately determine cost and performance. There are several approaches, such as high-level synthesis for data-oriented components (discussed above), and synthesizable RTL code generation from a high-level instruction set architecture specification, as discussed in Chapter 6.

Many developers use different kinds of configurable or scripting approaches. In general any approach used for IP component development needs to feed directly into a standard RTL-to-GDSII flow. Accurate cost and performance numbers are essential. With the growing need for high-level models, a number of these approaches also provide SystemC model generation. The SystemC model is used to integrate the IP into a single application subsystem design flow and subsystem model. Thus, for IP component development, SystemC code is in some cases an input. For an integration of this IP into a subsystem on a chip, there is an increasing need for SystemC models as an output.

In developing single application subsystems, models for key-processing elements typically need to be cycle accurate in order to allow accurate performance evaluation. Interconnects, memory subsystems, and standard peripheral models can be more abstract. Timing annotated and blocking interfaces are usually sufficient in this case.

The goal of single application subsystem analysis is to determine if the processing elements meet the performance targets, and to measure the latency and throughput requirements for accessing the memory (or other) subsystem. As discussed, today's products lead to programmable solutions for wide applicability and in-field flexibility. In general we can think of two kinds of programmable solutions: applications processors and specialized processors (the latter being an ASIP as discussed in Chapter 6). Specialized processors tend to be dedicated to specific tasks in the design. Typically the software running on the applications processor is not as timing critical as that on the specialized processors. For the purpose of the subsystem development in SystemC the application processor is mainly used to configure register values such that the rest of the system runs correctly. The performance of the applications processor and its' capacity to run the software stack is not evaluated in this kind of subsystem mode – this is often evaluated separately. At the subsystem level it is assumed that the interconnect and memory subsystem is designed to give the applications processor a fair share of memory bandwidth. The validation of this assumption is one of the main purposes of the analysis performed at the full multi-application platform level.

The timing and performance of the specialized processors are critical for each application subsystem. For processing-intensive loops, every cycle may count. The user needs to know if the specialized processor will be able to meet tight application-timing constraints. For example, in video with a 30 or 60 frame per second performance target, the average decode time for a full frame must be 1/30 or 1/60 of a second, with some buffering to handle jitter. To verify this, it is essential to use cycle-accurate instruction set simulators to accurately determine the number of cycles required per frame.

Single application subsystems typically have interfaces to both local private and global shared memory. The sizing of local memory and capacity of the interconnect to it must be determined to meet performance targets. A timing-annotated memory subsystem model may be sufficient for first-order analysis. However, if sophisticated pipelining with out-of-order execution takes place in the local memory subsystem, a cycle-accurate model with non-blocking transactions may be required.

At the end of this design step, we have a good idea of how many specialized processor cores (ASIPs) and of what types are required to run the different applications at the target speed. We also will have an idea of minimum and maximum memory latency and throughput requirements for the ASIPs and hardware accelerators.

The next step, multi-application platform analysis, analyzes how the concurrent execution of the different applications impacts each other. Global shared resources – memory and interconnect – have a significant impact on the performance of each application subsystems. In this step the designers analyze the bandwidth and latency of access to global interconnect and memory under various dynamic work-case scenarios. For example, for a car entertainment system, one person may be listening to a CD, while someone else tries to find another radio channel. Users expect that the CD playback is not disturbed by the radio search, but switching radio stations typically causes different traffic patterns than listening to a radio station. The analysis of worst case scenarios for any steady-state situation is now insufficient.

In multi-application platforms it is not reasonable to sum up the worst case scenario for each application and add some bandwidth margin that leads to over-engineering the product, making it more expensive than it needs to be. It is also extremely risky to assume that without any further dynamic analysis the system will meet the latency and throughput requirements of every application in every situation. A corollary to Murphy's law predicts that a new scenario will come along that will invalidate all the assumptions made through static analysis. In addition, in many cases the traffic, even in steady-state execution for one application is highly dynamic. When we mix transaction traffic from various applications we do not get a continuous stream of transactions with similar throughput patterns. Many applications have intervals of high traffic (bursty) followed by intervals of very low or no traffic during normal data processing. Thus further dynamic, simulation-based, analysis is needed.

Another challenge for system developers lies in the need to reduce energy consumption. Components in modern embedded systems are powered down when not used. When powered back up, the traffic load on global resources is extremely dynamic and may reach peak levels (for example, to refresh dynamic data and instruction memory or to refill the cache). This will affect the available bandwidth for other applications running in a steady state, and may impact their performance. The goal in this scenario is that the performance of all active applications must not be impacted by powering up components on demand.

A common approach today is to run SystemC simulations to analyze the latency and throughput performance for applications under various scenarios. For this analysis, cycle-accurate interconnect and memory models are essential. It is critical for this analysis step that contention of the shared resource is visible to the developer. This allows the developer to choose the right configuration with the minimum cost to meet performance targets so that that memory and interconnect resources are efficiently used.

In this phase of analysis, the models of the application subsystems do not need to be cycle accurate. It is only important to simulate with the right traffic patterns for the interfaces that access global resources. The functionality of the subsystem can be abstracted. For example, there may be cases where global resources are only accessed during SoC power-up and initialization. In other scenarios, there may be a constant stream of transactions to and from global memory. For such analysis, abstract component models generating reasonable traffic patterns should suffice.

For such analysis, traffic generators, rather than detailed models of hardware, may be used reflecting the bandwidth and latency requirements of the application subsystems. This gives higher simulation speed and it may be less complex to build such models. Sometimes real silicon can be used to generate real traffic profiles and the generator is really just a profile reader. As well as real profiles, various artificial and benchmark profiles can be created to exercise corner cases and stretch cases. From this analysis, the design team can define the global interconnect and memory subsystem architecture, and gain confidence that all applications will execute correctly and concurrently.

Multi-chip Development

In contrast to SoC design analysis, at the multi-chip level of the system, the analysis does not to be as accurate. Since many chips are pin-limited, communication in these cases must be implemented with fewer pins than optimal (via multiplexing), and this lowers the bandwidth of the inter-chip communication significantly from what is possible on-chip. Inter-chip communication must also be more robust – it cannot rely on fully synchronous communications at the board and backplane level, and instead turn to standard busing schemes in order that different vendor chips will communicate properly. Ethernet, PCI, or FireWire are the most commonly used inter-chip bus protocols for consumer or industrial products. The automotive industry standardized on CAN, LIN, and more recently on FlexRay for communication between devices inside the car.

There is a delay factor of 10–100X for latency for inter-chip communication when compared to on-chip interconnect. Systems cannot rely on low latencies and cycle-accurate timing behavior at this level – latency is not fixed and delay has significant variation. Communication protocols for inter-chip communication are defined to allow processing blocks to be insensitive to such variance or jitter. This is typically done using intermediate data storage buffers. Since latencies are the equivalent of many tens or hundreds of clock cycles, these models need not be clock-cycle accurate for analysis of systems at this level.

For SystemC models at this level, the focus is on the power of processing elements with respect to overall throughput and buffer size modeling. Instruction-accurate processor models are sufficient. Clock cycle-accurate delays for interconnects arbitrating memory accesses within the chip do not need to be modeled at this level. The impact of interconnect arbitration inside the chip is one or two orders of magnitude lower than the impact of arbitration of the communication link outside of the chip. The modeling of the processor pipeline or of memory accesses is also typically not required. Timing annotated models or even untimed models using blocking interfaces and loosely timed (LT) modeling styles is accurate enough. Models and platforms of this type will be discussed further in Chapter 5.

Chip and device-level system simulation gives design teams information about the sizing of memories and buffers and overall processing needs for processing units. It also confirms that the intended architecture supports the system throughput requirements. They may indicate the sensitivity of system performance to various sizes of buffers and choices of processing capability. These may be trends more than absolute performance numbers. Such simulations also may identify flaws that cause deadlock or buffer overflows.

3.1.2.5 Software System Specification

SystemC models are also used for modeling and analysis of software-specific issues. As with device level hardware analysis, the timing requirements are typically much more relaxed than for SoC architecture analysis. Here we describe how people are using SystemC to describe software-specific behavior.

Multi-threading and OS Level Simulations

The most common use case for SystemC, when using multi-threading or operating systems (OS), utilizes untimed or timing annotated SystemC models of the hardware. These models are functionally accurate (instruction accurate), which permits execution of the same software executable on the SystemC model as on the final product hardware (often object code compatible, requiring no recompilation). Typically, software developers carry out software bring-up on the model in order to greatly reduce the time spent on software bring-up after silicon and system hardware is finally available. There may be some fine-tuning of software that is necessary after final hardware becomes available.

Since a SystemC-based system model has greater visibility and controllability than hardware, some developers may continue to use the virtual system prototype or platform even when hardware is available, to debug software and hardware problems. Performance requirements for a virtual platform model during software bring-up, when using an OS, are similar to those required for multi-chip or device-level simulation. Generating timing information for an OS system call typically takes 50–500 clock cycles in hardware. Impact of pipeline delays for memory access and detailed interconnect arbitration is usually one to two orders of magnitude lower than the impact of thread switching in software. So abstract SystemC models that use abstract models of interconnect and processor pipelines are sufficient for software functional validation and gross software timing behavior at the task switching level.

Another use model is to simulate OS behavior directly in SystemC. SystemC threads are used to emulate and dynamically schedule user tasks and threads. This allows a quick analysis of different thread-scheduling algorithms without modeling the full hardware of the end product. The SystemC model mimics the OS behavior in simulating tasks, data communication, and task switching between the various applications. Application software itself is either a SystemC model or C/C++ routines using special APIs to register tasks and access system resources.

The new SystemC feature required for this kind of modeling is the dynamic creation of SystemC threads. Traditionally, SystemC threads used to implement concurrent processes need to be statically created during model elaboration and could not be dynamically spawned. However, OSCI added new constructs in SystemC 2.1 to permit spawning processes dynamically to allow emulation of OS threads. In this kind of model, the software and model is compiled to run natively on the simulation host rather than to run on a specific target processor model. This speeds up simulation, but is less accurate than when the software executable is simulated on a processor ISS model. However, the approach emulates OS behavior accurately enough to allow the designer to explore a variety of scheduling algorithms and communications protocols and make reasonable choices for controlling the target OS.

To emulate the timing behavior of a multi-threaded system, the user can annotate the behavior of each dynamic thread with timing estimates. This permits users to determine if there is enough capacity to accommodate OS overhead, and the amount of system slack. This may allow them to trade off use of an OS with multi-threading, vs. a multiprocessor approach where each processor is dedicated to a specific task.

The example below shows a basic task thread infrastructure BaseTask. Two tasks are shown, implementing a communications thread, which receives values from and sends values to other threads, as well as an autonomous thread. The class BaseTask is built on top of the OSCI dynamic fork-join thread functionality described in the SystemC 2.1 examples.

```
// base thread infrastructure example
class BaseTask : public sc_task_module
{
public:
 typedef sc_task_module     base_type;

BaseTask(sc_module_name name)
: base_type(name)
, processing_delay("processing_delay",10) // specifies processing delay
, wait_delay("wait_delay",100) // wait until next thread starts
, iterations("iterations",10)  // Number of iterations
, verbose("verbose",false)     // verbose flag
{
}

 virtual ~BaseTask() {
}

protected:
 unsigned int processing_delay;
```

```
 unsigned int wait_delay;
 unsigned int iterations;
 bool                  verbose;
};
```

A class inheriting from BaseTask will define a processing delay, a wait delay, the number of times the task is executed when the process is spawned, and control over messaging. From this generic behavior the user may develop specific task classes; for example, the AutonomousTask below, that does not communicate with other tasks.

```
// Basic Task example
class AutonomousTask : public BaseTask
{
public:
 typedef BaseTask        base_type;
 AutonomousTask(sc_module_name name)
 : base_type(name)
 , m_counter(0)
 {
 }

 // specifies the task behavior
 virtual void task() {

// start the execution
while (true) {
while (m_counter < iterations) {
task_consume(processing_delay);
task_wait(wait_delay);

        m_counter++;
}

 // initialize
initialize();

 // suspend the thread
task_suspend();
}
}

 private:
unsigned int    m_counter;
};
```

The CommunicatingTask example below sends and receives data from other threads. It binds the thread to base communication channels using **p_put** and **p_get** in the constructor.

```
class CommunicatingTask : public BaseTask
{
public:
 typedef BaseTask        base_type;
 typedef unsigned int        data_type;
 typedef sc_task_blocking_put_port<data_type>  put_port_type;
 typedef sc_task_blocking_get_port<data_type>  get_port_type;

 put_port_type p_put;
 get_port_type p_get;

 FeedForwardTask(sc_module_name name)
 : base_type(name,start_behavior)
 , p_put("p_put")
 , p_get("p_get")
 {
 }

 virtual void task() {
 while (true) {

 for ( int i = 0; i < iterations; i++ ) {
         // read input
         m_data = p_get->get();
 }

 // execute the task functionally with annotated delay
 task_consume(processing_delay);

 // write output
 p_put->put( m_data );
 }

 // suspend the tasks
 task_suspend();
 }
private:
 data_type     m_data;
 };
```

Both static and dynamic thread creation can be mixed in one SystemC simulation. This is useful for describing multi-core systems running multi-core operating systems. The multi-core operating systems distribute threads to various cores. Resource virtualization enables the system to run software while being unaware of the underlying hardware, although resources such as memory and processors can be shared. Of course, sharing has a performance impact, because threads must wait for access. In this kind of system, a model allows designers to analyze the impact of more cores, and more sophisticated virtualization algorithms in the OS.

3.1.2.6 TLM 2.0

OSCI TLM 2.0 standardized transaction-level modeling (TLM) in SystemC. The reason is that by utilizing the communications mechanism introduced in SystemC 2.0, using ports, interfaces, and channels, users could implement different transaction-level interfaces. Axiomatically, since the data formats, control formats, parameters, and specific function call syntax and names are not identical for different models from different sources, then the components are not interoperable. This is true even if the users are all writing models and interfaces for exactly the same bus protocol, since there was no standardization of names, parameters, types, and methods. For example, for an Advanced Microcontroller Bus Architecture (AMBA) Advanced High-performance Bus (AHB), one user might call a burst write method "write_burst" and another user "Burst_WRITE"; one user might order the parameters for address and write data {address, data} and another might order them {data, address} in the methods. Even though the users are implementing the exact same transaction semantics for the exact same bus protocol, their method prototypes would be incompatible and thus their models would not be interoperable.

TLM2.0 fixes this issue by standardizing the interface method calls. It defines a set of transport calls, which have to be used to be compliant. The goal is to enable interoperability between high-level component models, which can then be plugged into any TLM 2.0 compliant system model. This reduces the necessity of using adaptors or wrappers for model integration. It also makes models from different sources available for integration. As well as the standardized interfaces, the TLM 2.0 standard defines a set of modeling styles, a generic payload as well as more than 150 rules for the expected behavior of TLM 2.0 compliant models in the simulation.

Some of these rules clarify the semantics of a transaction. A large part of the rules specify restrictions to guarantee safe and reliable communication between components. Most of them ensure that no additional transactions are accidentally created and that no transactions are accidentally overwritten along a communication path. Besides standardizing on TLM 2.0, OSCI also defined a protocol specification, called the base protocol, which is one incarnation of the TLM 2.0 standard. For the base protocol specification, OSCI provides example models as a reference.

TLM 2.0 Modeling Styles

TLM 2.0 defines two modeling styles: loosely timed (LT) and approximately timed (AT). The LT modeling style is mainly targeted for system and instruction-accurate platform models, where the need for cycle accuracy is not necessary and functional validation only requires correct ordering of transactions. This means, for example, that while emulating the hardware it is not important that specific transactions occur at exactly the same time as in the real hardware. Since correct use of the LT style can speed simulation up significantly, especially when combined with other techniques, this is especially useful for software development models, often called virtual platforms or virtual prototypes (see Chapter 5).

Some timing information may be available in LT models; but arbitration of shared resources and the impact of resource conflicts and contention on the system performance is not modeled and thus not taken into account. For LT modeling TLM 2.0 defines two timing points, the beginning and end time of a transaction as shown in Figure 3.3. This modeling style is simple, but if used for hardware where concurrent accesses to resources have a significant impact on the behavior of the system, it could lead to significant errors in the model. Analysis of the impact of concurrent contending access to resources is not possible in LT. Note that the modeling style does not specify a level of abstraction. For example, systems without delays due to concurrent resource access can be modeled with 100% timing accuracy.

In addition to the LT modeling style, TLM 2.0 defines the approximately timed (AT) modeling style. The goal of AT modeling is accurate modeling of resource contention and arbitration. It is used for systems where correct functioning has a

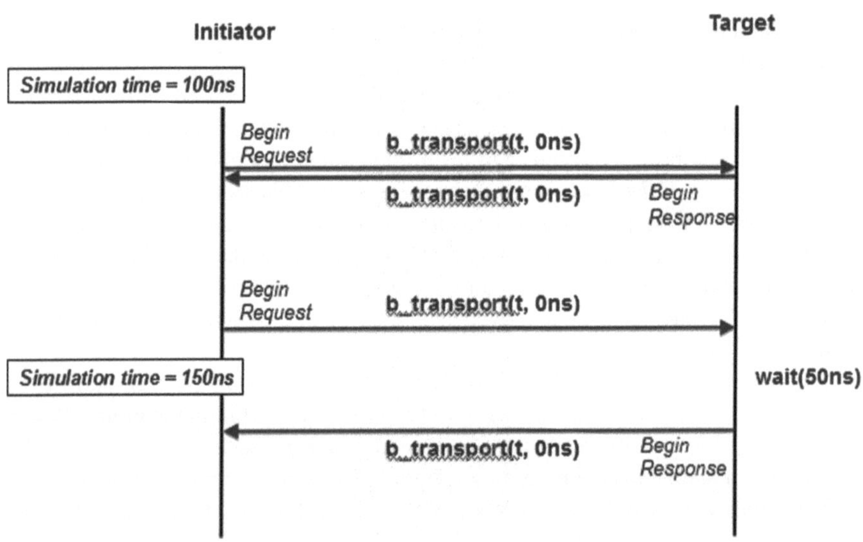

Fig. 3.3 LT model with blocking transport call

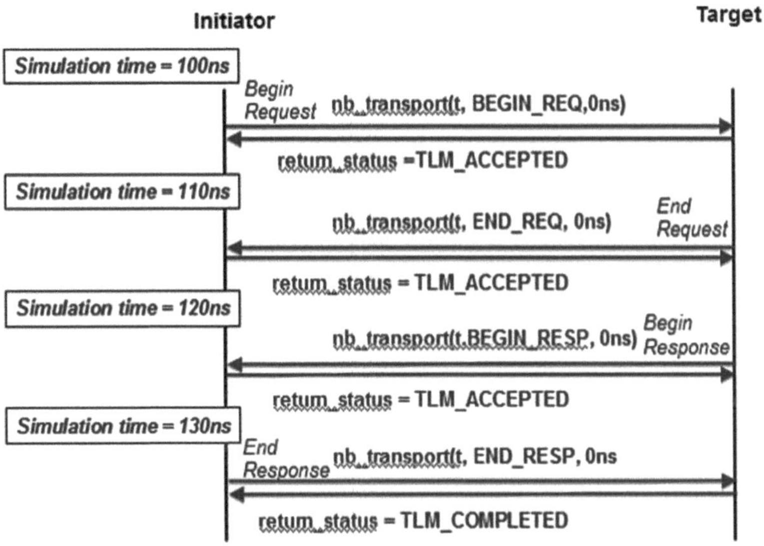

Fig. 3.4 AT modeling style with non-blocking transport calls

strong dependence on timing, such as those with stringent and detailed real-time requirements. These are often found in the dataplane rather than the control plane of embedded systems.

In AT, timing dependencies between components can be modeled very accurately. The AT style defines four timing points for a transaction: begin request, end request, begin response, and end response as shown in Figure 3.4. These act as synchronization points for models. AT can also be extended to additional timing points if necessary. Thus, systems with delays due to concurrent resource access conflicts can be modeled accurately. TLM 2.0 allows users to mix both LT-and AT-modeling styles together within one simulation. The goal of AT modeling style is to allow cycle-accurate modeling in TLM 2.0. It is possible, if all model developers agree on the exact meaning of the phases, and issue the transactions at the correct points in the overall simulation model.

The recommendation of the OSCI TLM 2.0 working group is to use LT modeling for software application development, software performance analysis, and for hardware architectural analysis for some kinds of systems (e.g., not particularly real-time critical control plane systems). AT modeling should be used for hardware performance verification and hardware functional verification.

TLM 2.0 Transport Calls

Beside the two modeling styles, TLM 2.0 defines two interfaces, a blocking and a non-blocking transport call. The blocking transport call has a reference to a generic payload and a time parameter (b_transport(TRANS&, sc_time&)). In the time parameter the user annotates how much time is consumed to transmit and process the data. The target either waits for the annotated time and then annotates its own delay to the transport call, or it can also return immediately, adds its time to the delay and annotates the results to the delay. The standard defines that the caller module or initiator does not continue any processing while issuing a blocking transport call. It waits until the transport call returns and then continues execution. Blocking transport calls can call SystemC wait for an event to occur, or enough time to pass to emulate timing behavior. This means that blocking calls can only be called from an SC_THREAD or an SC_CTHREAD process and not from an SC_METHOD process.

```
// blocking transport interface class uses generic payload
   as a template
template < typename TRANS = tlm_generic_payload >

class tlm_blocking_transport_if :
   public virtual sc_core::sc_interface {
public:
virtual void b_transport ( TRANS& trans ,
    sc_core::sc_time& t ) = 0;
};
```

In addition to the blocking interface, TLM 2.0 defines a non-blocking interface. Two types of non-blocking transport calls exist, a forward path nb_transport_fw(TRANS&, PHASE&, sc_time&) and the backward path nb_transport_bw(TRANS&, PHASE&, sc_time&). They indicate if the transport call comes from the initiator of the call or from the targeted component back to the initiator. They are referred to as the forward and backward paths, respectively. For both directions the transport calls return immediately, so that the caller module continues its processing, while the callee responds. This mimics, for example, continuous software execution on a processor, while the memory accesses are processed by the memory subsystem in parallel. Non-blocking calls can be called from any of the SystemC processes.

Beside the transport object and the timing annotation parameter, which are used in both the blocking and non-blocking interfaces, the non-blocking transport calls have an additional phase parameter. The phase parameter represents the four default phases of the TLM 2.0 protocol. As described above and shown in Figure 3.4, the phases indicate beginning of a request (BEGIN_REQ) and end of a request (END_REQ) as well as beginning of a response (BEGIN_RESP) and end of a response (END_RESP).

The blocking transport call is used mainly with the LT modeling style and the non-blocking transport call is used mainly with the AT modeling style. However, this is not a fixed rule, and either call can be used with either style; TLM 2.0 also allows mixing the two different transport calls in one simulation. It defines specific rules to control the usage and ensure interoperability between the two interfaces.

```
// enumeration time for return value of nb_transport callse
enum tlm_sync_enum { TLM_ACCEPTED, TLM_UPDATED, TLM_COMPLETED };

//
template < typename TRANS = tlm_generic_payload,
 typename PHASE = tlm_phase>

class tlm_fw_nonblocking_transport_if : public virtual
   sc_core::sc_interface {
public:
virtual tlm_sync_enum nb_transport( TRANS& trans,
                                     PHASE& phase,
                                     sc_core::sc_time& t ) = 0;
};
```

TLM 2.0 DMI and Debug Interfaces

Beside the transport interfaces there are also a direct memory interface (DMI) and a debug transport interface defined. The debug interface is intended to retrieve data from the models without delays or side effects in the model. The debug transport interface has only a forward interface. This makes sense, because the debug calls from the initiator need to return immediately and have no side effects on the simulation. Additional phases or a backward path is not required.

```
// debug interface is templatized with the generic payload
template <typename TRANS = tlm_generic_payload>
class tlm_transport_dbg_if : public virtual sc_core::sc_interface
{
  public:
  virtual unsigned int transport_dbg(TRANS& trans) = 0;
};
```

DMI is intended to speed up memory accesses for the LT modeling style by bypassing the sockets and the interconnect models and directly access the targets. In this case the targets are typically memory or memory-mapped registers in peripherals or accelerators. The DMI has both forward (initiator-to-target) and backward (target-to-initiator) interfaces. Some people use the DMI interface for tightly coupled memory.

TLM 2.0 Generic Payload

TLM 2.0 also defines a generic data format (generic payload (GP)) for the transport object, shown in Figure 3.5. It contains data and address information as well as typical attributes for memory-mapped buses such as control commands and several mode flags. In addition it has a response status field, so that the callee can

Attribute	Type	Default Value	Values

Command	tlm_command	TLM_IGNORE_COMMAND	enum tlm_command {
			TLM_READ_COMMAND,
			TLM_WRITE_COMMAND,
			TLM_IGNORE_COMMAND
			};

Address	uint64	0	
Data pointer	unsigned char*	0	
Data length	unsigned int	0	
Byte enable pointer	unsigned char*	0	
Byte enable length	unsigned int	0	
Streaming width	unsigned int	0	
DMI hint	bool	False	

Response status	tlm_response_status	TLM_INCOMPLETE_RESPONSE	enum tlm_response_status {
			TLM_OK_RESPONSE = 1,
			TLM_INCOMPLETE_RESPONSE = 0,
			TLM_GENERIC_ERROR_RESPONSE = –1,
			TLM_ADDRESS_ERROR_RESPONSE = –2,
			TLM_COMMAND_ERROR_RESPONSE = –3,
			TLM_BURST_ERROR_RESPONSE = –4,
			TLM_BYTE_ENABLE_ERROR_RESPONSE = –5
			};

Extensions	(tlm_extension_base*)[ ]		

Fig. 3.5 TLM 2.0 generic payload

communicate status information to the caller and vice versa. TLM 2.0 specifies that the GP object can be extended. To do this, an extension flag is set to indicate that the user is using the options to extend the transport object. For the base protocol the extensions need to be "transparent." That means that the callee does not have to respond to the specific items in the extended protocol. It does mean that the callee must make sure that the data object is forwarded correctly to the next downstream component. The caller cannot rely on the fact that every component understands the extended protocol, as long as it passes it on. If the extensions cannot be transparent, it is recommended to define a new traits class and not use the base protocol.

Traits class is a C++ specific construct. It allows you to group various template parameters into one traits class and then templatize the template with that traits class. So it relieves the user from having to know the detailed parameter list.

For the TLM 2.0 protocol this means that the initiator/target sockets are parameterized with a protocol traits class that defines the types used by the forward and backward interfaces. This is the payload type and the phase type. Each protocol traits class is associated with a specific interconnect protocol. The default protocol type defined for TLM 2.0 is the class tlm_base_protocol_types as shown below.

```
// The base protocol traits class:
struct tlm_base_protocol_types
              { typedef tlm_generic_payload   tlm_payload_type;
               typedef tlm_phase              tlm_phase_type; };
```

As an example to show the usage of the traits class you will see below the combined forward interface. The backward interface can be found in the TLM2.0 LRM.

```
// Forward interface
template< typename TYPES = tlm_base_protocol_types >
class tlm_fw_transport_if
 : public virtual tlm_fw_nonblocking_transport_if
                         < typename TYPES::tlm_payload_type ,
                          typename TYPES::tlm_phase_type> ,
  public virtual tlm_blocking_transport_if<
                          typename TYPES::tlm_payload_type> ,
  public virtual tlm_fw_direct_mem_if<
                          typename TYPES::tlm_payload_type> ,
  public virtual tlm_transport_dbg_if<
                          typename TYPES::tlm_payload_type>
{};
```

Temporal Decoupling and Global Quantum

The main purpose of temporal decoupling is to help increase simulation speed. With processor instruction set simulators at the instruction-accurate level reaching several hundred MIPS in simulation speed, and the potential to bypass interconnect models using DMI, the next bottleneck in a SystemC simulation is the thread context switch time in the kernel.

To speed up simulation, the goal is to avoid context switching that occurs for an SC_METHOD when the process reaches the end (run to completion), and in an SC_(C)THREAD when the process hits a wait statement. With enough modules and threads, a typical SystemC simulation model will have many context switches per second.

The temporal decoupling mechanism decouples initiator threads (for example, processors or key processing blocks) from the rest of the simulation, so that individual initiators can run ahead of the global simulation time – potentially hundreds, thousands, or even millions of cycles ahead of global time. They maintain an internal local time and execute without giving control back to the SystemC kernel – in fact, the objective is to run as long as possible, until either a specified interval has elapsed, or it is necessary to synchronize with some other models via an interface.

This significantly reduces the amount of context switching and increases simulation speed. It also enables processor models to use just-in-time (JIT) compiled code techniques which translate target instructions to simulation host equivalents and replace interpreted code with compiled code. TLM 2.0 specifies a quantum keeper (tlm_quantumkeeper), which controls the local time during the simulation. A set of functions specify the interface to the quantum keeper.

People developing initiator components (such as processor ISS wrappers) use the same concepts to control the local and the global simulation time. TLM 2.0 specifies functions to increment the local time with a specified delay to mimic timing delays inside the initiator. The user can get the current and the local time as well as check if synchronization is required and then synchronize with the rest of the system. The local time is defined as the offset to the global time. Any query to the simulation kernel reveals information about the global time before the quantum keeper starts to manage the local time.

TLM 2.0 also specifies a global quantum. The global quantum is the maximum amount of time an initiator can run ahead of the rest of the simulation. This ensures that each initiator model gets time to execute and keep up with the rest of the system – i.e., some amount of elementary fairness and to ensure that every initiator gets a time slice.

TLM 2.0 Rules

The TLM 2.0 protocol is targeted to systems with multiple callers (initiators) and callees (targets). Most of the systems have more than two components in the communication path. The initiator initiates the communication and the target responds to the communication requests. A typical design has several initiators (EInitiator) and targets (ETarget), which are connected via interconnect(s) (see Figure 3.6). The interconnect (EInterconnect) serves as both a target and initiator.

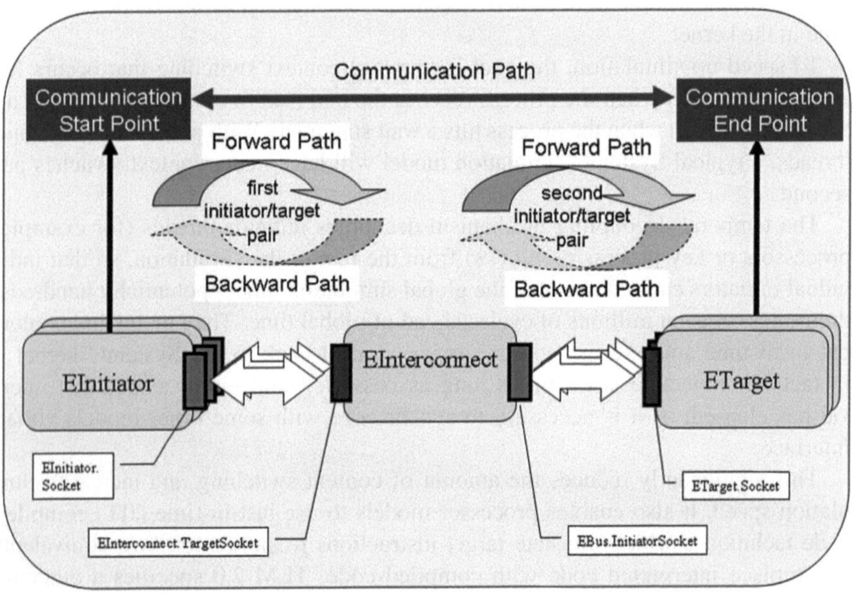

Fig. 3.6 System model

The communication path is the path between a communication start point and a communication end point. A communication path typically contains more than one initiator/target pair. In the simple example model in Figure 3.6 there are two initiator/ target pairs in one communication path (EInitiatorSocket → EInterconnect TargetSocket and EInterconnect InitiatorSocket → ETargetSocket). Real system models contain interconnect hierarchies, multiple different interconnects connected via bridges, multiple subsystems, and/or cache hierarchies. Thus in a system model, there are typically a lot more initiator/target pairs making up one communication path.

For the LT modeling style, only the forward path exists and the transport call between each initiator and target is typically blocking. For the AT modeling style the forward and the backward paths exist and the transport calls between the initiator and target are typically non-blocking.

However, this does not mean that the first initiator/target pair needs to wait for the second pair to finish before it can finish its local communication, in either style. Thus a large amount of concurrent processing occurs in the system model. To control the communication and make sure that transactions are not overwritten or lost in the system, TLM 2.0 defines more than 150 rules. Several rules specify the intended behavior of the communication. Several of them contain provisions to prevent components from malfunctioning in the system.

For example, there are several rules defined to avoid access conflicts for the GP. They regulate the ownership of the GP at any point in time during a simulation. They also define what the user needs to do to enable safe reuse of any transport object. The rules are designed to prevent incorrect communication behavior in the system as well as accidental duplications or destruction of transactions in the system.

The base protocol has been defined to illustrate one example of a TLM 2.0 compliance protocol for both the LT and AT modeling styles. For the LT specification, blocking interfaces are assumed and for the AT model non-blocking interfaces are assumed. The set of rules for the base protocol provides a detailed specification of the expected behavior. The goal of the rules is to clarify the semantics of specific constructs. They also specify restrictions on the protocol to ensure safe communication. The base protocol allows extensions of the generic payload as well as the phases. The intention is to specify the base protocol such that it covers a large set of interconnect protocols typical of current systems.

There are more than 150 rules defined in the TLM 2.0 library reference manual. Most of the rules define the semantics of the specification. About 50 of the rules apply restrictions to the models, required to achieve secure interoperability between models. In the following discussion, we will focus on those rules that impose restrictions on the user. See the TLM 2.0 documentation [8] for more details.

- **Expected phase transition:** A set of rules specifying expected behavior for each phase transition as well as disallowed transitions. In this way, every user has the same view of what states and what state transitions are allowed and which ones are not. Specifically, phase transition rules specified for the base protocol give a good understanding of what is required to guarantee safe communication, without either losing transactions or needing the creation of additional copies of transactions. The goal, while defining the phases and the transitions, was to provide enough flexibility to mimic various bus protocols, without making it too complex for every component to support the protocol. With the required two-phase protocol for the LT models and the four-phase protocol for AT models a large number of protocols can already be modeled. TLM 2.0 allows the extension of the generic payload as well as the number of phases. For the base protocol it is expected that the BEGIN_REQ and END_REQ are the first two phases and initialize the communication of one data token. BEGIN_RESP and END_RESP are the last two phases and end the communication of one data token. The base protocol also restricts the number of transactions with a BEGIN_REQ phase "in flight" from any initiator to any target. This requires that the END_REQ phase from the target has to be received by the initiator, before it can send another BEGIN_REQ. This prevents register overflow in the target of each initiator/target pair. The target can control the speed at which it responds to the initiator and this prevents more transactions than can be handled arriving at the target socket.

- **Initialization and restricted access to generic payload:** A set of rules addressing the initialization and accessibility of the generic payload. For efficiency reasons, the generic payload is not really transmitted through the high-level system model. References to one global copy of the generic payload from each component in a data path are used to mimic the transmission of data through the system. Beside the speed benefits of not copying values, this mechanism also keeps the communication between components flexible. Various transmission schemes controlling what data is actually transmitted to each component can be easily exploited without changing the model. The disadvantage is that all components can access and modify the generic payload at any time in a simulation. Implementing a mechanism inside the generic payload that identifies and controls which component accesses what data field at what time is too expensive in terms of simulation performance. The rules define which component type, initiator, interconnect, and target can modify which attribute in the generic payload. They also define the default values during the creation of the generic payload to prevent incorrect behavior in the system.
- **Memory management:** A set of rules addressing the expected mechanisms for memory management, visible to all components in the system. TLM 2.0 allows implementing specific memory management components or controlling their reuse directly in the initiator. The goal is to ensure proper memory management and avoid memory leaks. It also enables a safe generic payload reuse mechanism. The protocol specifies that a component can use the reference to the generic payload object beyond the actual transport call. So if an initiator or memory manager does not know that a component still has a reference to the generic payload and either deletes or reuses the object for another transaction, significant side effects can be created when this component uses the reference.
- **Timing:** A set of rules addressing the semantics of timing annotation specified in the transport calls. They also make sure that the simulation time progresses during a simulation to prevent any malfunction due to time stalling.

- **Transaction path:** TLM 2.0 rules also define the transaction path. They specify that every transaction has to take the same route through the system in the forward and in the backward path. This allows the detection of incorrectly routed transactions.
- **DMI and debug interface:** DMI and debug interfaces use different attributes for their transactions. There are a set of rules that define the initialization values as well as the restrictions as to which component can access which attribute.

What if the Model Is Not TLM2.0 Compliant?

In this section we provide several examples of rule violations and their consequences to the system behavior. Detecting interoperability issues is difficult, because more than one component is involved. If those components come from different sources, debugging becomes even harder. In addition interoperability issues are often not immediately visible. The developer only sees the consequences, such as a performance drop on a communication link or wrong pixel values for video processing. Typically the malfunctioning behavior is visible only much later in the simulation. The user might have to backtrack from the incorrect functionality over multiple hundreds to thousands of cycles to the original source. Interoperability issues require monitoring of transactions on the communication paths between the initiator/target pairs. It is often not possible to check for interoperability issues inside the initiator or target model, because the simulation time has already progressed, which impacts the results. Many rules are defined end-to-end, so many observation points along the path need to be taken into account, which are not visible from any particular component.

One of the access rules specifies that the target is not allowed to modify the address attribute of the generic payload. If in a TLM 2.0 model, virtual address translation is implemented in the interconnect, then the real and the virtual addresses must be stored. In the backward path the address in the transaction might be used to identify the corresponding real address from which the transaction originated. In the case that the target modifies the address field in the transaction, the interconnect cannot translate the address back to the real address correctly anymore. The consequence to the system simulation is that the response cannot be completed. This results in a deadlock, in which the initiator waits for the transaction to be completed before it can send the next. In other rarer cases, the transaction gets sent back to the wrong initiator, which can cause various unfortunate side effects in the initiator. One, for example, is that correct transaction information from a different communication path is overwritten with incorrect values. That typically happens at a time when the initiator expects transaction values to be stable. The time at which the correct values are overwritten depends on when the SystemC kernel schedules specific processes. So the user could potentially observe correct behavior for many transactions, until the problem suddenly becomes visible.

For the TLM 2.0 base protocol, TLM 2.0 restricts the number of begin-request phases allowed in-flight between one initiator/target pair to just one. So the target must send an end-request first, before the initiator can send the next begin-request. This allows protection of the targets from memory or register overwrites. It allows the target to decide if it must respond immediately to the begin-request, but then it has to be ready to receive another transaction. Or it can process the value in the transaction first, before it sends the end-request and thus does not yet have to deal with another transaction. Note that the rules are not relative to the full communications path; they are specified per initiator and target pair. This way features such as out-of-order execution and parallel thread processing can be modeled. In this case the intermediate target model can wait for multiple begin transactions before it forwards the transaction to the next target, while it changes the order in which they are sent.

Ensuring these interoperability restrictions and observing problems at the starting point of a communication path are trivial. In shared interconnect systems, perhaps with multiple interconnect hierarchies, this rule becomes more and more difficult to enforce or check. Now multiple transactions can be in flight inside the system. The interconnect models have to ensure that for all of the initiator sockets to all of the targets, this rule is obeyed. For each communication path, a number of transactions may be in-flight. For some systems, this means multiple10s–100s of parallel transactions may be in flight at any point in time during a simulation. In this case enforcement or rule checking needs to happen for every single transaction at every point in time to avoid malfunctioning system behavior.

In general, all of the examples discussed show that rule violations cause interoperability problems. This in turn can create major system behavior failure. Those system failures may be immediately visible in the form of system crashes, obvious wrong behavior or deadlock situations. Some others might be only visible to the system experts, because of inconsistencies with the expected behavior or performance results. It is these which are most dangerous to overall system correctness, because they may go unnoticed.

Why are these interoperability issues so difficult to find? In the following section we summarize a debugging effort and trouble-shooting approach, in the case that the model cannot be automatically checked for TLM 2.0 compliance.

3.1.2.7 TLM Compliance Checking

Manual Debugging of Interoperability Issues

This section describes the debugging effort involved to find interoperability problems. Finding interoperability issues during component development is often not possible, if there is no automated checking available. Typically the testbenches are developed by the same person who developed the component model, so if the interface specification is not correctly implemented, it is typically incorrect in both the testbench and the model. Thus, the first problem is that interoperability issues are only visible when models from different developers are integrated. In the following discussion we describe a debugging example to show the complexity and the effort involved in finding interoperability issues in a typical virtual platform.

The first problem is the complexity of the models. A typical virtual platform may contain 20–40 individual component models. It contains hardware as well as software components. All have their own debugging infrastructure. For example, each processor model is typically integrated with its own software debugger. Typically no one person has the expertise and knowledge about every component in a virtual platform. So either the model is well documented or multiple people are involved in debugging a virtual platform.

A top-down approach is used to narrow down the location of the problem. To be able to quickly identify problems in the system, the user monitors performance parameters as well as reference traces. The reference traces are typically system outputs, but sometimes intermediate results are also available. If performance results are inconsistent with expectations, or the outputs differ from the expected reference trace, interactive debugging starts.

To illustrate the different debugging steps to find interoperability problems a multimedia subsystem is used. Typically virtual platform models have application-specific viewers, which display the results. For video applications this could be a frame viewer, which displays the video sequence. While looking at the video sequence, the user sees that the individual frames in a video sequence are incorrect. Several lines in a frame are still intact, but lines are shifted randomly throughout the frame, or other visual artifacts are apparent. Because the lines of a picture are recognizable and some picture segments are intact, the problem appears to be incorrect address calculation for the pixel address. Because the picture is intact, most probably the problem occurs after decoding and before displaying the frame.

The next step is to look at the software which is responsible for triggering frame processing as well as the calculation of the pixel address. The user sets breakpoints in the source code relevant for frame processing. While stepping through the source code line by line, the user can observe values and value changes for specific register or memory locations in the system, either using physical references or in source-oriented debuggers, using variable displays. In most debuggers the user can set watchpoints to identify when values change. The simulation stops when a specified value occurs on a watchpoint. We set a watchpoint on the register value, which starts the processing of the next frame. If the value is 1, frame processing is supposed to be triggered. In our case the frame synchronization software works fine. Registers are set with the correct values and the frame processing in the hardware component starts. We no longer assume that the software application might have caused the problem and move to the next debugging step – looking into the hardware models.

For component debugging, ESL tools typically support system debugging as well as source code debugging for components. In our example multiple hardware accelerators participate in video processing. Most of the components move pixel information in and out of local and global memories. Many of the virtual platform development tools allow the display of all memory regions in the system. Some also allow tracing the read and write transactions to and from memory or between components with message sequence charts (MSCs) or other transaction specific viewers. Most ESL debugging tools are also SystemC aware. Being able to set breakpoints for individual SystemC instances, instead of the SystemC module source code, significantly simplifies debugging. The debugger stops the simulation only for that specific instance, instead of for every instance of that module in the simulation, when the source code is executed. It is typical that multiple frames are processed concurrently in a pipelined fashion. This makes debugging much harder because the user needs to know which frame is being processed in which component at any point in time.

In our example we identified that the processing is correct until the transactions reach the frame buffer. We can see that the register which stores the starting addresses contains some unexpected values. Now we trace back to the location where the next starting address is generated. This is also done in software. Here we observe the generated frame addresses. In this scenario, the generated addresses seem to be correct, but the address used to store pixels in the frame buffer was incorrect. So we have narrowed the problem down to the communication between the frame buffer and the processor, which calculates the start address for the next frame buffer. Now we need to have a much closer look at the TLM 2.0 socket interface of the processor. Because the processor is connected, along with other processors and DMA controllers, to a shared interconnect, we have to deal with multiple transactions from various sources in flight inside the interconnect model. The model used a four-phase AT base protocol, so we look at the transaction traces between the processor and the frame buffer.

Eventually we find out that we have an interoperability problem. To avoid the implementation of the detailed interrupt driven synchronization process between the hardware and the processor, the model developer used the TLM 2.0 END_REQ transaction phase as an indicator from the frame buffer to the processor that the current frame has been processed and the next frame address can be provided. Unfortunately the interface controller, which sends the transaction to the interconnect according to the protocol had a bug.

The TLM 2.0 protocol specifies that the initiator waits for the end-request (END_REQ) phase, before it sends the next begin-request (BEGIN_REQ). This guarantees that only one transaction is in flight to one specific target. In our case, the interface controller did not wait for the END_REQ transaction, but generated a new BEGIN_REQ transaction with a new frame buffer address in some corner case situations. What we thought was just a shift of lines inside one frame, was actually several frames compressed into one picture. The processor provided the next frame address too early and had thus overwritten the value in the register which holds the start address many times, before the start address was used. This created the effect that the start addresses were not nicely spaced, so that the addresses between them can store a full frame, but rather, the start addresses were much closer together, so that pixels of one frame were overwritten with pixels from the next frame. Thus, the problem that we thought was just an incorrect calculation for the pixel address was in fact a synchronization problem of the frame address generation of the processor with the rest of the system.

This real life example shows that debugging interoperability issues can be a very long and tedious debugging step. In this scenario, we also described only the successful debugging steps, and left out the cases which we investigated some hypothetical causes, but everything turned out to be correct. To be able to debug interoperability issues, the user needs detailed knowledge about the application, the hardware and software implementation as well as the synchronization mechanism for data communication in order to identify the problem. Successful debugging also assumes that detailed documentation of the model is available to identify expected values for specific scenarios. Typically multiple people need to be involved to support this effort.

Automated Compliance Checking

In the previous section we described what problems occur and the effort involved when interoperability issues occur between components. Now we describe an automated way to find those issues – using a commercial ESL tool to illustrate the concepts. The JEDA TLM 2.0 Compliance Checker detects rule violations automatically and reports to the user the time the violation occurs, which components are involved, and the transactions involved in the violation. If the user chooses certain tool controls, they can also identify the full payload information as well as additional status information about the violating transaction in a verbose mode. Pinpointing the time problems occur, the components involved and the violation reduces debugging time significantly. Removing interoperability problems lets the user focus on the internal functionality of the components.

Beside interoperability problems, the JEDA TLM 2.0 Compliance Checker also checks for routing problems. Typically multiple TLM 2.0 checker components are instantiated in a system. They monitor all transactions in the system in parallel, and follow the forward as well as the backward path of a communications path. Thus the tool has the visibility of the full system, and can later also be extended with user-defined system level checks. At the time of writing, the checker automatically checks for performance violations. Figure 3.7 illustrates how the JEDA TLM 2.0

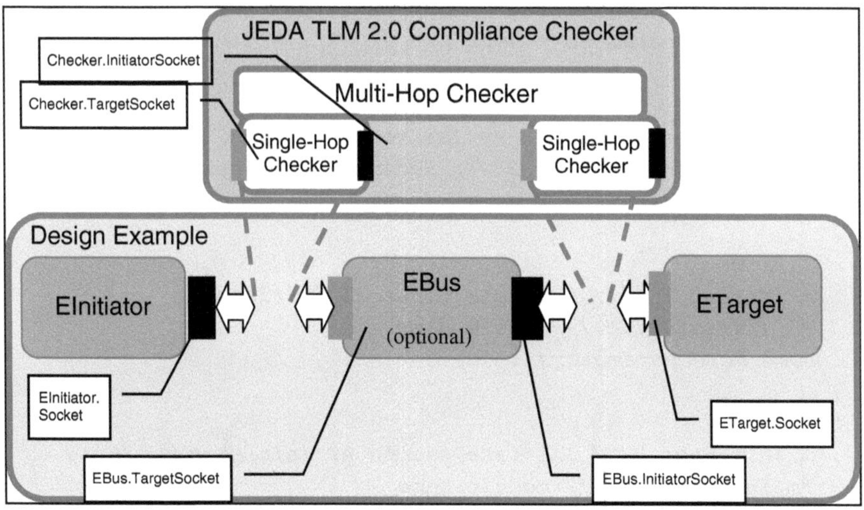

Fig. 3.7 TLM 2.0 compliance checker integration

Compliance Checker is integrated into a system. Unfortunately the TLM 2.0 analysis port is not sufficient to collect enough information, so such a checker needs to be inserted into the system.

The source code for the checker integration in a system of two initiators and two targets connected via one interconnect is described below.

```
// Checker integration example

// Include Checker header file
#include "jtc_checker.h"

// Define Checker modules
jtc_checker::jtc_t_socket<> initiator_checker1;
jtc_checker::jtc_t_socket<> initiator_checker2;
jtc_checker::jtc_t_socket<> target_checker1;
jtc_checker::jtc_t_socket<> target_checker2;

example_system_top::example_system_top
( sc_core::sc_module_name name
)
  : sc_core::sc_module /// Initiate models
  ( name
  )
  , m_bus /// Initiate Simple Bus
  ( "m_bus"
  )

  , m_at_target_4_phase_1 // Initiate first AT target
  ( "m_at_target_4_phase_1" // Sirst target name
... some more parameters here ...
  )

  , m_at_target_4_phase_2 // Initiate second AT target
  ( "m_at_target_4_phase_2" // Second target name
... some more parameters here ...
  )

  , m_initiator_1 // Initiate first AT initiator
  ( "m_initiator_1" // module name
... some more parameters...
  )

  , m_initiator_2 // Initiate second AT initiator
  ( "m_initiator_2" // module name
```

```
... some more parameters ...
 )
 , initiator_checker1 // Initiate checker between 1 initiator
 ( "initiator_checker1"  // and interconnect
 )
 , initiator_checker2 // Initiate checker between 2 initiator
 ( "initiator_checker2" // and interconnect
 )
 , target_checker1 // Initiate checker between 1 target
 ( "target_checker1"  // and interconnect
 )
 , target_checker2 // Initiate checker between 2 target
 ( "target_checker2" // and interconnect
 )

{
// Bind the checker between the initiators and the bus
initiator_checker1(m_initiator_1.initiator_socket);
initiator_checker1(m_bus.target_socket[0]);
initiator_checker2(m_initiator_2.initiator_socket);
initiator_checker2(m_bus.target_socket[1]);
// Bind the checker between the bus and the initiator
target_checker1(m_bus.initiator_socket[0]);
target_checker1(m_at_target_4_phase_1.m_memory_socket);
target_checker2(m_bus.initiator_socket[1]);
target_checker2(m_at_target_4_phase_2.m_memory_socket);
;}
```

The JEDA TLM 2.0 Checker is not only useful at the end of the development cycle, when the components are integrated into a system or virtual platform, but also during model development. When combined with other commercial tools, such as the JEDA TLM 2.0 Traffic Profile generator and the TLM 2.0 Smart target, the user can build a testbench for unit testing more quickly. During the simulation run, the checker provides detailed information in case a rule violation occurs. At the end of each simulation run, the checker generates a report about how many rules have been violated, which ones have been triggered or not triggered and which ones have been switched off by the user. This report allows the assessment of the completeness of the tests.

Figure 3.8 shows an example report. The detailed error report shows the time and the checker component which detected the violation, a detailed error message, and the expected behavior. Further down you can find the detailed information about the values for each TLM 2.0 attribute.

The summary report at the bottom of Figure 3.8 provides information on which rule has been violated with reference numbers corresponding to the rule number in

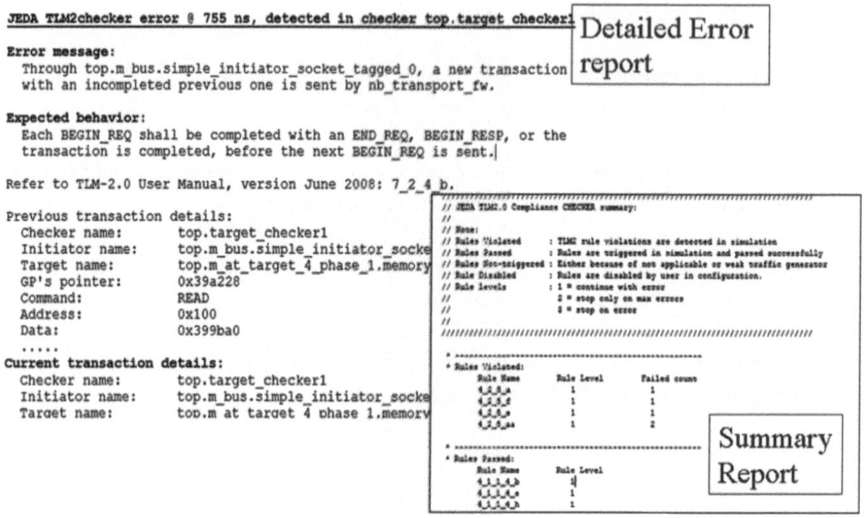

Fig. 3.8 TLM 2.0 checker report

the TLM 2.0 user manual. The rule level indicates the severity of the rule violation. Finally the information about how often the violation occurred within the simulation is provided. Besides identifying the rules which passed or failed, the checker also identifies the rules which are not triggered. This either identifies a hole in the tests, or it specifies a rule that is not relevant for that modeling style.

Typical average project delay due to protocol violations is at least 3–5 hours per protocol bug. Overall time can be lengthened by many days and even weeks. If many people in different locations are involved in the development of models, or to integrate those models into one system, the delay can grow significantly. The problems caused by interoperability issues identified with the JEDA TLM 2.0 Checker range from memory leaks, to deadlocks in the system, to system crashes and wrong performance numbers. The most problematic interoperability issues are those which alter the behavior of the system and cause system malfunctions or biased performance results. Even if TLM 2.0 models do not directly translate into hardware, interoperability issues, which are not detected, can cause system malfunctions not only in the model, but also in final silicon.

3.1.3 Architectural Models

3.1.3.1 Modeling IP

IP models are usually created in one of two ways: hand-crafted models most often written in C, C++, or SystemC; or generated models, possibly in any of these three language variants, as an output of an automated IP configuration and generation flow. Such models may exist for any IP component; as of 2009, models exist for many processors (ISSs at either the instruction-accurate transaction level, the cycle-accurate transaction level, or both), memories, bus bridges and other interconnect fabric, and a host of accelerators and peripheral devices.

When models exist in C or C++, an increasingly common practice is to also offer these models wrapped in a SystemC wrapper, potentially using the OSCI TLM 1.0 or TLM 2.0 standards, or a proprietary transaction-level modeling standard that can usually be adapted to an OSCI TLM standard in some fashion and can also be adapted to a variety of proprietary virtual platform modeling tools as are described in Chapters 5, 6, 7, and 8. Since TLM 2.0 was only introduced by OSCI in a definitive way in June of 2008, it may be that many models are not yet available in TLM 2.0 but an ad hoc user integration is usually possible.

Tensilica, for example, offers its processor ISS models in a variety of forms for different use models. These are described in Chapter 6, but include

- Standalone ISS which can be run in both cycle accurate and turboXim or fast-functional mode, or indeed, toggling dynamically between them, as the user requires. This can be used for accurate profiling of user software on a specific processor configuration, software verification, and performance and energy estimation.
- A proprietary transaction-level system modeling environment, XTMP, which predated the availability of OSCI SystemC by some years.
- A SystemC transaction-level system modeling environment, XTSC, which predated the availability of OSCI SystemC TLM 2.0 by several years.

Because these system modeling environments predated OSCI TLM 2.0, they use a proprietary notion of transaction, but these have been found possible to adapt both to TLM 2.0 and to a number of third-party ESL modeling tools with their own notions of transactions. This is very common in the IP modeling area. More details are available on these models in [9].

Many ISS models in the modern era are written to support instruction-accurate, compiled-code, just-in-time compilation, fast-functional execution for software verification and that fits in well with OSCI TLM 2.0 notions of time quantum execution. One reference to this style is [10].

3.1.3.2 System Models for Architectural Exploration

As discussed above, various IP models can be incorporated into system models for architectural exploration. This is discussed in more detail in Chapter 5, 6, 7, and 8. Architectural design space exploration with enough fidelity to allow good architectural tradeoffs will require cycle accurate or at least very accurate "cycle approximate" or AT models. This will then allow decisions made on various architectural configuration choices to be likely to lead to a superior architecture. A very simple example of a system model that uses underlying cycle-accurate models in

SystemC is the following simple XTSC model for a Tensilica processor with a simple memory attached is the following:

```
// Step 1: One include file for each component
#include <xtsc/xtsc_core.h>
#include <xtsc/xtsc_memory.h>

// Step 2: Add "using namespace" statements to simplify coding
using namespace std;
using namespace sc_core;
using namespace log4xtensa;
using namespace xtsc;
using namespace xtsc_component;

// Step 3: Get a logger for use as needed
static TextLogger& logger = TextLogger::getInstance("sc_main");
int sc_main(int argc, char *argv[]) {
 try {
 // Step 4: Initialize XTSC
 xtsc_initialize("../TextLogger.txt");
 // Step 5: Configuration parameters for an Xtensa core
 xtsc_core_parms core_parms(CONFIG_NAME, XTENSA_REGISTRY, "");
 core_parms.extract_parms(argc, argv, "core0");

 // Step 6: Construct the core
 xtsc_core core0("core0", core_parms);

 // Step 7: Configuration parameters for a PIF memory
 xtsc_memory_parms memory_parms(core0, "pif");
 memory_parms.extract_parms(argc, argv, "core0_pif");

 // Step 8: Construct the PIF memory
 xtsc_memory core0_pif("core0_pif", memory_parms);

 // Step 9: Connect the PIF memory
 core0_pif.connect(core0, "pif");

 // Step 10: Load the target program
 core0.load_program("target/hello.out");

 // Step 11: Set-up debugging according to command line
                                             arguments, if any
 xtsc_core::setup_multicore_debug(argc, argv);
 XTSC_INFO(logger, "Starting SystemC simulation.");
```

```
 // Step 12: Start and run the simulation
 sc_start();

 XTSC_INFO(logger, "SystemC simulation ended.");

 // Step 13: Clean-up
 xtsc_finalize();
 }
 // Step 14: Catch and log exceptions
 catch (const exception& error) {
 ostringstream oss;
 oss << "std::exception caught: " << endl;
 oss << error.what() << endl;
 xtsc_log_multiline(logger, FATAL_LOG_LEVEL, oss.str(), 2);
 }

 return 0;
}
```

3.1.3.3 System Models for Software Development

For software development, the fast-functional compiled code instruction-accurate ISS models are used, often bypassing the modeling of interconnect and using a variety of fast access methods to things like memory models, as was introduced in discussing SystemC TLM 2.0 earlier. In the case of the XTSC model above, the XTSC environment provides simple capabilities to let a user switch from one mode to another. If the SystemC model was compiled into a host executable called "mysoc," then using the command "mysoc –SimTurbo=true" would run the example in an instruction-accurate mode, up to 100 × faster than cycle-accurate model.

Chapter 5 discusses virtual platforms from the software development point of view.

3.2 Formal Models

3.2.1 Property Languages

In the previous sections of this chapter, we explored several languages that enable functionality to be defined. All of these languages are in the class of languages called procedural or imperative. What this means is that the lines of the model get executed in order, unless something causes it to branch to another location. In this way a set of steps can be defined that get the model to the expected result.

There is another class of languages called declarative languages that expresses the functionality without defining its control flow. In that regard they describe what the program *should do* rather than *how it will do* something. While declarative languages are not normally used for the description of hardware today (that is they are not in the primary implementation flow), they are an important way to specify functionality for verification purposes.

3.2.1.1 Uses of Declarative Languages

The declarative languages being used in the EDA industry are primarily the property specification language (PSL) and SystemVerilog assertions (SVA). These languages share a very similar subset which is used by most tools and will be the only parts of the language that will be discussed here. In both languages an aspect of functionality is defined in a property.

Accellera defined a property to be [11]

A collection of logical and temporal relationships between and among subordinate Boolean expressions, sequential expressions, and other properties that in aggregate represent a set of behavior (that is, a path).

In a typical design flow, there are several ways that properties can be used.

Property checking: Formal verification tools attempt to prove that a property holds true under all conditions. That means that whatever state a design may be in, and whatever inputs are supplied, that the property will always be true. If it is not then it means that a condition can be found wherein the design does not obey the property. As in all cases, this means that either the property is wrong or the design is wrong. This process is called formal property checking or formal verification. Formal tools will generally provide a counter example to illustrate how the property can be made to fail. This is highly useful in being able to debug the situation.

Interface checking: When a block is being implemented it is usually not appropriate to consider that all possible inputs can happen at any time. We expect them to follow reasonable sets of rules or behaviors that are associated with the functionality that is to be performed. In doing so a designer makes simplifying assumptions about the interface.

But what if those assumptions are not correct or if a block gets reused in another design which does not abide by those sets of assumptions? Properties are often used to define these assumptions, called constraints, so that if any input is sent into a block for which it was not designed, it will immediately flag an error, rather than attempt to fumble through and make a mess that could be very difficult and time consuming to debug. There have been some high-profile examples where these kinds of properties would have prevented catastrophes, such as the case of the Ariane 5 rocket [12]. In this case, software reuse from the Ariane 4 rocket was the culprit. The code for Ariane 4 was not subjected to the same acceleration levels as Ariane 5 was capable of and the code had been optimized based on the levels that were expected to be seen. With Ariane 5 this caused an overflow to occur and resulted in the guidance system of the rocket failing. Recognizing that its trajectory was in error, the rocket self-destructed. These interface constraints can also be used by formal property checkers to identify inputs that do not need to be considered when attempting to prove the property.

Assertions: While the ideal situation would be for all properties to be proven in a formal verification tool, this is not always possible. Formal tools have limits in terms of the design size or system complexity that they can handle. There are many innovative ways around these limitations, such as setting an initial state or limiting the depth of the number of clocks that are considered before a failure can occur. These conditions are usually called partial or incomplete proofs. When this happens, it may be necessary to continue checking those properties in a simulation environment. When used in this way they are usually called assertions.

3.2.1.2 Completeness

One of the problems associated with property set definition is knowing if the property set is complete. Put another way, are there any valid functionalities that have not been described by a property, or for that matter invalid ones that should not happen? In the simulation world a set of heuristics have been developed that attempts to find holes in the verification effort. This is normally referred to as coverage and will be discussed in the next chapter.

However, when attempting to perform formal verification the standard of completeness is set somewhat higher than for simulation. Simulation can never tell you that you are done with verification, but formal methodologies get a lot closer to that goal. When a user deploys formal tools, they want a much higher level of confidence than simulation is able to provide. That means that completeness is a much more important aspect of formal verification methodologies.

It should be noted that these languages were designed for use at the RTL level. The biggest difference that this makes is in terms of timing. At the RTL level, everything is synchronized to a clock. At the transaction level, there is no concept of a clock and thus events need to trigger the properties. This is somewhat awkward in PSL and SVA but not impossible [13]. Probably more important is the fact that at the transaction level things can happen simultaneously. There are no constructs in PSL or SVA to handle concurrency and no real work-around.

Another difference is that many of the things that properties would be applied to are no longer steady signals that can be sampled, but events or other ephemeral activities. One recommendation is that these be converted into signals. As an example, a transaction can be converted into a Boolean signal that is true while the transaction is in progress and false otherwise. Similarly, events can be turned into a pulse on a signal or a variable. If the width of a pulse is given a defined period, then it is also possible to have it contain a count of the number of occurrences within the defined period of time.

The downside to this approach is that it requires the transactions to consume time and that means they have to be suspendable. This in turn will slow down their execution considerably. There is a further problem in terms of potential overlap. Each transaction will have a start point and an end point. Assume that the transaction consumes no time, all we can say is that the end point will have occurred after the start point. But if another transaction happens notionally at the same time, then its start and end points are not known in relation to the first transaction. They could have been before the first transaction, during the period of the first transaction, after it, or somewhere in between as illustrated in Figure 3.9.

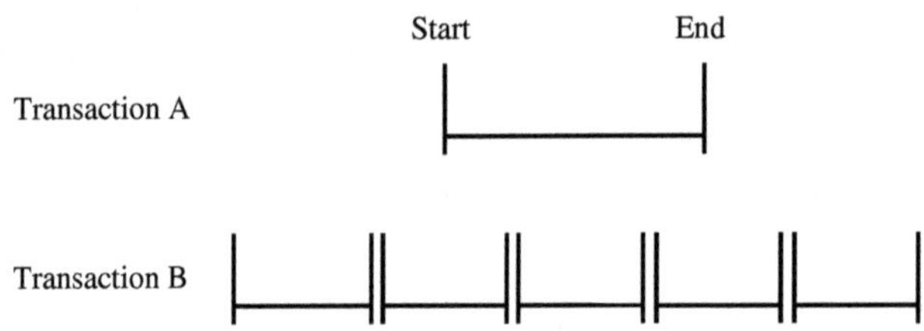

Fig. 3.9 Five possible relationships between two transactions

One solution to this has been proposed [14] that would extend the current languages with the constructs necessary. At the time of publication, no material work had been performed either in the PSL or SVA language committees.

Having painted a somewhat negative picture, there are still places where transaction-level assertions can still be applied. It does not matter what abstraction of system is being discussed, they all contain two main parts, a dataplane and a control plane which are closely linked together.

The control plane is essentially a state machine that makes transitions based on the inputs it receives. While not driven by clocks at the transaction level, it is still very possible to base this on events that are associated with the state transitions. The problems come when multiple communicating state machines need to be verified, and for this we will have to wait for the industry to come up with better solutions than the ones standardized today.

It is also possible to do some level of verification in the dataplane as well. Consider the humble FIFO. While it would normally be defined at the RTL level, a property such as the one described below performs a data integrity check on the FIFO without reference to the level of abstraction being verified.

```
data_integrity : assert forall dvalue in {0:'SIZE} :
always ( (push && data_in == dvalue) ->
eventually! (pop && data_out == dvalue));
```

The example basically says that after data has entered the FIFO, then sooner or later that data should come out of the FIFO.

One EDA company, namely JEDA technologies, has been attempting to create a native assertion language for SystemC [15]. It is based on SVA for the low-level clocked properties, but extends the language in three distinct areas for higher level properties. The three extensions are direct utilization of events, mutex operations, and high-level statistical functions.

Similar to the Ecker proposal [14], this identifies a begin and end to each transaction. A new set of operators then work on the function callbacks triggered by the transaction events. The operators include the ability to check if one sequence is included within another, or if it occurs outside of it, plus a number of others that deal with sequencing.

For the high-level properties, JEDA notes that coming up with a predefined set of primitives is difficult because the types of checks required are likely to be application dependent. They thus implemented a basic set of C++ classes that can be modified or extended by the user to fit their needs. Two of the defined classes are queuing models while the third provides a basic functional coverage capability.

References

1. http://www.mathworks.com/
2. Frank S. Eory. "A Core-Based System-to-Silicon Design Methodology," IEEE Design and Test of Computers, vol. 14, no. 4, pp. 36–41, Oct.–Dec. 1997, doi:10.1109/54.632879.
3. http://ptolemy.eecs.berkeley.edu/
4. IEEE Std 1666^{TM}-2005 Standard SystemC® Language Reference Manual, http://standards.ieee.org/getieee/1666/download/1666-2005.pdf
5. Thorsten Grötker, Stan Liao, Grant Martin, Stuart Swan. System Design with SystemC, Springer, 2002.
6. David Black, Jack Donovan. SystemC from the Ground Up, Springer 2004.
7. Srikanth Jadcherla, Janick Bergeron, Yoshio Inoue, David Flynn. Verification Methodology Manual for Low Power. Synopsys 2009.
8. OSCI TLM2 library, dated June 6 2008, http://www.systemc.org/members/download_files/check_file?agreement=tlm_2-0_080606
9. Sterling Augustine, Marc Gauthier, Steve Leibson, Peter Macliesh, Grant Martin, Dror Maydan, Nenad Nedeljkovic, Bob Wilson. "Generation and Use of an ASIP Software Tool Chain", Chapter 7 in Wolfgang Ecker, Wolfgang Müller and Rainer Dömer, Hardware-Dependent Software: Principles and Practice, Springer, 2009.
10. Achim Nohl, Gunnar Braun, Andreas Hoffmann, Oliver Schliebusch, Heinrich Meyr, Rainer Leupers. "A Universal Technique for Fast and Flexible Instruction Set Architecture Simulation", DAC 2002, pp. 22–27.
11. Accellera PSL standard version 1.1 2004.
12. Ariane 5 flight 501. From Wikipedia http://en.wikipedia.org/wiki/Ariane_5_Flight_501
13. Dave Whipp. Transaction Level Assertions in an Interface Definition Language. DesignCon 2008.
14. Wolfgang Ecker, Volkan Esen, Thomas Steininger, Michael Velten, Michael Hull. Implementation of a Transaction Level Assertion Framework in SystemC. DATE 2007: 894–899.
15. Atsushi Kasuya, Tesh Tesfaye. Verification Methodologies in a TLM-to-RTL Design Flow. DAC 2007, pp.199–204.

Chapter 4
Testbench Models

Verification has become as expensive and as important as design, and as such models and languages for verification are emerging that are highly directed toward their intended function. Most of what we think of as verification today is implementation verification taking place at the RTL level of abstraction or physical verification taking place at even lower levels of abstraction. Very little of the total verification effort goes into design verification. To put that another way, we spend most of our time verifying that we have implemented something correctly, rather than determining that we have specified the right thing, or that the collection of pieces that are being assembled are capable of performing the right function within the constraints imposed by the specification.

ESL verification attempts to ensure that the functional specification is right and complete, and looks not just at the hardware piece of the system but also at the software running on the hardware. It attempts to ensure that the architectural decisions that are being taken are correct. In short, verification should concentrate on the design choices that are being made at that point in the design flow. For ESL that means verifying system level functionality, performance, and providing the reference models that are to be used for the implementation verification.

In this chapter we will introduce the major languages that are in use today and how they are used to create reusable verification models. This includes some of the methodologies that define frameworks for how these models play together to make complete testbenches.

A testbench is not a single model or even a collection of models. It is a group of diverse models that, in conjunction with a set of tools, enables an act of verification to be performed. It is not possible, with the technology that exists today, to perform complete verification, and thus the goal is one of minimizing the risk of a serious bug escape while keeping to various budget constraints, such as manpower, size of server farms, or elapsed time. In this chapter we will examine some of the more common models that make up a verification environment, but it should be understood that there are many ways to put these together to help perform the task of verification.

No two companies have exactly the same needs or amount of money that they can dedicate to the task, and thus there is no one right solution to the problem. There are many good verification books [1] on the market that deal with particular

B. Bailey, G. Martin, *ESL Models and their Application*, Embedded Systems, DOI 10.1007/978-1-4419-0965-7_4, © Springer Science+Business Media, LLC 2010

languages [2] or methodologies [3, 4] and this chapter is not meant to replace those in any way. By reading this chapter it is hoped that you will get a better understanding of the possibilities that exist for an ESL verification flow and the languages, tools, and methodologies that are the most likely to help you accomplish your goals. Then you will know which other books may provide the necessary details to implement a complete flow. You will also understand some of the problems that exist today in accomplishing ESL verification as most of the tools and languages that exist today were designed to be used at the RT level.

There are two principal means of performing verification. These are generally called static and dynamic methods, or more commonly formal verification and simulation, respectively. However, the boundary between these two types has become blurred in the past few years, as we see verification tools and methodologies combine both techniques into a somewhat unified flow. While there are still some major challenges in this combination, it is a trend that is likely to continue.

4.1 Testbench Basics

No discussion about verification is possible without first looking at the definition of verification. While this was discussed in detail in *ESL Design and Verification* [5], it will be summarized here. Verification is defined by the IEEE as follows:

Confirmation by examination and provision of objective evidence that specified requirements have been fulfilled.

In layman's terms, this can be stated as making sure the implementation is correct and ensuring that it conforms to the requirements recorded in the specification. This is in contrast to validation, which is similarly defined by the IEEE as follows:

Confirmation by examination and provision of objective evidence that the particular requirements for a specific intended use are fulfilled.

This is very similar to the definition for verification in that they are each concerned with demonstrating that design requirements have are fulfilled. However, validation is about ensuring that the specification is correct and that it captures the product design requirements, whereas verification is ensuring that the implementation matches the specification.

In most of the verification methodologies in use today that are based on RTL models, it is almost impossible to perform validation in a virtual environment. This is because the size of the systems coupled with the level of detail contained in the models that are available means that verification environments run so slowly; nothing more than somewhat superficial tests can be performed. Because of this, many companies rely on emulation, prototyping, or even the first sample run of chips to perform many of the validation tasks. In addition, some other verification tasks such as system performance evaluation are performed on these late-stage evaluation methods.

ESL with its higher abstraction models promises to change this and also make validation something that is performed in the early stages of the development process, rather than at the end. Having said that, many of the verification tools and methodologies in the market today have not yet been modified or enhanced to take advantage of higher abstraction models. This does not mean that it is not possible to do abstract verification with them, just that they are not ideal and is an area for which we can expect to see progress in the future.

Going back to the definition for verification, there are four key phrases in that definition:

- *Confirmation by examination:* This quite simply says that if you do not look, then it has not been verified. Many older verification methodologies, such as directed testing, used a small set of observations to define that a test passed. If the correct values were seen in certain registers or memory elements, then it was assumed that everything happened correctly throughout the entire design. While this may have been a good indicator for relatively simple designs, the designs of today have many things happening in parallel, and what is going on in one part of the design may have adverse functional or performance impacts on another part. It thus becomes more important than in the past to ensure that the entire block or system is being observed, even if a test is only intended to exercise smaller portions of it.
- *Provision of objective evidence:* Objective evidence basically means something that can be trusted to provide the results against which the implementation is being compared. This is the essence of verification, in that it is fundamentally the comparison of two models. If those models are independently derived and they agree, then there is a good chance that they are both correct. A disagreement indicates a problem with the design, the comparison model, or a problem with the specification such that something was missing or ambiguous leading to two different interpretations. A model developed during an ESL flow may be used as the objective evidence against which an implementation is compared. However, if that model were also used as the input to a behavioral synthesis flow, then it cannot be used as a verification model, except to ensure that the tool or the decisions made did not introduce any error. This then becomes an equivalence checking process rather than a verification that the specification has been implemented correctly. This will be discussed further in Chapter 9.
- *Specified requirements:* As already stated, it is not possible to verify everything so we must decide on what is to be verified. While EDA marketing departments love to quote figures that say the number of things that can happen in a design is of the same order of magnitude as the number of atoms in the Universe, this is not true. However, the numbers of unique behaviors are large and it is not possible to do everything. So in that regard they are partially right in that the number may as well be that large. Choices must be made. Priorities have to be established. This is normally done as part of the planning process for verification. It is also the most subjective aspect of verification, which makes it difficult to be precise about actual progress. The verification plan often contains a coverage model that will help to indicate progress as the verification tasks are performed.
- *Have been fulfilled:* This phrase is the one that ties everything together. Before we can indicate that a specified requirement has been fulfilled, we must tie it to

an actual *confirmation by examination* against *objective evidence*. In other words, hitting a coverage point is not an act of verification, and if an act of verification was not performed, then the results obtained are not trustworthy. The easiest way to think about this goes back to the days when code coverage was the principle means of measuring progress. When deadlines were drawing close, engineers would look at the coverage details to find out which lines of code were not being executed and devise a stream of stimulus that would ensure the line was executed. They did not spend the extra time to work out how to ensure the line of code was correct. This managed to quickly increase the coverage numbers, but no additional verification was actually performed. While modern verification methodologies ensure that more verification is being performed, it is still a problem. Many tests do not propagate an error effect to a place that is being monitored and thus they remain hidden. The objective of tools, such as constrained random test-pattern generation, is to obtain coverage, not to ensure that error effects are propagated.

4.1.1 Testbench Components

There are a number of models within a testbench that are always present, and some that are specific to particular verification methodologies. In some cases, a model or part of a model may be incorporated into another piece of the testbench making it look as if it does not exist. For example, consider a verification strategy called directed test. With this strategy, a stream of stimulus is composed that will guide the design through desired functionalities and produce some result. That result needs to be compared against the expected results from running that test. In practice there are several ways in which that could be built as a testbench.

The first method, which was the predominant method in the past, would create a single file that contained the stimulus, the results checker and an implied comparison model. Implied model means that an engineer has sat down and determined on paper, or some other method, what the correct results should be and directly coded them into the checks that are made on the outputs. While it may seem that a comparison model does not exist, it exists in its implied form, distributed across the sum of all of the directed tests that were created.

The problem with this methodology is that when a change is made in the design specification, which happens quite frequently, the implied model in all of the tests has to be updated – creating a maintenance nightmare. In addition, working out what the correct outputs should be is non-trivial, especially with the complexity of the systems being designed today. It may have been somewhat practical when just a few gates were being considered, but not for systems containing many concurrently executing components that are communicating with each other. In many cases, the expected results were obtained by eyeballing the results produced and if they appeared to be right, establishing them as the new "golden" vectors against which further runs would be compared.

The models that are always required in a testbench are the plan, a comparison model, a progress model, and an input constraints model. Each of these will be discussed briefly below and then later in the chapter some modeling techniques available for each will be discussed.

4.1.1.1 Verification Plan

For many people, this is not yet a model, but a free-form text document. This is changing as more adaptive verification methodologies are being introduced. The plan defines a number of things, but primarily the "what, how, and when" of things that are to be verified, and how to record those acts of verification. The verification plan is often the only way in which multiple verification strategies can be brought together effectively without it causing large amounts of duplicated effort and to ensure that all of the right verification components are being created so that they are ready at the appropriate times. A good verification plan is built upon experience and knowledge of both the verification methodology and the design itself. What would be the right verification plan for one design and company may not apply to others. The verification plan will be discussed in Section 4.2.

4.1.1.2 Comparison Model

This is the most obvious model associated with verification and is always present, no matter what strategy is used. We have already discussed the way in which this model may be implied and distributed in a directed test scheme. In more modern verification strategies, including modern directed testing, the comparison model is a complete and stand-alone model, often being the system-level model that was developed in the early stages of an ESL flow.

Indirectly associated with the comparison model is another model, which is the checker. While this is often independent from the comparison model, they work together closely. The role of the checker is to ensure that the implementation and the comparison models match. This is made more difficult by the fact that timing and in some cases sequence of events may be different between the two models, as well as differences in data representations. Consider the case of a router that has to deal with traffic overload. In the specification it may state that the router may drop any packet with a priority below a certain threshold. What happens if the abstract model and implementation model drop different packets? We would now have two different output streams being generated by the models and a comparator would never be able to successfully match the results. Both may be correct interpretations of the specification – just different. In fact an abstract model usually contains many different implementation possibilities because that model has not yet been constrained by later architectural or micro-architectural decisions.

One common form in which the comparison model and checker come together is as an assertion. Assertions are primarily used for localized checking, such as within a FIFO. These checkers ensure that the correct temporal actions take place within the block and that no data gets lost, or becomes out of order, for example. Assertions use a modeling approach called properties, which are also the primary comparison model for formal verification methods.

In this chapter, we are talking about ESL verification. This creates a bit of a conundrum when discussing the comparison model. In many verification strategies, the model used to verify an implementation was the system-level model of the design. But what was used to verify that system model? Does it imply that a third model has to exist – the comparison model for the system model? The answer to this question lies in the overall methodology that is established for the ESL flow. There is one truth we cannot get away from and that is that verification cannot be performed without two independent models.

In an ESL flow using high-level synthesis, the implementation model is derived from the system model. The verification model is the model used to verify the system model, which is the same model that will be used to verify the implementation, or any other act of verification that needs to be performed. We thus continue to compare the two models throughout the verification flow. This also presumes that the system model has been written in such a way that this becomes possible.

If, on the other hand, the implementation is hand-crafted RTL, which is based on the system model, then they are not independent models. They share a common interpretation of the specification. In those cases the implementation model is a third model and thus this flow actually requires more total resources than one using synthesis.

In the rest of this chapter, we will be assuming that behavioral synthesis is used and thus the verification environment created will be deployed for all stages of verification and not just for verifying the system model.

Comparison models will be discussed in more detail in Section 4.3.

4.1.1.3 Progress Model

While the verification plan talks about what and how verification will be performed, it is necessary to measure progress. For dynamic verification, this is often in the form of a functional or structural coverage model that measures the completeness of stimulus that has been applied. There are many forms of coverage models including structural metrics which range from line coverage to state-transition coverage, ad-hoc metrics such as bug rates and densities, functional coverage which is derived from the specification, and assertion coverage, which is being defined by an Accellera committee [6].

Within the formal world, a coverage model attempts to define the completeness of the checkers that have been defined and thus identify the holes that are yet to be adequately verified.

Coverage will be explored in more detail in Section 4.4

4.1.1.4 Input Constraints Model

The task of verification is bad enough without saying that a block must be verified under all theoretical conditions. An input constraints model defines what constitutes valid inputs, or the subset of erroneous conditions that are to be tested for. The constraints may also define sequences, such that particular scenarios may be defined. This is probably the model with the maximum level of diversity across various verification methodologies. It can range from an independent model all the way through to the use of a system model to constrain the inputs that are generated.

The constraint model is the principle input to the constrained random stimulus generation process that attempts to generate as many input sequences as it can that conform to the input constraints. In a formal world it is used to simplify the proving that a property holds true under all conditions that conform to the given constraints.

Constraints will be discussed in more detail in Section 4.5

4.1.2 Verification Methodologies

The verification methodologies used for RTL design methodologies are based on the V diagram as depicted in Figure 4.1.

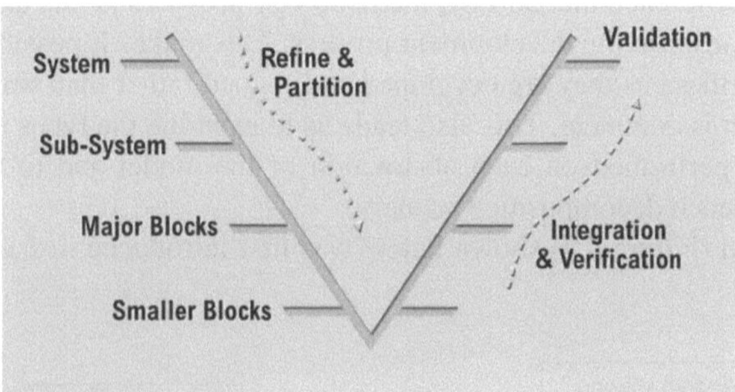

Fig. 4.1 Classic V diagram

On the left-hand side of the V is a view of the traditional design process. It is seen as the decomposition of the system level specification into sub-systems, which are further decomposed until pieces that are manageable by a single engineer or small team of engineers are created. These pieces are then implemented at the RTL level. In most cases, formalized models do not exist for the upper layers of the design process, except perhaps for some algorithmic models that were developed, usually in MATLAB or similar development tool. Once the RTL models have been implemented, the verification process starts as depicted on the right-hand side of the V. The low-level blocks are first verified and when most of the problems have been resolved they are integrated into the subsystems. The subsystems are verified and then integrated into a complete system, where verification may again be performed and eventually it may be possible to perform some very limited amounts of validation.

While this verification methodology has been in use for a long time, it is generally agreed that it is inefficient and increasingly ineffective. The inefficiencies come from the fact that all verification has to be performed on the RTL models. While this is appropriate for some amounts of verification, it is not appropriate for verifying high-level functionality as it contains too much detail. The second major area of inefficiency comes from the fact that testbenches used for block-level verification are often not usable for verifying the integrated blocks, thus requiring new testbenches to be developed at each stage of the integration process. There are many other aspects to the inefficiencies that this methodology incurs, but they will not all be listed here. While verification teams have made huge advances in the technologies they use, the complexity of the designs has also been increasing. The results that were demonstrated by the Collett International Studies [7] in 2002 and 2004 indicated that the numbers of functional problems that were escaping verification and being manifested in the chips themselves was increasing. While it is not known if those trends have continued, there is a general consensus in the industry that the verification process is somewhat out of control and a big desire to see more rational proportions of the total time and effort being allocated to verification. The only way to achieve this is through fundamentally different methodologies, and ESL makes some of those possible.

The major difference that an ESL methodology provides is that abstract models are available earlier in the development process. This makes it possible to perform verification on these as they are developed and refined rather than waiting until the implementation is complete. This also leads us to examine the types of verification that are to be performed on each abstraction of the model and to think about a top-down testbench decomposition scenario.

The diagram (Figure 4.2) shown below was first introduced in *ESL Design and Verification* [5].

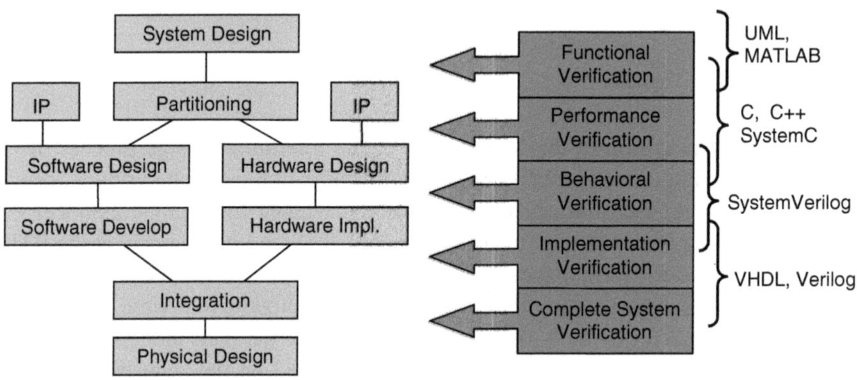

Fig. 4.2 Facets of design and verification

This diagram reflects the types of verification that are possible as we progress through an idealized design flow. When the system has been designed, it should be possible to perform functional verification on this model. In some cases it may also be possible to perform validation on this model. This is especially true for consumer devices where the user-operational model needs to be tested to see if a typical user feels that the device is performing in the ways in which they want to see or expect.

These may be in the form of mock-ups, or user interfaces in front of the functional models, or in other forms. Complete functional models can also be built that enable the system to be comprehended and to assist in making the partitioning choices, which comes next.

Once the system model is complete and partitioning is being considered, the fundamental design decisions being made are the types of resources that will be used to perform computation and communications. As these decisions are made, approximate timing becomes known and this enables performance verification to be conducted. Within the pieces that are to be implemented as hardware, further refinement yields a behavioral model of the block on which fairly detailed verification can be performed. Finally, after implementation has been completed, the RTL blocks can be verified. It should be quite clear that the scale of verification diminishes as we progress down through these layers, such that the size of the design portion being verified diminishes at each step. This does not mean that the complexity of the verification task diminishes. As additional detail is added, more aspects of the block have to be verified, and the additional detail slows down the verification tools. This top-down approach of incremental verification is a complete reversal of the original V-diagram methodology which started small and attempted to end big. The other major difference is that integration verification, as defined before, is almost completely eliminated since the purpose of this verification in the original methodology was to discover problems with the interfaces between blocks or inconsistencies in the specification when it was partitioned. Given that these have all been verified before block implementation started, it is very unlikely that any of these problems will exist. We can thus see the new methodology becoming as depicted in Figure 4.3.

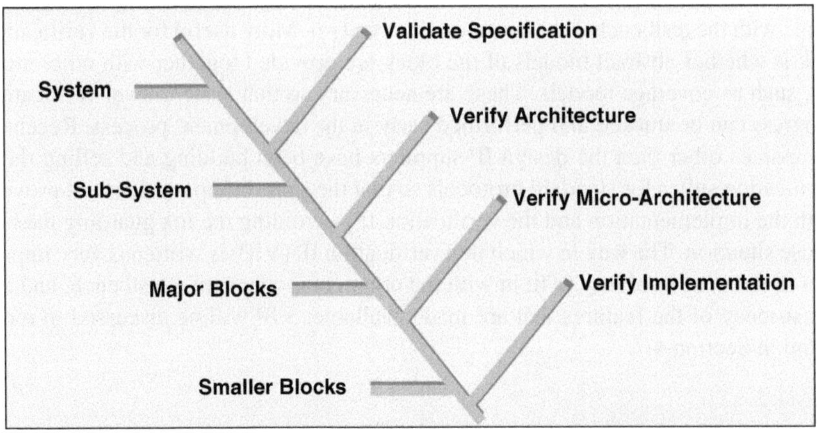

Fig. 4.3 Modified V diagram

We can also think about this in taxonomy terms. At the highest level the description is essentially untimed. Data may well be floating point; little thought may have been applied to concurrency and internal communication is point to point over dedicated channels. It is pure functional verification.

As the system architecture gets created, we start to see partial ordering in the timing domain, data types refined and course-gained concurrency specified. The partitioning creates the need for communications and so this now needs to be considered. Performance verification can be started at this level, and this includes both the timing and power domains.

Once the major blocks have been identified, the micro-architecture for them is considered. Timing is refined to the numbers of clocks and data types have turned into fixed point with defined bit widths. While course-grained concurrency was inserted in the last step, finer gained concurrency gets created here and the detailed choices about communications are made.

Finally as we consider the smaller blocks, sub-clock timing is under consideration as well as a host of physical factors. This is not considered to be in the realm of ESL verification.

It should be noted from this that if a single reference model is to be used across all stages of verification, then it must be constructed in such a way that additional information and refinements can be made in this model without destroying the aspects of it that have already been verified. While we may not have the ideal answer today, the solutions that exist are getting closer to this ideal.

4.1.3 Verification IP

Few designs start from a blank sheet of paper. They are often incremental designs based on an established platform, or are an assembly of many pieces of IP that were developed either internally or from third-party companies. This has dramatically decreased the amount of time that design takes. Most of that design IP will also come with the testbenches that were used to verify it. More useful for the verification task is whether abstract models of the block are provided together with other models, such as coverage models. These are necessary so that system-level verification progress can be tracked and performed early in the development process. Recently, companies other than the design IP suppliers have been building and selling these verification suites for standard protocols so that the same company does not provide both the implementation and the verification IP – avoiding the fox guarding the hen house situation. The way in which this verification IP (VIP) is written is very important in terms of its ability to fit in with the other components of a testbench, and the consistency of the features that are made available. VIP will be discussed in more detail in Section 4.7.

4.2 Verification Plan

One of the most important and difficult questions for any verification team to answer is "Are we there yet?" In order to be able to answer that question, there must be a strategy as to how to tackle the verification problem and how to keep track of progress. At the heart of this is a verification plan. In the past, this was a document, often created using whiteboards, sticky pads, and a group of people sitting around a table [8] in an attempt to provide some kind of order to the verification tasks and models that needed to be written. While it is good to have a plan, this method had a number of downsides including the difficulty of maintaining the document and back annotating information into it as the verification task proceeded. It was also found in some cases that appropriate verification strategies were not known early on in a new design and would only become apparent as architectural choices were made. This meant that the plan either had to be delayed until the hardware design was well on its way, or it had to be continuously updated to take into account decisions that had been made. This rarely happened, so the main value in developing it became the thought process that led up to it, rather than the document itself.

Constrained random verification produces so much coverage data that information overload is common. At the same time different team members need to be able to access different kinds of information. As an example, managers may only want to see the overall progress that has been made, whereas a verification engineer, assigned to a particular part of a design, wants detailed information about their assignment so that they can decide where to focus attention.

What is needed is an "executable" verification plan. In this context executable means that it is a dynamic document that is adaptable, extensible, and can be analyzed by a collection of tools, such as report generators. It also has to be able to consume data from various places, such as simulation results. We are beginning to see some highly creative solutions emerge that use the verification plan as their starting point, such as the ActiveDesign [9] product from Jasper Design Automation. This creates a semantic cross-reference between the specification and the implementation such that functional searches can be made based on queries. Most of the solutions that have been created take on a spreadsheet format as their user interface. This provides a familiar environment to many people.

To illustrate the main aspects of a verification plan, the VMM Planner [10] from Synopsys will be used. VMM [11] is a methodology based on SystemVerilog for the creation of testbench models that enables efficient integration and reuse of those models. Other similar methodologies exist such as OVM [12] and GamePlan [13] from Jasper that is distributed for free.

VMM Planner is an active document that coordinates the verification process. It contains the features of the design that are to be verified, the metrics that define completion, and automated links into the verification tools for data collection. The plan can also be used to keep some non-functional requirements and track progress against them. As an example, VMM Planner currently allows the user to enter power budgets. Actual power figures are then extracted from tools and annotated into the plan so that it can be seen how power is being consumed and where it may be out of line with the requirements. We can logically expect that over time, the verification plan may become a more generalized executable requirements document, where any number of attributes about a system can be stored and analyzed.

A verification plan is primarily composed of three main types of data, namely the verification requirements, the verification environment, and the verification implementation plan. The verification requirements list all of the features that are to be verified and in many cases this can be prioritized. Features are normally decomposed in a hierarchical manner such that major features are defined first and then these are broken down into the individual pieces that are necessary to support these major features. It is the leaves of this tree structure that become the primary verification targets. This is shown in the left part of Figure 4.4.

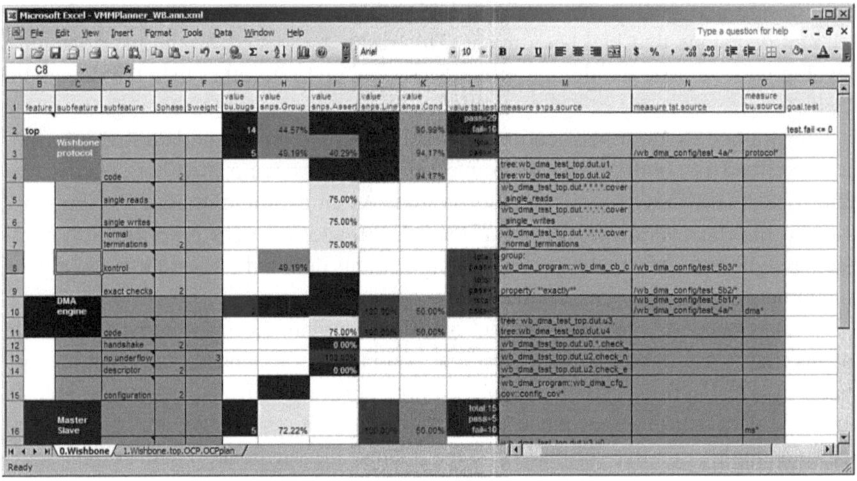

Fig. 4.4 VMM Planner screenshot

In Figure 4.4 we can see the "top" feature is broken down into the Wishbone protocol, DMA engine, Master Slave, and possibly more. Each of these major blocks are then broken down into sub-features, such as single reads and writes. Note that after the sub-feature has been listed, it then details what phase of verification will target this feature and the associated priority of the feature. It should be noted that in the example provided above, the decomposition is somewhat structural in nature. This is because it is targeting an RTL block. When used in an ESL environment, it can be expected to be based more on functional decomposition since the structure may not yet be known.

Moving from left to right in Figure 4.4, we then see the associated coverage data. At the top of each feature section, we can see a row associated with code coverage, then a row for the functional coverage for each individual sub-feature. This data is aggregated back up the hierarchy so that both detailed data can be seen or summary results. A tool such as VMM Planner can handle many types of coverage data, or the user can define their own. Examples of user-defined metrics may be bug rates, code stability, run times for regression tests, or many other operational metrics. At present VMM Planner accepts coverage data directly from the Synopsys range of verification tools. An Accellera committee [14] is also defining a unified coverage

API that will enable data to be collected from a number of sources (such as different simulation vendors) and analyzed in the users tool of choice.

The data on the right defines the test-case attributes such that features can be associated with specific coverage metrics, or other information can be attached, such as references to specifications. The coverage model itself will be described in Section 4.4. Also important are the goals. It is not always required that 100% achievement of the goals be attained at early stages in the verification effort. The target level for these goals may change as a project progresses.

When a set of tests have been run, the data can be examined in several different ways. For a glimpse at the current status, the Unified Report Generator produces an html page that can be viewed by anyone with intranet access to the data repository. An example is shown in Figure 4.5.

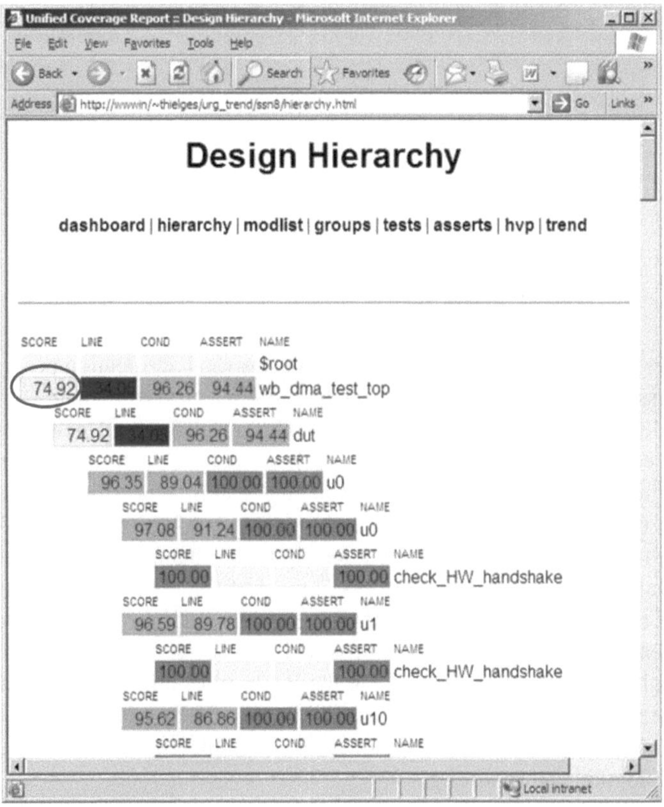

Fig. 4.5 Web status report

After several runs, progress reports can also be generated based on the historical progression of data as shown in Figure 4.6.

VMM Planner comes with several filters, such that, for example, the results will only show progress of activities scheduled for a particular phase of verification.

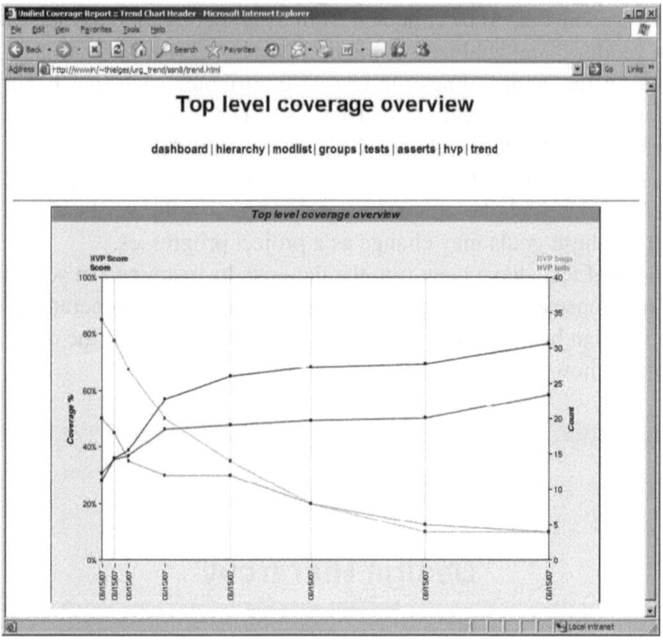

Fig. 4.6 Progress report

Thus rather than see what may look like a very low total verification progress score, the progress to the specific milestone can be seen in a clearer manner.

What do you do when you find a bug? This invalidates all existing coverage data and requires a regression run to refill the database with new coverage data. However, in the event that this was the "last bug" and time is critical, it may be safe to assume that most of the results obtained in the past will still be valid and that only those tests that would be directly affected by the change need to be re-run. This kind of analysis is still a manual process in most of the verification plans that exist today, and thus the intelligence behind them remains with the users. The continued development of tools such as ActiveDesign will provide some degree of automation, even though today it is targeted more at the RT level of abstraction. While a lot of manual intervention is required, the executable verification plans take a lot of the drudgery out of the task, and eliminate a lot of simple errors that can be made in data consolidation and reporting.

One area of active thought is how to compose coverage data across a project. For example, if verification is performed on a block, how is this data mapped into a complete chip-level view? Similarly if a block has been verified using formal verification tools, how is this integrated with the results from dynamic verification? Another area of active thought is in the use of abstract models for part of the verification process. How can we intelligently combine this data in meaningful ways? There are no definitive answers to these questions today, but the good news is that the industry is now taking coverage seriously and at least is forming the right set of questions.

One thing that these executable verification plans do not contain today (but is usually contained in a written plan) is the way in which the verification environment is constructed. What are the components and how do they relate to each other? If we think about the constrained random verification process, the stimulus generated is dependent on the constraints and these can be manipulated to increase the chance that solutions will be found in particular areas. Thus if coverage is lacking in certain areas, then it would be nice for the verification plan to be able to know how to manipulate the constraints automatically. This is perhaps a development that we can look forward to in the future.

It should thus be clear that verification plans are perhaps one of the least evolved of the verification components, but also the one at the heart of the next level of automation and efficiency. One alternative way that we could see this develop is characterized by some of the intelligent testbench approaches, such as those by Mentor's inFact [15] and Breker's Trek [16]. These are graph-based verification approaches that directly tie together the plan with the generation and coverage operations. However, they still need to consider prioritization and other issues such that the verification engineer cannot be completely taken out of the process.

4.3 Comparison Model

When talking about ESL verification we have to be very careful that we do not end up in a recursive argument about models. At the RT level, it has become an accepted fact that two models are required – the implementation and the model against which it is to be compared. Each of these models must have been independently derived from the specification for there to be value in this comparison. Now if we think about a migration up to the ESL level, we have to think about where those models came from. For the implementation model there are two possibilities. Either the RTL implementation was generated by a synthesis program, such as the one described in Chapter 9 or it was hand crafted after the system-level model had been verified.

Let us start by considering the synthesis route. The synthesis process is not infallible. This does not mean that we expect the synthesis tool to have bugs, although it might, but that synthesis is driven by the options and parameters that were set in the synthesis tool. These will affect the micro-architecture created that will, as a minimum, affect timing. It is also possible in some synthesis tools that the order of events will change and these can cause alternative, but equally valid, functionalities to be executed. Sequential equivalence checking is emerging as a technology that will enable us to find these differences in an efficient manner.

For most people today, they will prefer to perform simulation runs on the synthesis results and check that the functionality is the same. This is not functional verification; this is determining that no errors have been introduced during the synthesis process. We are in effect doing equivalence checking by simulation. None of this tells us that the input description works.

However, if we have done a good job in the verification of the model that is fed into the synthesis flow, then no additional functional verification should be necessary on the synthesized output. We should be able to use the same testbench and reference model for verification at either level of abstraction and expect to get the same results. Structural coverage figures for the generated RTL may be lower if IP blocks were used and not all of the available functionality of those block utilized.

If a manual transformation path is used, the abstract design model cannot be used as the reference model for verification purposes. This is because we are deriving the implementation from this model and they may both have common problems regarding interpretation of the specification. The only use for this model is to check that the implementation matches the intent. There is thus no alternative but to have three models: the abstract design model, the implementation model, and the reference model. This is unfortunate, but necessary in what is not truly an ESL flow, but an added step in the product development process, probably used to help with the understanding of a specification.

Now let us turn our attention to the verification model. When we are creating the testbench, we need to create all of the component pieces: the stimulus generator, the reference model, the checkers, scoreboard, coverage model, etc. These need to be verified. Does this mean that we require another model? Who is verifying the verifier? No – we do not need another model. At the end of the day it will be compared against the implementation, at whatever level of abstraction it exists in. The challenge is that we must have one testbench and reference model capable of verifying all levels of abstraction that the implementation may exist in.

4.3.1 Testbench Languages

There are two languages designed for the construction of testbenches that are both IEEE standards. These are SystemVerilog (IEEE 1800) which is supported by all of the major EDA vendors and *e* (IEEE 1647) which is supported by Cadence. The *e* language has existed for a longer period of time, but no reliable figures exist to show which has more active users. In addition, a number of people still write their testbenches in C or C++, but these languages do not have the built-in capabilities for constrained random stimulus generation or for the construction of complex assertions. A verification class library was also created for SystemC, but this did not have the same level of capabilities as the dedicated languages and today is not being developed or used by many people in the industry. It is thus expected that the dedicated languages will become the languages of choice in the future.

Neither of the dedicated testbench languages, however, are the language of choice for the implementation of the reference model or the system model. The community clearly favors either C, C++, or SystemC for this role. This means that the verification environment will almost certainly be mixed language as shown in Figure 4.7. In this diagram, the main testbench components are shown as being implemented in SystemVerilog. SystemC is being used for the reference model although it may also contain native C or C++ components or models coming from

MATLAB/Simulink. The implementation model could itself be in multiple languages, such as SystemVerilog, SystemC, or C++ for the behavioral model and Verilog or VHDL for the RTL model.

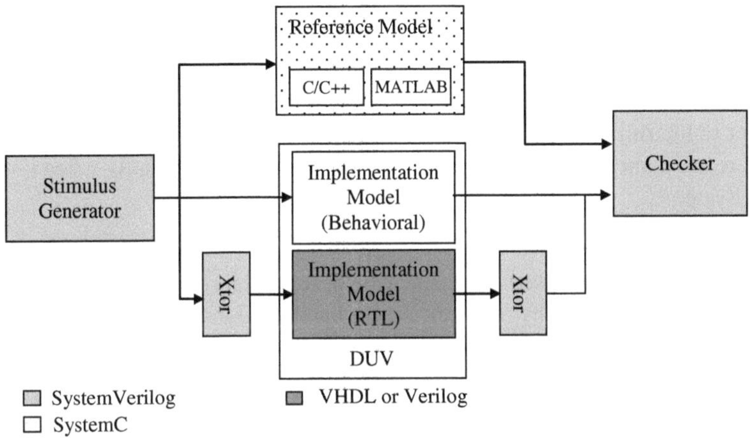

Fig. 4.7 Mixed language environment

It thus becomes clear that the construction of the testbenches, particularly at the module interfaces, must be written in such a way that this kind of language intermixing becomes possible. This is not as easy as it sounds and is primarily concerned with control of threads of execution. The easiest way to think about this is in a mixed SystemVerilog/SystemC environment. Both of these are complete simulators which means that they contain event schedulers to deal with issues such as timing and concurrency.

If one environment hands control over to the other, such as passing a transaction across one of these boundaries, the calling simulator expects to regain control without any time advancement. But what if the other language cannot complete the request in zero time? This creates a problem which has been solved many times in the past but with severe performance penalties. While those penalties may not have been large enough to cause problems with, for example, a gate level to analog simulator connection, they would not be viable at the speeds with which an ESL model needs to operate. We thus need better ways to accomplish this.

SystemVerilog contains a foreign language interface called Direct Programming Interface (DPI). This allows it to connect to a C model in a very efficient manner. This is a bi-directional interface in that it allows SystemVerilog to call C and for C to call SystemVerilog. Once a task has been declared and made accessible to the other side, they look exactly like a native call. Thus for example if we had

```
Res1 = call_function_1(param1);
Res2 = call_function_2(param1);
```

there is no way to tell if each of these calls is a native or foreign model call.

This interface does not really get around the problem with time advancement. It is expected that when a call is made from one side to the other, it will complete in zero time. While SystemVerilog tasks, which consume time, can be exported into a C environment, this does not provide a generalized fix to the problem.

Before we become too pessimistic, it should be remembered that a large number of testbenches are written as untimed models. They rely more on active synchronization with the design. Thus the intermixing of languages is a viable way to go for the creation of verification environments. The major EDA vendors are also making it easier to intermix them if you use all of their tools.

There are some additional limitations in this interface, especially when connecting to SystemC.

- SystemC is a C++ based language which supports class methods. Since DPI cannot traverse hierarchy it cannot access handles to instances or objects.
- No methods for synchronization are provided and thus the only way to accomplish it is through value passing.

Because of these limitations the interface between the two language worlds needs to be crafted carefully, and this is best done at the transaction level. This also makes it easier to handle the mixed abstraction situation since transactor models can be added to perform the conversion to signal interfaces for RTL models.

If FIFOs are used in the interfaces, they allow for simple time separation between the two worlds. Each side can then block the simulation advancement based on the FIFO being full or empty and thus achieve rudimentary synchronization. A simple form of this is shown in Figure 4.8.

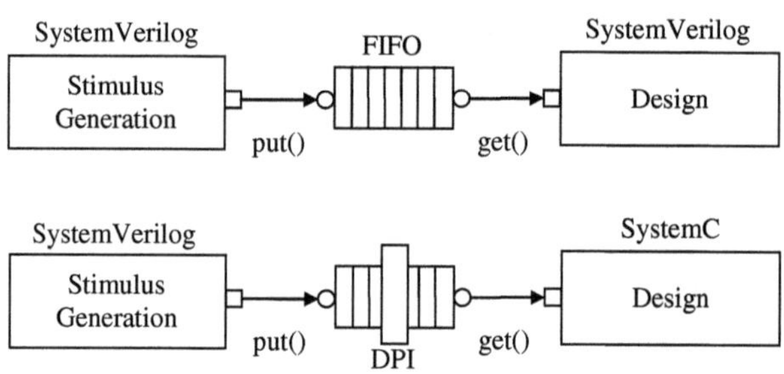

Fig. 4.8 Using FIFO interfaces

With this method the interfaces stay the same independent of the language that is being used on the other side of the interface.

In a paper at DVCon 2009 [17], Mentor described a new layered interface that they are working on, which they are calling DPI-C++. DPI-C++ provides a means for passing class object arguments "by-reference" via import and export DPI calls.

- Import classes are based, owned, and managed on the C++ side of the interface. SystemVerilog proxy class objects are used to call methods on C++ objects from within SystemVerilog code.
- Export classes are based, owned, and managed on the SV side of the interface. C++ proxy class objects are used to call methods on SV objects from within C++ code.

On top of DPI-C++ they define another layer that they call DPI-SC to more directly support the integration of SystemVerilog with SystemC. The main feature added at this level is that SystemC can consume time, and thus manages to overcome all of the limitation described previously.

Given the importance of multi-language interfaces, such as the one described here, we can expect to see a lot of progress in this area very shortly. To say that complete ESL flows depend on it may be a little melodramatic, but without it, the creation of a reference model that can be used at all stages in the ESL flow is cumbersome at best.

4.4 Progress Model

The principle form of progress model is a coverage model. Some of the more important ones will be described briefly below. The book *ESL Design and Verification* [5] discussed the general form of coverage models, particularly related to the construction of functional coverage models.

4.4.1 Ad Hoc Metrics

Some of these are very subjective measures, such as bug rates or error densities. While these may not be seen as qualified means of defining completion, they can be very powerful indicators of a smoothly running engineering process, or when things go awry, they can be early warning indicators. While we may not wish to acknowledge it, time is also a highly important progress metric. In some cases this is the most important of them all.

4.4.2 Structural Metrics

Structural metrics are derived from the structure of the component or system that is being tested. For an RTL model, this means the RTL code. But there is no reason why these metrics cannot equally be applied to higher abstractions as well. There are many types of structural metrics, but the one that most people are familiar with are the code metrics such as line, branch, and expression coverage. Other types of structural metric include state/transition coverage and path coverage. Some of these metrics may not be available for abstract models in your simulator of choice as they require a higher level of design analysis, particularly those involving state.

4.4.3 Functional Metrics

Functional metrics are derived from the specification and are related to functions that should be present in the implementation. There are two primary metrics in this class, functional coverage and assertion coverage.

Functional coverage is actually a bit of a misnomer. What it does is to define indicators of the presence of functionality. So, for example, if we think of a processor and if we see the op-code for an "ADD" instruction in the instruction register, then we conclude with functional coverage that the processor has indeed performed an addition instruction. In other words it assumes that something else – the testbench – is checking to see that the ADD op-code was put in the instruction register correctly and that the processor did indeed perform the desired function with no undesired consequences. There is nothing to check that this did indeed happen.

Assertions define actual functionality that must be present, or not present in the design. An assertion, usually written in a declarative language such as PSL or SVA, is a partial reference model with a built-in checker. It can thus be thought of as a mini testbench itself, although it does not have the ability to create stimulus. Assertions can also be used by formal tools, where stimulus is not required. Formal tools will tell you if there is any way that the assertion can be made to fail or if it is valid under all possible conditions.

So while assertions are actual acts of verification, assertion coverage is a lot more difficult to pin down. The simulation vendors have not yet agreed upon a consistent way to measure this, although the Accellera UCIS committee is working on that problem.

But the story does not end there. Consider for a moment the formal verification world. If all of the assertions written are proven to hold true, then is verification complete? To answer this we have to know if there are any functionalities that have not been included in the set of assertions, or put another way does the distributed functionality of the asserts exactly match the functionality as described by the design? Unfortunately there is no commercially accepted way to answer that question today although a large amount of research effort has been expended.

4.4.4 Coverage Metrics in SystemC

SystemC does not have built-in functional coverage constructs and while readily available code coverage metrics, such as GNU COVerage tool gcov (http://gcc.gnu.org/onlinedocs/gcc/Gcov.html), can be gathered directly on the C++ code, this does not really provide an adequate way to measure verification progress. These coverage tools include the SystemC kernel and all of the other code that is not part of the model in their calculations. Several ways to overcome this have been proposed.

All coverage methods basically rely on the insertion of monitor points into the design. These monitor points feed data into a mechanism for storing, sorting, and analyzing the data. It is clearly possible to embed these directly into SystemC code (or any other code), but this has many drawbacks. The main problems are that this is slow and tedious. It requires modification of the code and this may lead to problems being injected. Also, the injected code now becomes something that will become part of the debug and if, for example, monitor points were injected for functional coverage purposes, this will skew the results for code coverage collected at the same time as the model now includes the instrumentation code. Commercial simulators do all of this behind the scenes and ensure that the user has no knowledge of the instrumentation when they are looking at the model.

4.4.4.1 Simple Code Coverage

A simple approach was introduced by [18]. Here they automated the introduction of the monitor code for code coverage. The augmented code was compiled and linked with the SystemC kernel and their coverage libraries. This was then executed to obtain the coverage information. The collected data was analyzed and reported back in terms of the original source code. While they did not report the impact of the instrumentation of simulation performance, it is the kind of thing that can be quickly and easily implemented and does provide some degree of confidence in the verification process. However, few companies rely on code coverage today as it omits many aspects of the complete verification problem.

4.4.4.2 SystemC Verification Library

In 2002 Cadence release a verification library for SystemC called SystemC verification library (SCV) based on the DAI Testbuilder. This included a transaction-based functional coverage capability. It consists of three sets of calls, namely database creation, stream creation, and transactor creation statements.

The database creation is fairly simple as it defines a text file where coverage data is to be stored. The stream creation statements are used to define an object which is to be collected. This often does not correspond exactly to a single signal but may aggregate many signals and events. One way to think about this is a bus over which transactions will flow. The list of transactions that may appear on that stream are defined. For a bus, it may include read, write. Then the code can be annotated to record events on these transactions, such as begin_transaction and end_transaction. The library also enables a hierarchy of transactions to be defined, such that a begin_transaction can be recorded as being a child of another transaction.

This requires a lot of effort on the part of the user. Coware has used SCV in the past as a means of recording transactions in their platform creator/platform architect products. This included software instrumentation and memory analysis.

In [19], the authors propose the adoption of a simple fault model that should be applied to SystemC descriptions. That fault model is basically a bit-level failure. So each variable, constant, signal, or port is considered to be a vector of bits and each bit can be stuck at either one or zero. Similarly, a condition which is based on the

Boolean result of an expression can be stuck, meaning that a path will either be taken or not taken. The model assumes that the elemental operators, such as +, −, >, work correctly and thus errors only exist on the inputs or outputs of the operators. This is somewhat different from a mutation analysis approach [20] which also includes the possibilities that an incorrect operator may have been used. The authors state that they believe this to be adequate for the RTL aspects of the language, but when modeling at a higher level, some additions are necessary to take into account things like events and channels. For events, they include two faults. The first is to change the eventual parameter of the notify method and the second is to block the event notification. Parameter modification itself is already handled by the bit faults. For channels they say that they are similar to ports and signals and can be faulted by modifying the data managed by the methods.

The advantages they claim for this fault model is that it covers all of the major classes of code coverage including statement, branch, condition, and block. While they also put path coverage in the list, they put in a disclaimer saying that it only defines a finite subset of paths. However, given that their approach includes notions of propagation it does as they say *cover the subset of paths that activate and propagate the injected faults from the inputs to the outputs of a SystemC design within a given time limit.*

Rather than attempting to measure functional coverage directly in the SystemC code, Cisco [21] decided to use the SystemVerilog functional coverage capabilities. To do this they relied heavily on the DPI interface previously described. Below are snippets from their paper.

They started by defining coverage groups that contain cover points for the variables and expressions needing to be covered. The following example shows the definition of a SystemVerilog coverage group, used to cover the variable “enable.” Since the clocking event is not specified in the coverage group, coverage collection is triggered using the sample() method.

```
`include "scsv_fcov.svh"
module example_assert();
covergroup ENABLE_CG;
    coverpoint enable{
        bins enable_1 = {1};
    }
endgroup
ENABLE_CG ENABLE_CGI = new;
endmodule : example_assert
```

The SystemC reference model instantiates scsv_fcov, which defines DPI calls to trigger the SystemVerilog functional coverage group. As shown in the code sample below, fcov->sample is called from SystemC with the value of the variable to be sampled. This call to sample ENABLE_SIG in turn calls ENABLE_CGI.sample() in SystemVerilog. After several simulation runs, the results are merged to see the functional coverage achieved.

```
#include "scsv_fcov.h"
SC_MODULE (example_module) {
    sc_in<bool> clk;
    scsv_fcov *fcov;
    SC_CTOR(example_module)
    {
        Fcov = new scsv_fcov("fcov");
        SC_METHOD(capture_rtl_intf);
        Sensitive << clk;
    }
    Void capture_rtl_intf() {
        If(enable == SC_LOGIC_1)
fcov->sample(ENABLE_SIG,enable.to_bool());
    }
};
```

4.4.5 Coverage Metrics in SystemVerilog

Unlike SystemC, SystemVerilog has dedicated constructs for the definition of functional coverage. There are three principle constructs: covergroup, coverpoint, and cross. We already saw one example of covergroup and coverpoint in the mixed language example above.

A functional coverage group defines how data that is to be used for coverage measurement is obtained and how events are counted. As an example, consider the following code snippet:

```
covergroup cov1 @(posedge clk);
  Addr: coverpoint addr
  {  bins low = { [0:'h10] };
    bins mid = { 17, 18, 19 };
    bins high = { ['h14:'hFF] };
  }
endgroup: cov1
```

A covergroup called cov1 is defined. The data will be sampled on the positive edge of the clock clk and the captured data will come from the coverpoint addr. Rather than looking at every possible value for addr, three bins have been defined that cluster the low, middle, and high values that we could expect to see. Coverage measurement will include how many times we have seen addr fall into each of the value bins. While this example used a clock to trigger the data capture, it could be any event within an ESL verification model.

The cross keyword keeps track of simultaneous conditions of multiple coverpoints or covergroups. So, for example, if we had two variables a and b, each of which could take the value of 0 and 1, then the cross of a and b would define four coverpoints, looking for a and b to take the values [00], [01], [10], and [11].

In addition, assertions can be used for control-oriented coverage. For example, we could say

```
sequence seq1;
  @(posedge clk) fifo_full |=> ##[1:]fifo_empty;
endsequence
cover property(seq1)
```

Here we have defined a sequence that looks for the fifo to be full and then at some point in the future for the fifo to become empty. By adding the cover property to this, we will keep track of how many times this sequence actually happens.

4.5 Input Constraints

Constraint languages are one of the newer and still evolving aspects of a modern testbench. Before getting into the details of constraints, let us take a quick look at why we need them. When directed tests were the primary means of verification, constraint languages were not necessary. It soon became apparent that some of these directed tests could be made more powerful by adding some degree of randomization to them. For example, if we had a test that wrote to memory and then read back the results, we could write a program such as

```
for (i=0 to 2**32-1) memory[i] = i;
for (i=0 to 2**32-1) check memory[i] == i;
```

It would write the address as data into each and every location of memory and then check that each location has the right data in it. This test would most certainly find a lot of errors, especially with the address decode logic, but what if the data bus were actually shorted to the address bus such that this relationship between address and data would always hold true, but no other data value could be read?

While it would be quite easy to improve the program to correct the previous limitation, what other limitations exist? Randomization helps to find the problems that you never thought of looking for. Consider the previous example if we were to add randomization. Creating the data would be simple

```
for (i=0 to 2**32-1) memory[i] = random();
```

Now every location would have a random value written into it. The question now is how do we verify it? Without being able to check the values then all we have done is to consume simulation time. One way to get around this would be able to control the seed of the random number generator so that the sequence of numbers could be repeated. The following pseudo code would accomplish that

```
seed = random();
memory[0] = random(seed);
for (i=1 to 2**32) memory[i] = random();
check memory[0] == random(seed);
for (i=1 to 2**32) check memory[i] == random();
```

The first read and write operations set the seed for the random number generator and each iteration of the loop then calculates a random number based on that seed. Now we are likely to find more of the errors that may exist in our memory system. This very simple example has shown why we need to have checkers that can verify the data when randomization is used. It is no longer possible to perform verification without that ability.

The example is still performing the write and read to every memory location. This is probably overkill and we could accomplish just as effective verification by doing it for a random sampling of memory locations.

The use of completely random values is seldom that useful. If the memory map contains I/O functions, they need to be properly set up before they can be expected to perform any useful function. Similarly if error detection is included, the generation of data that does not take this into account is likely to produce 99.9% bad values. Constraints are used to ensure that the random values that are injected make sense and are, to the best possible extent, useful.

A constraint solver is exactly the same as trying to solve simultaneous equations. Given

```
2x + y = 12 and
x + y = 8
```

What values can x and y take? Clearly x = 4 and y = 4. When coupled with pseudo-random generation, it is likely that the solution will have a lot more degrees of freedom, such that there is not one right answer, but many that can then be randomly picked from.

```
constraint c1 { x > 2; y < 3; z > x+y; }
```

A simple example would be the generation of an address, in which only even numbers are valid, or the generation of a data packet which contains one or more check digits. In the case of the check digit, it may be that after the random selection of the data, the check digits are calculated and inserted. But we may not always want the check digit to be correct. It is likely that we do most of the time, but perhaps 5% of them to be wrong.

```
constraint parity { parity_correct ( 0:=5, 1:=95); }
```

Returning to the case of the address selection, we may also have a non-continuous range of addresses due to the construction of the address map. There is clearly no point in generating addresses for locations that do not exist.

```
constraint addr_1 { addr inside {[0:0ff],[2ff:7ff]}; }
```

The biggest limitation to specifying input constraints in SystemVerilog is that it does not have built-in constructs that enable sequential constraints to be defined. Properties can be defined that restrict the constraint solver, but these can be very difficult to write.

```
property p1; @(posedge clk) X |=> !X;
endproperty
constraint c1 { p1; }
```

In this example, it says that whatever value is chosen for signal X, on the next clock it must be inverted.

Below is a complete example for the generation of an Ethernet frame.

```
class ethernet_frame;
 rand bit [47:0] dest;
 rand bit [47:0] src;
 rand bit [15:0] type;
 rand byte payload[];
 bit [31:0] fcs;
  rand bit [31:0] fcs_corrupt;
 constraint basic {
  payload.size inside {[46:1500]};
 }
 constraint good_fr {
  fcs_corrupt == 0;
 }
endclass
```

Here we see that each of the fields is defined as being randomizable, except for the **fcs** bit which represents the corruption of the packet. The random generator would thus always generate good packets. A function that computes the CRC digits for the packet could then override this bit if it decided to generate illegal packets.

4.6 Verification IP

Verification IP is a collection of verification models that are packaged to enable both the verification of a block and the verification of that block when integrated into a larger system. Verification IP or VIP has existed for a long time although not in the sophisticated form that it is found in today. An early VIP example was called a bus functional model (BFM). Its job was to convert transaction-level information on one side of the model into corresponding pin-level activity on the other side of the model and similarly pin activity back into transaction-level activity. The name came from the most common example which was for a bus protocol. Read and write transactions would be fed into the model and the corresponding signals would be wiggled making the tests much quicker and easier to write. While the models have become a lot more sophisticated than these early BFM's, a large percentage of them are still related to interface protocols including USB, I2C, and Ethernet as well as processor interfaces such as AMBA AXI, AHB, and APB. Web sites, such as Design and Reuse [22], have sections dedicated to verification IP, where you can find the most widely used IPs and new IPs that are being offered by many companies.

4.6.1 VIP Components

Modern VIP contains many connected pieces as shown in Figure 4.9. This example [23] comes from Denali Software, but all suppliers of VIP, such as Synopsys and Cadence, have a similar structure. This structure would be very similar for all VIP that supports a protocol. Examples are a bus, such as AMBA from ARM or CoreConnect from IBM, or an I/O protocol such as SATA or USB.

At the heart of the VIP is the protocol FSM. This is what keeps track of the current state of the protocol and the valid states that can be transitioned into based

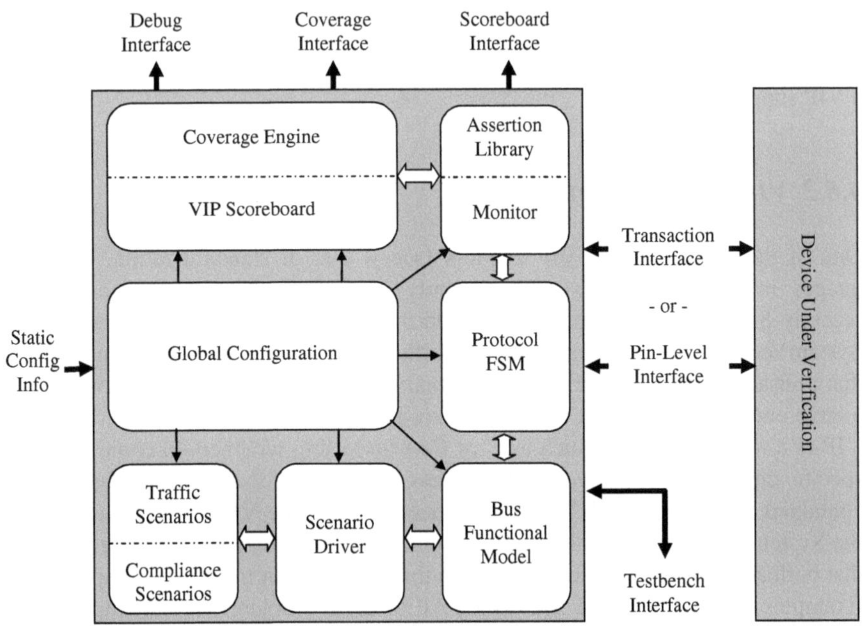

Fig. 4.9 Internals of a Denali VIP protocol block

on input that would either come from the design or the stimulus pattern generator. The connection to the design is provided through the BFM that is likely to have multiple abstraction levels supported, such as the transaction level and the pin level. Its purpose is to convert between the transaction level activity within the VIP and the pins of the design being verified. When data is being collected from the design, it too is converted back into transactions that may then be monitored and compared against the expected protocol, or by additional checks that may be implemented as assertions.

The scenario generator is the primary stimulus creation engine that understands useful ways to move through the protocol to achieve useful functions. In this way, scenarios are a combination of both directed test and constrained random test methodologies. The scenario ensures that useful paths are executed, but allows randomness in the transactions that are generated in certain aspects of each transaction. Often a library of scenarios may be provided that covers all of the essential elements of the protocol, such as specified operations, corner cases, stress tests.

The coverage engine is a block that tracks progress in the verification effort. It may keep track of which assertions have fired, the coverage of the state machine and transitions that have been taken. An external interface is also required so that coverage from this block can be combined with other pieces of the complete system.

The scoreboard provides some of the high-level checking, such as data integrity and temporal checks. Examples of its checking may include dropped packet detection, response times, and out of order verification. These checks are likely to be protocol specific.

For a more in-depth evaluation of VIP, the reader should consult Chapter 7 of [23]. This book also has information about selecting IP providers or how to become a VIP supplier yourself.

4.6.2 VIP Standardization

One of the problems that VIP vendors face is that of standardization. The languages in use by the design community include Verilog, VHDL, and more recently SystemVerilog and SystemC. On the testbench side there is *e*, Vera, and SystemVerilog. To complicate matters further, there is no consistency in the way that transactions are defined. This could make it difficult to write a piece of VIP for a large enough market. Part of this problem is overcome by writing the core of the VIP in a neutral language such as C or C++ and using wrappers to connect it into specific environments. However, this means that much of the advanced capabilities contained in the modern HVLs are not accessible to the VIP writer. Even within the SystemVerilog community, multiple class library extensions have been created that both add capabilities, but at the same time bring about more non-compatibility. Examples of this are VMM and OVM. All of these problems are slowing the rate at which VIP is being made available. This is why Accellera [24] started a new standardization activity to bring about some order to this chaos.

4.7 Conclusions

ESL verification is still in its infancy. Most of the tools and languages being used were intended to be used for RTL implementation verification and have not received the necessary extensions and modification to support higher levels of abstraction. Part of this problem is due to the rate at which RTL verification has been changing in recent years. Constrained random test generation is still new and many companies have not yet adopted this. Assertions are also new and the best ways to insert these into an RTL flow is still being actively discussed. The major EDA companies are thus too busy perfecting these new methodologies that are targeted at the existing large user base, rather than attempting to solve the emerging issues that today are found by only a few users. We thus have a chicken and egg situation. Until more people demand that the verification methodologies and tools be extended into higher levels of abstraction, the EDA companies will not see it as being economically viable. But because of the lack of adequate tools and languages, many companies will not embrace a full top-down ESL verification flow.

References

1. Bruce Wile, John C. Goss, Wolfgang Roesner. Comprehensive Functional Verification. The Complete Industry Cycle. Morgan Kaufman, Elsevier 2005
2. Thorsten Grötker, Stan Liao, Grant Martin, Stuart Swan. System Design with SystemC. Springer 2002
3. Harry D. Foster, Adam C. Kronlik, David J. Lacey. Assertion-Based Design. Springer 2003
4. Frank Ghenassia [Ed.]. Transaction-Level Modeling with SystemC. Springer 2005
5. Brian Bailey, Grant Martin, Andrew Piziali. ESL Design and Verification. A Prescription for Electronic System Level Methodology. Morgan Kaufman/Elsevier 2007
6. Accellera Unified Coverage Interoperability Standard Committee (UCIS) http://www.accellera.org/activities/ucis/
7. 2004/2002 IC/ASIC Functional Verification Study, Collett International Research
8. Peet James, Verification Plans. The Five-Day Verification Strategy for Modern Hardware Verification Languages. Kluwer 2004
9. Jasper Design Automation ActiveDesign™ with Behavioral Indexing™ website: http://www.jasper-da.com/products/ActiveDesign.htm
10. Practical usage of VMM applications in Verification Process. EDA Interoperability Forum 2008
11. VMM Central website. http://vmmcentral.org
12. Open Verification Methodology. http://www.ovmworld.org/white_papers.php
13. Jasper Design Automation GamePlan website. http://www.jasper-da.com/gameplan/
14. Accellera Unified Coverage Interoperability Standard. Website: http://www.accellera.org/activities/ucis
15. Intelligent Testbench Automation. Mentor website: http://www.mentor.com/products/fv/infact/
16. Breker website: http://www.brekersystems.com/
17. Arnab Saha, Doug Warmke, Hui Yin. Introducing DPI-C++. DVCon 2009
18. Daniel Große, Hernan Peraza1, Wolfgang Klingauf and Rolf Drechsler. Measuring the Quality of a SystemC Testbench by Using Code Coverage Techniques. Obtained from website: http://www.informatik.uni-bremen.de/agra/doc/konf/07fdl_tbquality.pdf
19. Wolfgang Müller, Wolfgang Rosenstiel, Jürgen Ruf. SystemC: Methodologies and Applications. Springer 2003

20. Wikipedia Entry for Mutation Testing. http://en.wikipedia.org/wiki/Mutation_testing
21. Srinath Atluri, Nimalan Siva, Anant Sakharkar, Rebecca Lipon. Migrating Complex Networking ASIC Verification Environment to SystemC and SystemVerilog. IC Design and Verification Journal March 2008. http://www.icjournal.com/articles/2008/20080318_cisco.htm
22. Design and Reuse website – Verification IP. http://www.design-reuse.com/vip/
23. Brian Bailey, Kathy Werner [Eds]. Intellectual Property for Electronic Systems: An Essential Introduction. IEC Press 2007.
24. Accellera website http://accellera.org

Chapter 5
Virtual Prototypes and Mixed Abstraction Modeling

In the preceding chapters, we talked about models and the strengths and weaknesses that they posses. In this chapter, and the chapters that follow, we start to see how those models are combined and applied to solve some of the problems associated with ESL. Unsurprisingly, most ESL methodologies that have been constructed start with a model that is generally called a system-level virtual prototype (SLVP). This is the closest thing that the electronics industry has to an executable specification.

An SLVP is a fully functional software model of a system, including the processors, memories, I/O, and user interface, and is capable of running unmodified production code, including the drivers, OS, or application. Speed is of the essence with these prototypes as they must run as close to real time as possible so that execution times for long operations are kept to a reasonable length. Other concerns that users may have about SLVPs are time of availability, accuracy, development cost, bring up-cost, debug insight, execution control, and system interfaces to the environment in which a system resides. We will talk about these later in this chapter.

Early SLVPs were proprietary and constructed for a particular purpose, such as hardware–software partitioning, architectural exploration, or to enable early integration and testing of software. Each of those posed different requirements on models where certain types of decisions required specific accuracy of models. For example, an SLVP designed for early software development may only make sense if it is available early enough prior to silicon or physical prototype or if it offers significant advantages like debug insight and execution control. We have already talked about the problems that this caused, primarily related to the inability to migrate models from one SLVP to another or promoting an effective third-party IP business. Thus these SLVPs were also targeted to one or a small number of industries such that the quantity of models required was limited.

While we tend to think of the SLVP as something new, the software community has been using similar concepts for quite a while. These earlier prototypes were not fully functional and often were less useful for the lower layers in the software stack. But the software community has shown that it is very willing to adopt these systems. For example, when Apple released the software development kit (SDK) for the iPhone [1], which includes a software simulator, it was downloaded 100,000

B. Bailey, G. Martin, *ESL Models and their Application*, Embedded Systems,
DOI 10.1007/978-1-4419-0965-7_5, © Springer Science+Business Media, LLC 2010

times in the first 4 days of its availability, and that number reached 250,000 [2] over the following 3 months.

SystemC and TLM 2.0 changed the industry since now it has a respectable way of integrating models together and promoting interoperability. What we see today are SLVPs that are in transition from being closed proprietary systems to open systems based on standards. This transition cannot happen overnight, not just because of the tool development that is required, but because of the large number of models that were created using the old languages. Models are one of the biggest roadblocks when it comes to the adoption of SLVPs, so enormous value is still attached to these legacy models.

In the next few chapters, we will see that there are still several different types of SLVP, each one targeting a specific task within an ESL subflow. Each of these in effect becomes the anchor for a subflow with, for example, very specific requirements on model accuracy, which has yet to be fully integrated into one continuous flow. It is thus important to decide what you want to accomplish with an SLVP before deciding if one or another approach is the right one.

In this chapter, we will be looking at an SLVP of the hardware system that is specifically intended to be used for the development and testing of software. It thus concentrates on processors, buses, memories, and abstract notions of hardware blocks. By performing this early integration, it enables architectural choices to be made in both the hardware and software architectures maximizing the utilization of resources, optimizing power consumption, and ensuring that the interface between the hardware and software is well constituted. Perhaps even more importantly, it enables the development and testing of software a lot earlier than has been possible in the past. Hardware/software co-simulation started this migration path, but this modeling technology makes it possible to start platform development at the same time as the hardware is being designed. As soon as this has been completed, the software team has a full binary compatible model on which they can start to develop and debug the software, shaving months off the schedule and in many cases taking software off the critical path in the development flow.

In addition, this chapter will demonstrate some of the process of change in the industry. Aspects of the SLVP described here preceded the introduction of SystemC and TLM 2.0 and were thus proprietary and closed. Synopsys has always been an adopter and promoter of standards and thus when a suitable standard emerged they knew the importance of following it and converting the systems that they had to be compliant to those standards. This is not an easy process. Legacy is an important part of this industry, not only in terms of tool support but also in the models that have been created. Synopsys has a huge library of models. To lose those models would have been a setback, not only to Synopsys but potentially to the industry adoption of SLVPs in general.

It should thus be noted that aspects of both the old and the new will be discussed in this chapter, but that Synopsys is dedicated to the future based on standards. At the time of the actual publication of this book, it is likely that more of the legacy capabilities will have been transferred into the SystemC/TLM features of the environment.

5.1 Introduction

Consider for a moment a modern automobile. Up until a few years ago, the electronics in them was limited to a radio and either a cassette or CD player, electronic ignition, anti-lock brakes, and possibly a few other places such as trip computers. Today, up to 40% of a vehicle's cost is attributable to electronics and the software that runs on it; 50 – 70% of that cost is associated with the software itself. For many people, this means that they spend more for the software in their car than all of the other software that they buy, either directly or indirectly.

The electronics in a car is composed of many subsystems, as shown in Figure 5.1, each one having a CPU of its own and communicating with many of the other subsystems in the car. Each of these subsystems is becoming increasingly complex such that it presents both an intra-system and inter-system challenge.

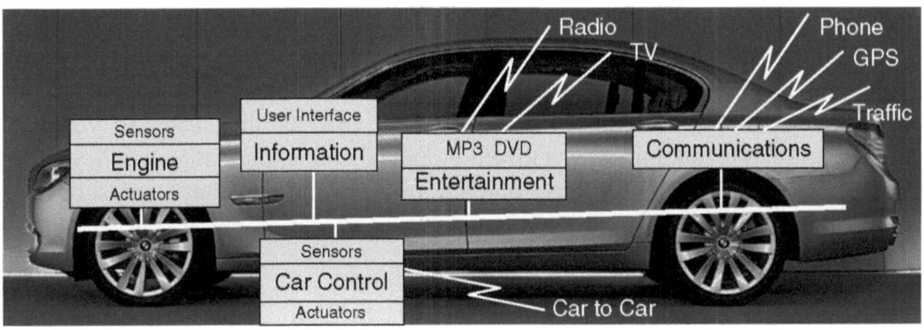

Fig. 5.1 Typical electronic components in a car

The challenges in a car also vary between each of the systems. While some of them, such as the infotainment systems, need to be flexible and user friendly, others must be highly reliable and of high quality, such as those used in safety-related aspects of the vehicle. In all of the systems, problems can result in increased warranty costs and can significantly affect the prestige of a manufacturer. Remember back just a few years to 2001 when BMW introduced the iDrive. This system was intended to simplify the control of infotainment, navigation, passenger comfort systems, and many other functions of the car. It replaced an array of individual controls with a general purpose computer running Windows CE and a single input device, the computer equivalent of a mouse. This excerpt [3], taken from one article, was typical:

> BMW's 2001 introduction of iDrive, its pioneering driver information/entertainment system, was arguably the biggest corporate disaster since Coca-Cola Co. decided to tinker with the formula for its eponymous beverage.

> To say that the automotive trade press and nearly every contributor to a Web discussion of the system hated iDrive is a huge understatement. How, one wonders, could anything in an automobile priced from $75,000 generate so much venom, especially considering the design objective and the fact that a driver didn't have to use it?

Given the turmoil that the automotive industry has been going through over the past couple of years, cars have become highly competitive and compete heavily in terms of safety, fuel efficiency, and new features. At the same time, they must remain cost-competitive and maintain high quality and reliability. Mistakes, like the iDrive, could force a weak manufacturer out of business today.

In the past, integration and debugging issues with software have been solved either by throwing more people at it or by the use of physical prototypes which are by definition late in the game and often very difficult to put together and control. Alternatively, long integration, testing, and rework cycles can be deployed, but this will make the vehicle late to market, which is no longer an option.

In this chapter, we will be demonstrating the technology solutions from Synopsys and specifically their Innovator product and the DesignWare System-Level library. Synopsys is a pioneer in this area, not only in the pure technology sense, where they were one of the major developers of the SystemC language and contributed significantly to the TLM 2.0 standard, but also in the development of products to address these problems. The Innovator product development started in National Semiconductor's Research Laboratories. In 1999, the technology was spun out into as company called Simulation Magic. They later changed their name to Virtio. At the heart of that technology was a mixed graphical and textual language they called Magic-C. Synopsys acquired this company in 2006 and has continued to develop and market the tools.

5.1.1 Historical Perspective

There has been an evolution in the approaches used to create SLVPs over the past 10 years or more. Historically, physical prototypes were the primary means of integrating and testing software. When processors were separate chips, most of them provided an in-circuit emulator (ICE) which would replace the processor in the system and offered increased visibility and controllability of the processor. Subsequent technology advancements made these very difficult to create and use, especially when

- processors became embedded into chips
- multiple processors started to become common
- systems became too complex
- the clock control of the system meant that it was no longer possible to stop and start processors without disrupting the state of the system.

Some attempts were made to create hardware systems where multiple hardware blocks, including processors, could be integrated by software. These worked reasonably well for very short sequences, but were not viable for real software since all of the vectors going into each block had to be saved and re-run after each time-slice was captured on the outputs. Emulators improved on this by providing hardware interconnect, and they also had the ability to freeze the whole system. Many emulator providers also created custom processor cards that would allow the processor to execute independently from the rest of the system. This meant that it could, for example, execute background debug code. Still, there were often problems getting bonded out versions of the processor cores that could be used in this manner. And of course, this could not help projects with new processor architectures that were not yet available from fabrication.

To help with the processor being integrated into larger chips, on-chip debug systems were incorporated as an extension to the JTAG interfaces, originally used for board level interconnect testing. The JTAG standard has seen many improvements and extensions which mean that it is still possible to perform debug on real systems. For problems found in the field, this may be a reasonable approach to isolating a problem. However, real systems tend to be non-deterministic and not highly controllable, making debug frustrating at times.

Thus the need for virtual prototypes has been growing. The earliest commercial success was the Seamless CVE product from Mentor Graphics. This successfully connected a processor ISS model to a logic simulator. Even though the ISS model could execute much faster than a full model of the processor, it did not take much in the way of hardware content to bring the software execution speeds back down to a few cycles per second. Seamless added concepts such as memory and IO virtualization that raised execution speeds for software that did not need large amounts of timed interaction with hardware. Many companies were successful using this for low-level software tasks, such as diagnostics, drivers, and boot code. Some even managed to perform limited OS level execution.

Another avenue of enhancements to the SLVP included co-emulation, where part of a system could be in software and part could be on an emulator. While both aspects of the system could now execute fast, the communications between them became a bottleneck. Transaction-level interfaces such as the Accellera standard co-emulation–modeling interface (SCE-MI) have been addressing these issues making it possible to create hybrid SLVPs.

To execute at speeds required for true software integration and test, all of the system has to be modeled in an abstract form. Execution speeds close to real time are required, and to achieve this, some compromises in accuracy have to be made. This is why a SLVP created for software execution may not be the right kind of prototype for low-level issues that require accurate timing or power measurements. However, SLVPs that have accurate timing may be too slow for software execution. When choosing model types, SLVP users have eight major concerns:

- Time of availability: Once the specifications for a specific design are frozen, the time it takes for a software execution platform and its associated development environment to become available directly determines how long software developers will have to wait before starting on the project.

- Execution speed: Ideally the chosen development method provides an accurate representation of how fast the real hardware will execute. For software regressions, execution that is faster than real time can be beneficial.
- Accuracy: The type of software being developed determines how accurate the development methods have to be to faithfully represent the actual target hardware. This is necessary to ensure that issues identified at the hardware/software boundary are not introduced by the development method itself.
- Production cost: The cost of a development method is comprised of both the actual cost of production and the overhead cost associated with bringing up a hardware/software design within it. The production cost determines how easily a development method can be replicated to all members of a software development team.
- Bring-up cost: Any activity that is expended on a development method that goes beyond what is absolute necessary to get to silicon can be considered overhead. The overhead for the creation of a SLVP must be small in comparison to the returns that will be obtained in order to make it a good investment.
- Debug insight: The ability to analyze the internals of a design, such as being able to access signals, registers, and the state of the hardware and software.
- Execution control: During debug, it is important to be able to stop the representation of the target hardware using assertions in the hardware or breakpoints in the software, especially for designs with multiple processors in which all components have to stop in a synchronized fashion.
- System interfaces: If the target design is an SoC, it is important to be able to connect the design under development to real-world interfaces. For example, if a USB interface is involved, it is important to connect the development method to real USB protocol stacks for verification and software development. Similarly, for network and wireless air interfaces, connection of the design representation to real-world software is important for software development.

During the late 1990s, many proprietary SLVPs emerged, such as those marketed from VaST, Virtio, and Virtutech that were processor agnostic, or vendor solutions such as those from ARM – AXYS that was a spin off from Rockwell Semiconductor/Conexant. It should be no surprise that there were many of these given the sometimes conflicting concerns for these SLVPs. Many semiconductor companies also had their own internal systems such as MOOSE (Motorola Objected Oriented Simulation Environment). Most of these used proprietary modeling systems and had their own APIs to connect them with external models. When SystemC and TLM 1.0 became available, they did support these, but this did little to solve interoperability problems and models were certainly not portable. Some vendors, such as CoWare, did release free communications libraries built upon these standards, but there was no industry uptake of them.

With the advent of SystemC and TLM 2.0, almost all vendors have implemented external compatibility to the same standard. While this is not yet capable of modeling all bus interconnection without significant manual effort, it has brought about a revolution in the acceptability of these SLVPs and made it possible for models to become portable. These APIs may still be slower than the internal ones developed by the vendors, and so in some cases a SLVP will consist of many of the models written and communicating with the proprietary languages and APIs and use the external standards when it is necessary to integrate with an external model. A good example of this is that few vendors will choose to model their processors using SystemC, since C or C++ provides all of the capabilities required and can execute much faster. They will, however, provide TLM 2.0 interfaces into the model so that it can be used within any SystemC compliant simulator. SystemC and TLM 2.0 thus provide the interconnect framework into which several types of models can be plugged and can communicate and synchronize with each other in an efficient manner while maintaining portability and performance.

This is also the case with the Innovator tool discussed here. We will look at both the proprietary and standard parts of the suite. It should also be noted that the open interface execution speeds are being improved over time as additional optimizations are put into the systems, making the speed difference between the proprietary and open interfaces marginal. In fact Imperas [4], a provider of high-performance models that integrate into virtual platforms, reports that they see only a 3% degradation in performance using the TLM interfaces as compared to their own proprietary interface [5]. We can thus expect to see many of the legacy components being faded out in the future.

5.1.2 Use Models

While in the previous section we talked about some of the characteristics of SLVPs, we did not talk about some of the real advantages that they provide over and above what real hardware can provide. These include early availability and determinism. In addition they also provide levels of controllability and observability that are unattainable with real hardware.

These advantages make abstract SLVPs useful in a number of ways. We will briefly discuss four of them, including early verification and validation, architectural analysis, software development, and finally debug.

5.1.2.1 Early Verification and Validation

Verification and validation are tasks that should be performed continually and iteratively throughout a design as discussed in Chapter 4. Until the emergence of the SLVP, there were no concrete models on which to perform either of these tasks. Thus verification was delayed until the RTL was complete and validation was delayed until product integration was complete. With a comprehensive SLVP, this fundamentally changes the design paradigm as validation can now be conducted at the beginning of the design cycle and then each successive refinement step can be verified against the original model. It thus becomes impossible for major errors to persist in a model since detection should happen close to the point at which they were injected.

Just as importantly, the software team now has access to the abstract hardware model well in advance of the RTL, which means that their development timelines can now be parallelized and the interfaces between these two worlds can be developed and tested together. Software architecture is as important, if not more so, than the hardware architecture, especially, as systems start to incorporate more and more processors. Creating the right amount of concurrency in the software, dealing with data transfers and communications, control synchronization, load balancing, and many other tasks need to be performed early in the software development, and it is difficult to change these aspects of software once the hardware has been fully developed.

The development of the SLVP thus has more value than just to perform architectural exploration or any of the other tasks performed at this point in the development flow. It becomes the verification model against which all stages of the development will be tested. The way in which the SLVP is constructed should thus consider the other environments in which it is to be used. This includes the connection to other simulators, such as an RTL simulator, but could also include connections to physical components such as emulators and physical prototyping systems. As an integral part of the verification environment, it also needs to fit into methodologies such as the Verification Methodology Manual (VMM) or the Open Verification Methodology (OVM).

Perhaps the biggest requirement for a SLVP is that it should be able to run production code in an unmodified form. This means that an actual OS, application, or third-party code that may only be available in binary form can be executed on the prototype. This is subtly different than saying that production code should run in an identical manner to the real hardware. While this may also be desirable, it may not be possible under all conditions since some code will depend on timing, which as stated has still not been defined at this point. However, a timing difference that causes a failure may actually imply a restriction in the system, so even this failure may be important to understand.

5.1.2.2 Architectural Analysis

Before we can make any decisions about the systems architecture, be it either the hardware or software architecture, we need a means of conducting experiments, receiving feedback, and testing choices. Design is the process of removing degrees of freedom in a specification until we reach an implementation. Until the emergence of abstract SLVPs, the primary means of making these decisions involved the use of spreadsheets and a large dose of experience based on previous designs. This has all changed, as it now becomes possible to execute large amounts of code or data streams in order to gather sufficient information from which informed decisions can be made. This kind of analysis cannot be expected to depend on accurate timing since the architecture of the solution is not yet known. While some degree of timing is necessary, speed of execution and accurate reflections of dataflow through the system are the most important attributes for this type of SLVP. It also needs to be possible to quickly modify the SLVP such that experimentation is encouraged.

5.1.2.3 Software Development

While software is sometimes developed concurrently with the hardware, it cannot be tested on the hardware until a prototype becomes available. For emulators and physical prototypes this means when the RTL coding is complete and has been verified, at least to the point that the software team will not be spending most of their time debugging hardware failures. This may be too late in the schedule, and resolving problems in the hardware–software interface can often result in compromises being made, or schedule slips for the hardware.

Increasingly, software from a previous generation of product must execute on new hardware, and it is important to find out early if there are any problems with the proposed hardware. When pieces of software do need to undergo modification, it is highly desirable that these can be done early such that compatibility testing can resume.

Developing and debugging software on a virtual platform also overcomes other limitations that have traditionally slowed down this process using physical targets. This is the extra controllability and observability that come from the virtual platform. In a SNUG 2008 paper [6], it was reported that this alone can provide a 2-5X productivity improvement over debugging on a physical target.

Software exists at many different levels, even though hardware engineers often tend to think of it in homogenized terms. Not only do we have the many layers of software, such as hardware-dependent software, which may include drivers, diagnostics, boot code, operating systems, board support packages, middleware and applications, but many of them can exist in multiple levels of abstraction. As some of the software languages such as C and C++ are being used more in the definition of hardware, engineers are discovering that manipulation of code can yield significantly different results when synthesized. That is because code contains architectural choices and it is not unusual for prototype code to look and perform very differently from production code. Specific examples of this will be seen in the next several chapters.

The execution of software requires high levels of performance from the SLVP above all other considerations, but for the lower levels of software, differing levels of timing accuracy are also necessary. It is also necessary that a SLVP have the look, feel, and integration into the environment that the software engineer chooses to use. This means that the tools they use must be consistent across all phases of the design, development, and test processes, and on the real hardware. It is thus essential that a SLVP be integrated into the same environments that software tools use and where that is not possible present a similar look and feel to their operations.

As power becomes a limiting factor in many of today's systems, power management hardware and software are becoming an increasingly vital part of a system. There have already been known examples where a chip failed because of power management problems. SLVPs can also be a great aid to understanding and building the correct power control systems.

5.1.2.4 Debug and Visibility

Any engineer who has tried to debug an intermittent problem on a physical system knows the frustrations of non-determinism. Sometimes this can come from the instability of the environment that it exists in, or sometimes just because of random effects within the hardware itself. A typical physical prototyping system is shown in Figure 5.2. Often these problems can be affected by environmental elements such as heat and light. While some of these problems may remain completely hidden in a virtual prototype, most functionality problems exist in any model or physical manifestation. However, we still want to remove the non-predictable elements in the environment, and thus a simulator provides a much more stable environment within which we can debug problems and have a faithful meaning of regression runs.

Fig. 5.2 Debugging a physical prototype

Another problem with the real world involves visibility and controllability, both essential elements for efficient debug. RTL simulators can provide infinite visibility, although they often produce so much data that problems can be hidden or hard to find – the needle in the haystack problem. An SLVP running at higher levels of abstraction is good at removing the low-level details, making it much easier to find the big picture problems. Remember from the definition of ESL, we want appropriate abstractions to aid in comprehension.

Several companies have produced special purpose debugging hardware that can be integrated into chips. They increase both the visibility and controllability of a system. The solutions range from banks of multiplexors that can feed signals to the external pins of a device, through to on-chip breakpoints, built-in assertions, and trace systems. While these help considerably, they also impart a production cost on each and every device that is made. This may be an expensive compromise when only a few of them will ever be used for the purposes of debug. For FPGAs, the debug logic can be removed after debug is complete, and so long as the FPGA has sufficient room can be a very viable solution. In some cases, where the largest FPGA in a family is not required in production, the next larger size than necessary can be used in the early stage of bring-up to make room for this on-chip debug capability.

While it would seem obvious to perform debug using a SLVP during the design phase, there are often times, even after a system has been produced and is running in the real world, when it makes sense to move a problem back onto the SLVP after it has been isolated. This again provides a more efficient and controllable environment in which to understand the problem and come up with a fix that can be properly verified along with all of the other system tests.

Another use of SLVPs is for verifying and debugging protected code. An example of this is with the Cell processor, where access to the memory and processors in the compute engines is purposely made non-accessible to the outside world such that encryption and other sensitive software can be executed without a user being able to break through the hardware security layer provided. This means that the only way to see what is happening is to execute the code on a virtual copy of the system where those restrictions are not enforced.

5.1.3 Technology

A SVLP is a collection of several different types of models. These include one or more processors on which software is to be executed, buses and other internal protocols to connect blocks together, memories, peripherals, and possibly blocks that are going to be implemented in custom hardware. Most SLVPs will also need to model parts of the environment in which the system is meant to operate. This brings in the need for virtual IO devices and user interface emulation as well.

5.1.3.1 Taxonomy

Two types of SLVP are described in this chapter. The first is correctly described as a functional transaction-level model, meaning it has little to no timing, except for the timing necessary to maintain a consistent ordering of events at the system level. The taxonomy diagram for this is shown in Figure 5.3. The complete key for these charts can be found in Chapter 1. In the SystemC/TLM 2.0 terminology, it would be described as either a programmer's view (PV) model or a loosely timed (LT) model. As such it has only a minimum of timing necessary to ensure the progress of time in a controlled manner.

This compares to a more accurate type of platform where the timing is more refined. This can be accomplished at several different levels of abstraction without

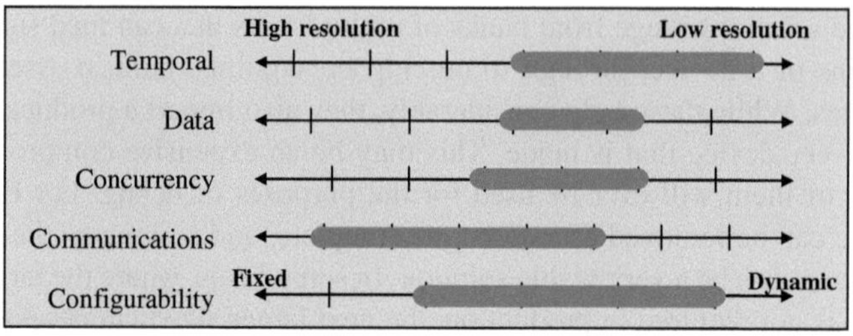

Fig. 5.3 Functional transaction level model

being completely cycle accurate. With cycle accuracy, the timing at clock cycle boundaries is expected to be identical to the final hardware. These intermediate SLVPs are called behavioral transaction level models or in the SystemC/TLM 2.0 terminology as an approximately timed (AT) model.

The principal difference between the two models is in the abstraction of timing. For example, the processor models, which are at the heart of almost all platforms, are modeled at the instruction-level of accuracy for the PV level. This means that they produce the correct functional results at the boundaries of each instruction, but they do not attempt to replicate the actual timing of the processor. All instructions are idealized in that they take one tick of the system clock. Using this approach, it is possible to model processors with performance on the order of 20 MIPS per GHz of the host machine. We will discuss the technology behind ISS models in more detail in Section 5.1.5. For the behavioral platforms, the processors would estimate the exact number of clock cycles that they expect would be consumed by the instruction. Some even go down to the exact clock cycle at which memory operations would take place.

In addition to relaxing the timing accuracy of the processors, other aspects are also relaxed, such as the interfaces. For both types of SLVPs, these are not modeled at the pin level, instead presenting a transaction interface for connecting to the bus model.

The bus may not be fully modeled either (Figure 5.4). For the functional SLVP, address decoding is done in an idealized manner and all components connect onto the bus using a simple fixed API. This API consists of just two calls – read and write

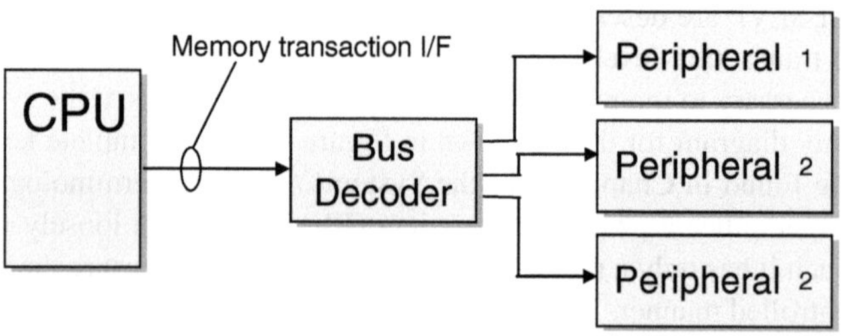

Fig. 5.4 Functional bus model

plus an interrupt mechanism. The models have no concept of bus phases or bursts, instead making all requests as atomic transactions. When multiple masters share the bus, simple arbitration schemes are used.

The peripheral models that connect to this functional bus share the same simplistic interface as described for the functional bus. The models, shown in Figure 5.5, have three principal layers, namely the register interface layer which interfaces the component hardware to the processor via the bus, the functional layer which provides the correct functional results in response to its interfaces, and the I/O interface layer which defines how the platform talks to the real world or other components.

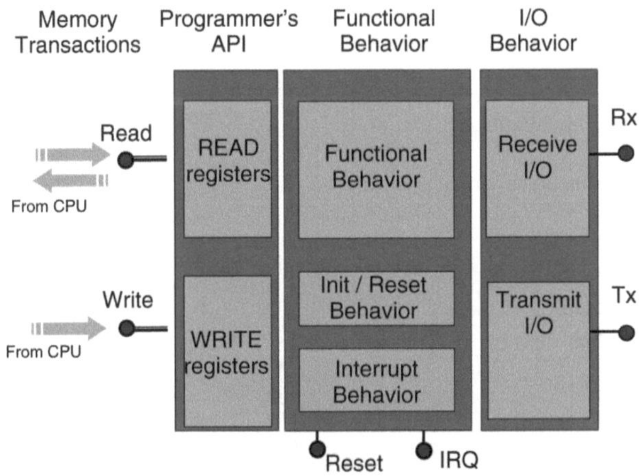

Fig. 5.5 Peripheral models

From these components it is possible to construct high-performance platforms (Figure 5.6). An example of the former would be the DesignWare™ model created by Synopsys for the OMAP platforms that are distributed by TI to their customers. These models enable early porting of software and help the customer understand and gain the maximum benefits from the hardware that TI will supply at some point in the future. In many cases, these models can be shipped 12 months before any real hardware becomes available.

Many of the model abstractions change when we migrate to a behavioral SLVP (Figure 5.7). For example, cycle-approximate processor models are deployed using PVT bus models and many of the peripherals and memories may be SystemC models coming from multiple suppliers. The principal use models for these SLVPs are architectural exploration, performance estimation, communications throughput and latency estimation, and memory architecture design. A platform such as this requires additional tool support such as the ability to configure peripherals and to collect and analyze performance data.

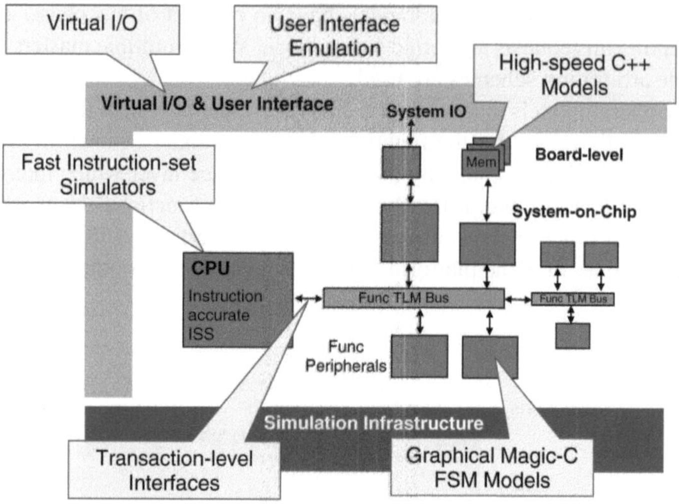

Fig. 5.6 Synopsys functional virtual platform

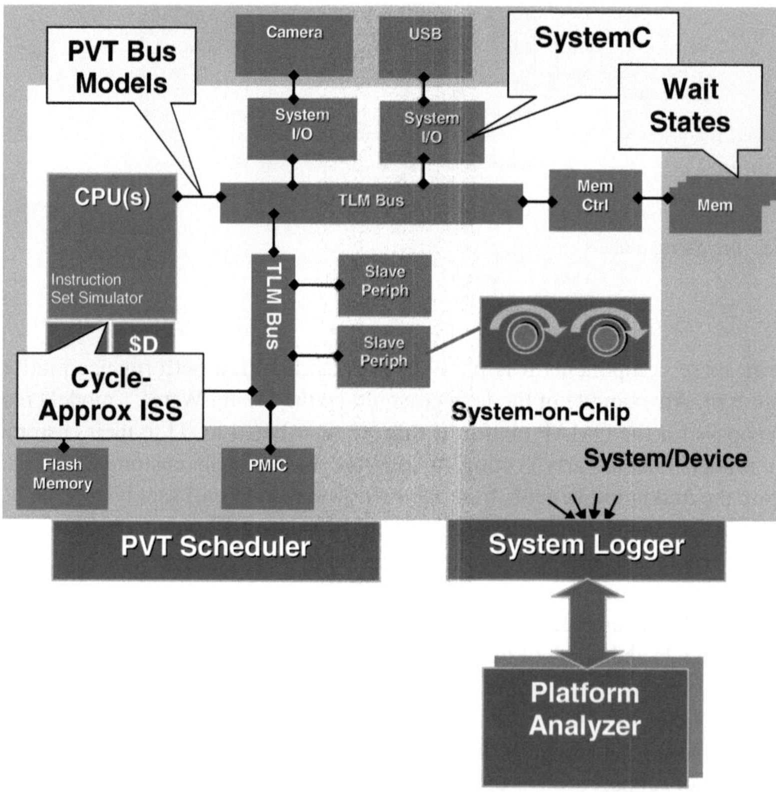

Fig. 5.7 Synopsys behavioral model

5.1.3.2 Time Advancement

In the functional prototype, all of the processor models execute one instruction per system tick. All memory accesses take zero time and are performed through instantaneous non-blocking API calls. Timers or other special constructs cause the system to progress in time. Timers are also used within the peripheral models to poll the virtual I/O interfaces. All of the timers can be tuned to calibrate real-time events, such as OS timers to match the host speed and thus ensure that the platform runs in "real time."

This is very different from the time model in use when SystemC is used. Here the more traditional event-based execution semantics are used, meaning that all models progress in lockstep with each other. However, having said this, the LT mode of operation within TLM 2.0 allows for exceptions to this which enable faster execution. This is because every time a model has to defer to the simulation kernel, a context switch happens, and this is an expensive operation causing the platform performance to degrade. Thus timing can be passed as an annotation rather than as an actual wait, and these annotations can be stored until it becomes necessary to actually perform the context switch. This mechanism was more fully described in Chapter 3.

5.1.4 Interfaces

While interfaces between design blocks are important, there are other interfaces, particularly those between a model and the tool that are equally important when attempting to run and debug a design. One example of a model to tool interface that was standardized as part of the TLM 2.0 standard is the memory interface. This allows one model to access the memory contents in another model without having to utilize the normal access mechanisms. Without such an interface, it is not possible to quickly load memory contents or to perform many other debug tasks.

Other interfaces that are equally necessary but not yet standardized are debug and control interfaces. Examples of these would be profiling and tracing interfaces, debugger interfaces, and command and control scripting. While it is possible to build these using standard SystemC, this requires the models be changed every time different levels of visibility are required. The user will thus want instrumentation to be performed behind the scenes in an efficient manner.

One final category of necessary interfaces for a SLVP is virtual and physical I/O interfaces. These allow the prototype to emulate hardware devices, such as displays and input devices, or to be able to utilize real peripherals, such as a USB or Ethernet connection. These will be discussed in greater detail in Section 5.3.5.

In a press release [7], Synopsys talked about the advances that they were making in their Confirma™ and CHIPit® product lines. This allows FPGA-based prototyping to be combined with their rapid prototyping boards. They then utilize the Accellera SCE-MI standard, which provides a high-performance transaction modeling interface targeted at emulator to software connections, to connect into a software world. It will clearly not be a major step before this is integrated into the SLVP [8] allowing a single unified prototyping environment to exist all the way from design conception to deployment in the field with the abilities to arbitrarily replace any design block with models at different levels of abstract, with an FPGA-mapped version or with real hardware.

5.1.5 Processor Models

Processors are at the heart of everything these days. While products used to be primarily hardware, with some software, products today are primarily software with hardware used as a generic platform for running the software and possibly with a small amount of custom hardware that is either a product differentiator or as an accelerator for software that cannot execute fast enough.

While early SoCs often contained custom processors, today there are a handful of companies that specialize in the development and verification of processors, which are then reused on SoCs. This has been the biggest success story in the IP and reuse category.

As discussed in Chapter 4, verification has become one of the limiting factors in the creation of ever more complex systems. It is important that we do not unnecessarily duplicate verification effort, and this is one of those areas where most users have, unknowingly in many cases, separated the facets of verification correctly. When we construct a system and perform verification on it, we are not trying to verify the processor. This has already been done at a greater level of detail and confidence than any single SoC verification program will be able to perform. The processor companies are betting their life on the quality of their product, and the processor is one of the last places we would expect to find a problem. (This may be less true the first time a new processor is used, but that is a special situation where both parties would be acting as partners in the SoC verification.)

Since we are not attempting to verify the processor, we do not need an actual model of the hardware, only a model that emulates the hardware. Nobody would try to use an RTL model of a processor to verify an SoC. It is too slow and would consume way too many simulation resources. Instead, most companies have become comfortable with the use of an ISS model of the processor. This is a model, normally written in C or C++ that interprets the instruction stream and ensures that the internal registers of the processor are updated correctly and that the correct operations are performed on the bus. Other than this, it does not have to operate in the same way as the real device. Only at the end of each instruction does the model attempt to faithfully represent that its state is the same as the hardware. These models execute much faster than an RTL model.

There are several levels of detail at which these ISS models can operate, and the terminology can sometimes get confusing. The most accurate models are called cycle-accurate models. This means that when we examine the model externally, there is no way to tell if it is an ISS or an RTL model. They behave the same under all conditions. It still does not mean that the internal state of the models is the same at every clock boundary, only from the outside can we tell no difference.

Most ISS models fall into the category of being cycle approximate. This means that there are occasions when we will see a difference in the timing behavior between the ISS and RTL models. For most models, the differences will only be seen under extreme or corner case conditions, so you will have to check with the model creator to find out just how accurate these models are. However, the functionality from the outside is still expected to be identical. For most models, they would also claim that instructions will always take the right number of clocks to execute. The timing variations are in the exact points at which bus activity happens.

Some processor models utilize statistics to define the timing of events within the processor. An example of this will be described more fully in Chapter 6.

Processor models that are not very accurate in terms of timing are often called instruction-accurate models. This means that they function correctly, but timing is likely to differ quite markedly from the real processor. For some it means that they have made no attempt to model timing at all. Many of the processor models that will be used in this chapter take one nominal clock tick to execute an instruction, no matter the complexity of that instruction. Of course, for every step in timing accuracy that we give up, we expect to get compensated by having higher execution speeds.

But this is nowhere near the end of the road in terms of gaining higher performance in the execution of the SW. So far, we have remained in the realm of processor models that interpret each instruction and maintain the same internal state as the actual device. But what if we can do without the model entirely? Most processors have a lot in common and just a few things that differentiate them. Differences may be in the way that status information is stored or in specialized instructions, or in the way in which instructions or data are cached. This is true not only of embedded processors but also of the processor in the computer on which a simulation is running.

Compiled code techniques [9 – 11] replace some or all of the SW that would run on the processor model by a specially compiled form of the code that is designed to execute on the host processor. This compilation could be done ahead of time, or sometimes based on the execution patterns that are observed in an ISS simulator (often called just-in-time compilation). For example, if the same sequence of instructions is seen to execute many times, then compiled code can be created dynamically to speed up this piece of the instruction stream. In one experiment conducted by Synopsys, they took their ISS model for an ARM7tdmi and compared it to a dynamic translation execution of the VPDHRYSTONE benchmark. The ISS came in at a very respectable 19.6 MIPS running on an Intel Centrino Duo CPU T7200 2 GHz. Using dynamic translation, it clocked in at 132.3 MIPS – a speed-up of almost seven times. Other processors have shown similar performance increases.

So if some of the software can be executed on the host processor, how far can we take this? This depends on the objective of the verification. If we are attempting to verify high-level application code, then do we really care about the values held in the registers of the processor? This is unnecessary intermediate detail and until something is written into memory, in a place that will cause a change in behavior of some other part of the system, then we do not care if it even exists. If we assume a standard layering of software, then these interesting activities could be associated with calls to a driver. So why not execute everything on the host until a driver is activated? If the program counter associated with entering the driver can be detected, then we can switch to execution on the target processor, and then revert to host execution once this is complete. In this way any OS call, or for that matter any interesting call, can be trapped and replaced by execution on the target. Using these techniques, Synopsys has seen a further two to three times performance improvement.

The right level of accuracy for the execution of the processor will always be dependent on the verification task that is being performed and may also be affected by the number and types of processor that exist within your simulation environment. In this and the following chapters, we will see processors at various levels of abstraction being utilized.

The level of detail of the processors can impact other parts of the system model, such as the bus. In one common usage model for the Synopsys platforms, each processor executes one instruction per clock tick. In addition, the processor may be allowed to execute multiple instructions without returning control to the system scheduler. During that period the master may have issued several bus requests. Then the next master runs for the same period of time and it too produces bus operations for the same bus. This continues for each of the masters.

At the end of the time-slice, we may have seen more transactions pass on that bus than would have been possible in a real system. The bus accesses have to be overlaid in a fair manner as shown in Figure 5.8 below.

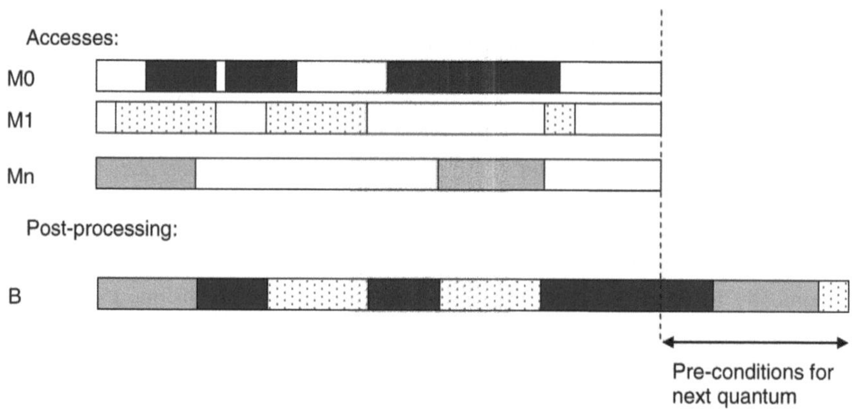

Fig. 5.8 Bus access overlays

The way in which the transactions are ordered on the bus needs to preserve a notion of order and fairness. In the scheme shown above, events that are clearly ahead of others in time are allocated first. When arbitrary choices exist, the masters are overlaid in a round robin scheme. So, for example, in Figure 5.8, Mn has an event that is before either of the other masters, so it gets the bus first. By the end of Mn access, both M0 and M1 have a transaction pending. M0 is arbitrarily picked first. By the end of that transaction, both M0 and M1 still have transactions pending, but the token moves on to M1 to ensure all processes get fair access.

Transactions that overrun the notion of the time quantum become the preconditions for the following time advancement. The overlaid transactions are used for calculating bus utilization, latency, and other bus-based statistics.

5.2 System Prototypes

5.2.1 Development Environments for Software Development

Even though this chapter is focused on virtualized prototypes, i.e., SLVPs executing in software, there are hybrid alternatives that combine execution in software simulation with hardware acceleration. There are three basic options: software-based techniques, hardware prototyping-based techniques and production silicon-based techniques. Each of these will be discussed briefly.

Within the software execution models, the first option is to develop virtual platforms using loosely timed (LT) transaction-level models (TLMs). With software development being the main target use model, virtual platforms can often be made available 4–6 weeks after specifications are frozen. Virtual platforms execute very fast, typically between 20 and 50 MIPS for complex platforms, and north of 100 MIPS for less complex designs. They offer excellent execution control and debug insight, with the ability to stop all components in simulation in a synchronized manner and provide full debug access to all aspects of the design.

The production cost for virtual platforms is fairly low, making it easy to replicate them for all members of the software team. Virtual platforms offer a variety of system interfaces, allowing them to be connected to USB, Ethernet, SATA, etc. However, the accuracy of loosely timed virtual platforms is limited to functionally and register-accurate representations. They will not reflect all timing aspects of the target hardware. Also, assembling a virtual platform from a library of TLM components or even developing new TLM models may be considered overhead, i.e., high bring-up cost by the hardware team if the models are not directly usable for verification as well.

Staying with the software execution models, the second option is to develop approximately timed (AT) virtual platforms. These models contain more timing detail than the LT models so that architecture exploration becomes possible. However, they require significantly more development effort which means that they become available later in the project and also execute the target hardware slower than LT virtual platforms, typically in the range of 100 KIPS – 10 MIPS.

The third option for software execution models is to develop cycle-accurate (CA) virtual platforms. For complex designs and components, cycle-accurate modeling can take just as much time as the actual RTL development and in many cases knowing cycle accuracy implies that the detailed design has been completed. As a result, the industry is moving away from specific C/C++ based cycle-accurate modeling and is instead adopting RTL for cycle-accurate software based-representations. As a result, cycle-accurate virtual platforms are available late during a project – when RTL is stable – and they often execute the target hardware slowly, in the range of 100 IPS – 100 KIPS. This is not the case for all processors – in particular, configurable, extensible processors as discussed in Chapter 6 have automated development tool flows that will generate appropriate cycle-accurate and instruction-accurate models as a deliverable from the processor generation flow. Models for particular processor configurations do not need to be developed by hand in this scenario.

Within the category of hardware-based development methods for software development, emulation and FPGA prototypes are the main alternatives. Generally emulation and FPGA prototypes are available much later in the design flow – after RTL is stable – which can take up to 70% of the time from requirements to tapeout. Given the substantial time it takes to produce engineering samples after tapeout, they still offer significant time advantages over using the actual silicon engineering samples.

Both methods reflect the RTL accurately but vary in execution speed. Emulation typically executes in the low MIPS range, while FPGA prototypes execute somewhat faster, in the range of 10s of MIPS. But this is only one of the trade-offs that exist between the two alternatives.

- The bring-up effort for emulation is typically less than it is for FPGA prototypes
- Emulators have much higher production costs than FPGA prototypes. This makes it more difficult to provide affordable replicates for software development teams.
- Debug insight and execution control are pretty good in emulation. FPGA prototypes, while not as good as emulators, still provide much better insight compared to the actual silicon engineering samples.
- Both emulation and FPGA prototypes offer real-world interfaces that connect to the target system environment although speed differences may create some issues.

Software development methods using real silicon usually mean the use of a previous version of silicon or the use of the engineering samples of this project once they become available. While previous chips are available at project start, they may not reflect the new target silicon accurately enough in areas in which new software is to be developed. Changing register interfaces and the lack of new hardware functions, only available in the new design, limit this method to high-level software development. While both types of actual silicon can be executed at real-time speed, engineering samples are available very late in the project allowing little parallelization of the hardware and software efforts. In addition, while on-chip instrumentation (OCI) and on-chip debug have improved over the years, they are costly due to their silicon overhead and do not offer the same level of control and visibility as users can find in software-based and hardware-assisted development methods for software development.

5.2.2 Hybrid Hardware–Software-Based Development Platforms

None of the development methods for embedded software development, in the context of hardware, come without disadvantages. As a result, hybrid solutions are emerging, which allow "System Prototyping" of systems and SoCs combining the advantages of software-based and hardware-assisted development methods, as illustrated in Figure 5.9.

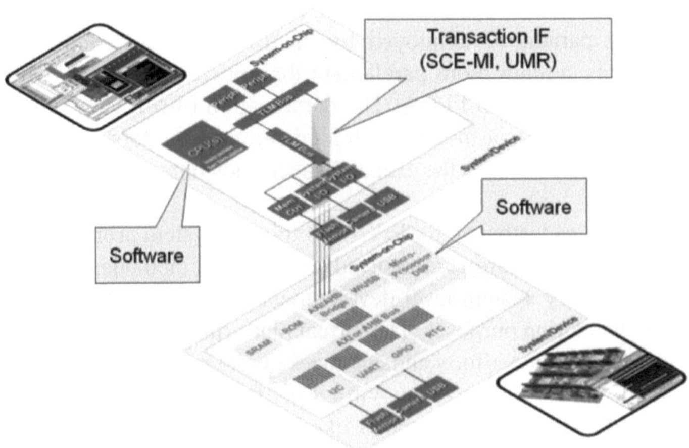

Fig. 5.9 Hybrid system prototyping

Hybrid system prototyping requires three technology components:

- Physical interfaces must be available to connect the actual hardware prototype to the workstation running the simulation. PCI Express is a common solution for this.
- Data must be transported using an agreed upon protocol between the software and hardware worlds. Some hardware-assisted technologies offer proprietary interfaces but the standard co-emulation modeling interface (SCE-MI) has become a standard in this domain [12]. It was developed under the umbrella of Accellera [13].
- For conversion from the transaction-level model to the transport interface, transactors are necessary to translate high-level protocols like AXI, OCP, and AMBA,

into the actual signals driving and observing the blocks that are executing in emulation or FPGA prototypes.

5.2.3 Hybrid System Prototyping Use Models

Hybrid system prototyping enables five use models:

1. RTL reuse and architecture verification: To reduce the bring-up effort for virtual platforms, existing RTL from previous projects can be mapped into FPGA prototypes avoiding the modeling effort of potentially complex IP blocks. The cycle-accurate execution in FPGA prototypes increases overall fidelity and allows the replacement of virtual models with RTL to verify that architecture decisions are correct. This provides fast execution of cycle-accurate models.
2. Accelerated software execution: Software typically runs on workstations and virtual processor models faster than in FPGA prototypes due to the different modeling paradigms employed by the two methods. Many accelerators and peripherals contain more parallelism than processor models which makes them execute faster on an FPGA. The combination of software running on the workstation on processor models with hardware accelerators and peripherals in FPGA prototypes combines the execution speed advantages of both development methods.
3. Virtual platform as testbench for FPGA prototype: System prototyping mitigates the late availability of hardware-assisted development methods, enabling the efficient reuse of early system-level development efforts for RTL verification and post silicon validation purposes. The virtual platform acts as a testbench for RTL, which avoids duplicate efforts and enhances model reuse.
4. System environment connections: Virtual platforms already provide real-world and virtual I/O for popular interfaces like USB, Ethernet, and SATA. Daughter cards in FPGA prototypes provide real-world I/O with interfaces to real-life streams, like the wireless physical interfaces. System prototyping with hybrids of virtual platforms and FPGA prototypes allow real-world stimulus to be used where most appropriate – at both the transaction and signal levels.
5. "Virtual ICE" connected to FPGA prototype: Software developers often dislike development boards on their desk and instead prefer a development environment that combines a keyboard, screen, and their familiar software debugger. Reuse of the virtual development environment allows better access to FPGA prototypes and decreases set-up time. System prototyping allows the FPGA prototype to be kept remote and this makes the development environment more familiar to the software developers.

While software-based, hardware-assisted, and real silicon development methods have found adoption in their own right, system prototyping allows mitigation of many of the disadvantages of these individual methods. Hybrid system prototyping may allow design teams to capitalize on the combined advantages of software- and hardware-based development methods.

5.3 Constructing a System-Level Virtual Prototype

There are two principle ways in which a SLVP can be created. The first is using freely available components such as a standard C compiler, the OSCI open source SystemC, and the text editor that comes bundled with an operating system. The total cost of this environment is either free or close to it, and it is possible to do it this way.

Having said that, this solution is far from ideal. Most companies choose to use one of the integrated development environments (IDEs) specifically designed for the purpose. An IDE provides many things that improve modeling productivity, such as browsers, graphical input tools, design-rule checkers, version management, analysis, and much more. In addition, having a library of existing models can reduce the total development time down to a fraction of what it would have been.

Figure 5.10 is a generic screenshot of the Innovator IDE from Synopsys that is being used to establish some of the parameters of the SLVP. On the left of the screen shot, we can see the principle browser for this IDE which is based on hierarchy. That can be either the design hierarchy or, as shown in this example, the parameter hierarchy. An alternate view is based on signals and connectivity. To the right of this is the main working window, which in this case is showing the top-level view

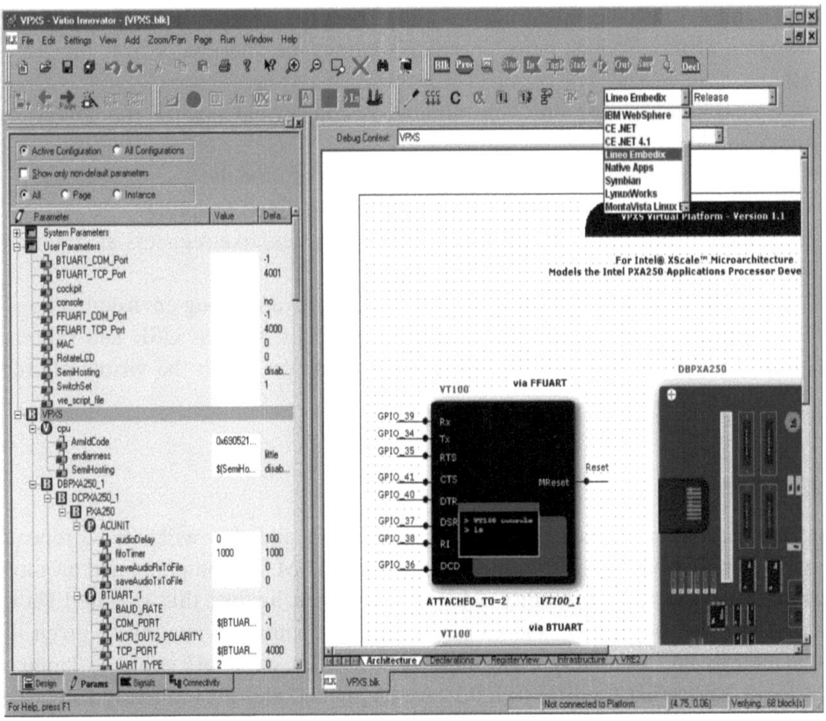

Fig. 5.10 The Innovator IDE

for one of the supported platforms. This window would also be used for building or modifying connectivity or for constructing Magic-C models.

5.3.1 Modeling Languages

Innovator provides three related environments in which models can be constructed. These are SystemC, Magic-C, and VRE C++. Each environment provides a different mix of attributes for operations such as editing, debugging, and visibility. They will each be discussed briefly. At the time of the writing of this book, tools like Synopsys Innovator are in a transition toward full SystemC support. As always in standardization, not all user requirements are standardized in parallel, leaving a mix of proprietary and standardized techniques. Today, SystemC TLM-2.0 APIs sufficiently standardize the simulation, enabling simulation of models from different sources without noticeable impact on speed. However, some other items are currently in the process of standardization (for example, in the OSCI SystemC CCI working group). Some of the capabilities provided in the Innovator VRE C++ environment may be subject to standardization such as

- Profiling – allowing non-intrusive collection of profiling information from inside models. Important characteristics like number of transactions, cache hits/misses, and number of executed instructions allow users to optimize the SoC architecture.
- Logging of transactions between models allows the analysis of SoC integration effects.
- Sideband signals used only for simulation control but not for the actual SoC functionality.
- Configuration interfaces allow parameterization of models.
- Command interfaces allowing flexible and user-defined scripting.
- Enhanced visualization of key model characteristics, like registers at simulation time.
- Debug interfaces connecting to software and hardware debug environments.
- Virtual and real I/O connecting standard interfaces, such as USB and Ethernet, directly to the protocol stacks of the workstation on which the virtual platform simulation resides.

5.3.1.1 SystemC

SystemC has already been described in Chapter 3 and this will not be repeated here. SystemC modeling within Innovator is fully TLM 2.0 compliant and promotes maximum re-use and portability. Models developed against this standard should be portable into any SLVP and can be assured of future compatibility given that the standards are ultimately controlled by the IEEE. SystemC is also a language that looks somewhat familiar to hardware engineers, and thus is likely to be their preferred modeling language. Innovator includes a built-in editor for SystemC and is now the Synopsys preferred language for all platform component modeling.

Models created by Synopsys incorporate two back door mechanisms that can help in the creation of system-level platforms. Both of these are a standard part of TLM 2.0 so will work in all simulators. The first of these is the direct memory interface (DMI) which allows a model to obtain a pointer to target memory. Thus if one model, for example a RAM model, contains a block of memory, another model, such as a processor, can access this memory directly once it has obtained the pointer. This completely bypasses the transport mechanism in TLM and enables high speed loading or unloading of memories. The second mechanism is the debug interface. This is a special set of calls within the transport mechanism that allow memory to be accessed without any timing or other side effects in the model. This is a more general access capability that can work with any memory device and not just those that support the DMI interface.

5.3.1.2 Magic-C

Magic-C is a proprietary modeling methodology that combines both graphical control constructs and textual input using standard ANSI C. This mix of graphics and text was thought to improve modeling productivity and to provide high levels of visibility and control. It was optimized for control-dominant peripherals and based on a language called specification and design language (SDL) [14] that has been in use for several decades in the telecommunications industry. It has found favor particularly among peripheral model developers due to the amount of control often contained within them.

Fundamentally, the language is a hierarchy of concurrently executing processes. These processes can be used to build finite state machines (FSMs). Thus the environment uses the concurrent, communicating FSM model of computation.

Figure 5.11 shows the Magic-C editing environment. At the top of the window are a set of buttons that relate to the graphical constructs of the language. These will be familiar to anyone who has worked with flowcharts. The main window shows the model. In here it can be seen that the symbols may contain actions. A model may contain multiple sheets and they are displayed as tabs along the bottom of the window. At the bottom of the window, a transcript panel can be used to locate errors which are then related back to the graphical diagram. Rule checks are performed as the model is edited, so it provides immediate feedback about problems.

Consider the state machine shown in Figure 5.12 as a simple example of using Magic-C.

This is a basic Mealy state machine. Each transition is labeled by the input that caused the transition and the output it produces. Mapping this into Magic-C is a straightforward process as shown in Figure 5.13.

At the beginning of the simulation, all state machines are started at the oval block shown at the top of this diagram. They execute until the first state is reached, namely start. Here we are looking for either signal A or C to be seen. If it is signal A then we go to state S1 with no change in output signals. The state machine will then wait for signal B, at which point it will set signal D and move to state S2. There is no exit from this state. Alternatively if signal C was observed from the start state,

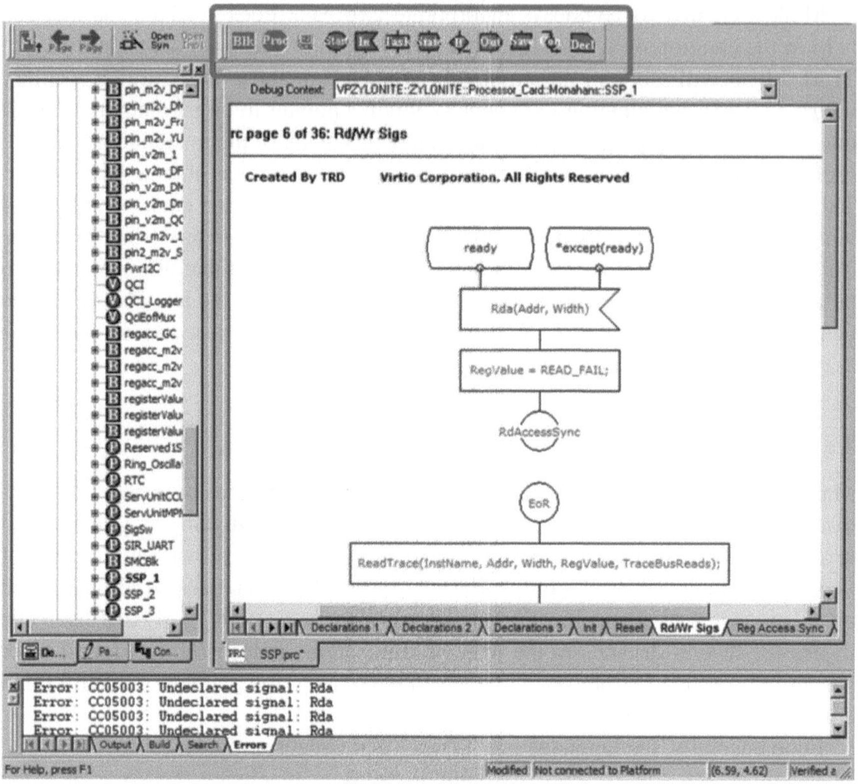

Fig. 5.11 Magic-C editor

Fig. 5.12 Simple Mealy state machine

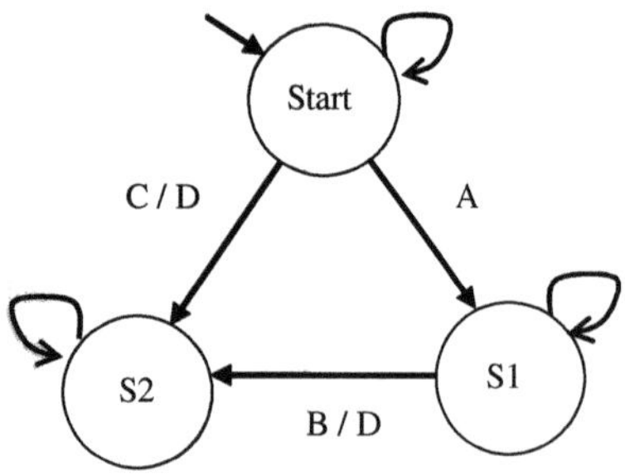

signal D would have been produced and it also would have transitioned to state S2. Asynchronous inputs such as interrupts can also be handled easily as well as constructs that remember the state before an operation so that it can be returned to.

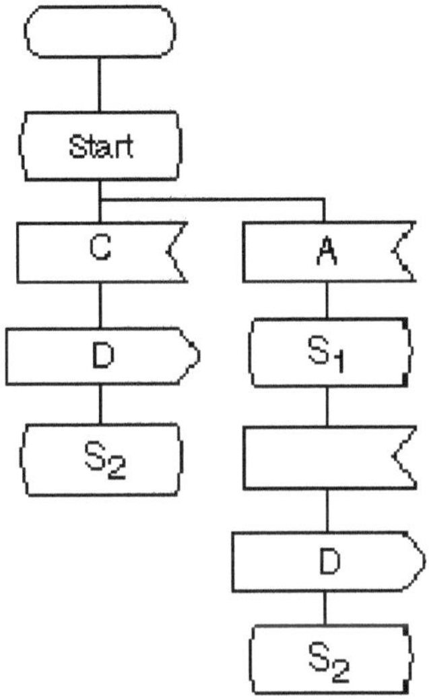

Fig. 5.13 Magic-C representation of state machine

While many of the examples provided in this chapter will include Magic-C, it should be remembered that Synopsys is now de-emphasizing this programming language in favor of the SystemC/TLM 2.0 methodology. Models in Magic-C can be exported as SystemC compliant models for use in SystemC compliant simulators.

5.3.1.3 VRE C++

VRE stands for virtual platform runtime environment, and is a scheduling and modeling infrastructure which is a Windows COM-like C++ component language. VRE C++ is an abstract modeling environment allowing faster execution that Magic-C models, but models are essentially black box, in that internal details cannot be seen during simulation. Each model is a dynamic linked library (DLL) which communicates through interface pointers. The environment itself will not be described here as it is one of the legacy environments that is being de-emphasized by Synopsys. Several components of VRE C++ were donated to OSCI in January 2007 [15]. VRE is still used as part of the scheduling environment for hybrid SLVPs composed of VRE and Magic-C models. When SystemC models exist, all VRE models are encapsulated and execute under the control of the SystemC kernel.

The scheduler is very simple: it follows a round-robin protocol for all of the VRE masters. Masters, such as a processor, are not kept in lock-step but are free to execute any number of instructions as controlled by the user. Thus the amount of skew between masters is controlled and context switching between them is minimized. The Magic-C models become a collective master in that when they get their opportunity to advance, they are all allowed to progress for a defined period of time. There is no centralized time keeping or global clock within the system.

When a VRE model requires data from a Magic-C model, that simulation domain will be advanced until the data becomes available or some other exception occurs. Asynchronous signals such as interrupts are recognized when the receiving block advances.

VRE also provides many of the command and control facilities for the IDE which acts as a command router. Much of this functionality has been preserved as the control infrastructure for the updated IDE.

5.3.2 *Model Creation*

SystemC and Magic-C models can be created within the Innovator tool. Within this tool shown in Figure 5.14 the memory structure that is to be exposed outside of the model is defined along with the interfaces into the model. For the memory structure, the address map is defined, placing both memory and registers within the defined space. For the registers, the bit fields are defined along with the restrictions on any bits, such as if they are read only.

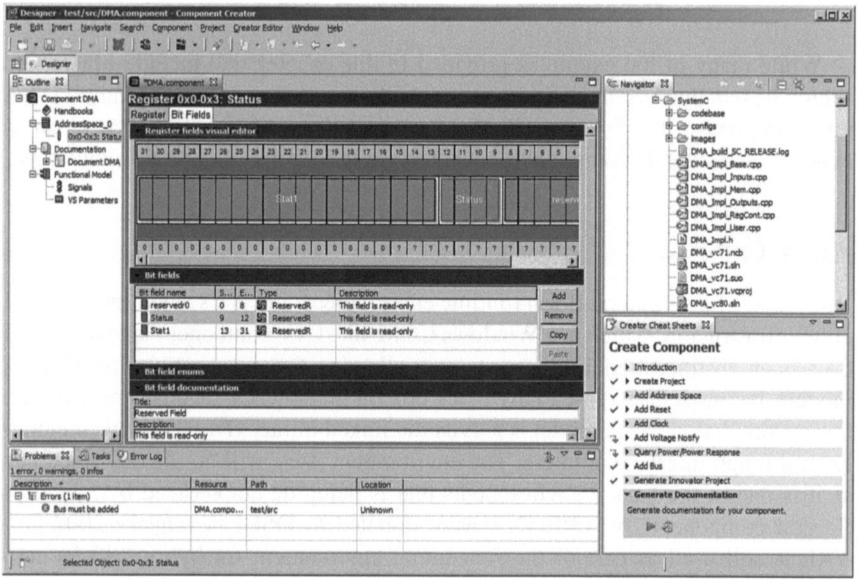

Fig. 5.14 Innovator model creation

Additionally the parameters that are to be settable within the IDE are defined.

From this information the headers and structure of the model are created along with a symbol that can be used for connecting it into the system model. The user is then expected to fill in the desired functionality for the rest of the model (Figure 5.15).

Wherever possible, information is derived from IP-XACT specifications.

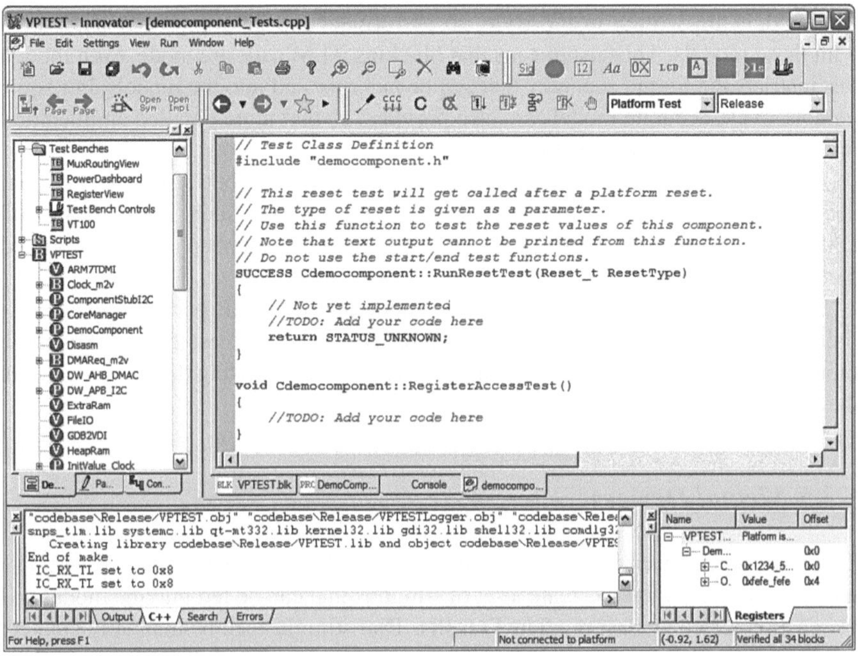

Fig. 5.15 Model template creation

5.3.3 Model Import

Many system architects develop their algorithms in C or C++ and an easy way to integrate these without extensive recoding is necessary. Innovator allows model import into all three language environments. For SystemC, an interactive import wizard is used. For Magic-C or VRE C++ a wrapper can be created, but this is a manual process and will not be described here.

SystemC models can also be imported into the Innovator environment. The import tool parses the source code and from this builds the necessary graphical symbols. The importation of a textual SystemC model into an IDE provides many advantages. Most of these are associated with being able to integrate or modify the model into a system without having to change the source code directly, which can be error prone and time consuming. It also enables things such as parameters to be modified by scripts, or other such mechanisms, such that a set of simulations can be run with various configurations and the best one identified.

5.3.4 Model Libraries

As already stated, Synopsys is a supplier of models as well as a tool supplier. The Synopsys DesignWare system-level library is a portfolio of transaction-level models written in SystemC that can be used in any SLVP that is also IEEE 1666 compliant.

The models are written in PV style, so performance is the primary consideration with these models. This also means that blocking TLM calls are made, even though this restricts the timing accuracy that can be incorporated into the models. For some models, cycle-accurate models are also provided so that more detailed architectural investigation can be performed.

The library is highly biased toward processor models but also contains connectivity cores, such as USB, bus infrastructure cores, such as interrupt and DMA controllers, peripherals, such as a UART and a number of pre-assembled platforms.

Synopsys has a significant business in supplying design IP, particularly connectivity IP. They supply high-level system models with all of these cores. Thus these models match with the capabilities that can be obtained by using the Synopsys IP in the final product.

5.3.5 Virtual Devices

There are several ways in which testbenches can be constructed for a SLVP. The more traditional way is to create a set of stimulus that can be fed into the model and its reaction tested against predicted results. That stimulus can be manually created, sampled, or randomly generated. This is the same basic methodology that is used for RTL verification today. Another way to verify and validate the model is to integrate it into the real world such that it can be used in the environment in which it is intended to operate once implemented. This concept relies on a technique called virtualization, and in its most basic form means utilizing the capabilities of the host system as if they were part of the model.

Virtualization can exist at many levels, and all virtual platforms use some of the most basic forms. The simplest is the virtualization of memory. When a block of memory is instantiated in a model, we do not actually create a memory out of the basic electronic building blocks. Instead we allocate a piece of the host system's memory to hold the contents of the memory in our model. When reads and writes are made to a memory model these are translated, either within the model or by the operating system, into the physical addresses in the host memory system.

This results in much faster execution of the model and efficient usage of resources since we are not trying to verify if the model of the memory element itself works; only that it provides the correct functionality to the system. In addition, if it were a dynamic memory that was being modeled, we would not have to implement any of the refresh logic, which not only adds complexity but also takes unnecessary simulation time. At this stage in the implementation process we assume that any dynamic memory element that is eventually used will be able to perform its function and that none of that needs to be verified at the system level.

Figure 5.16 shows some of the generic virtual I/O functions provided by Synopsys. As an example, if we consider the virtualization of the Ethernet, the I/O

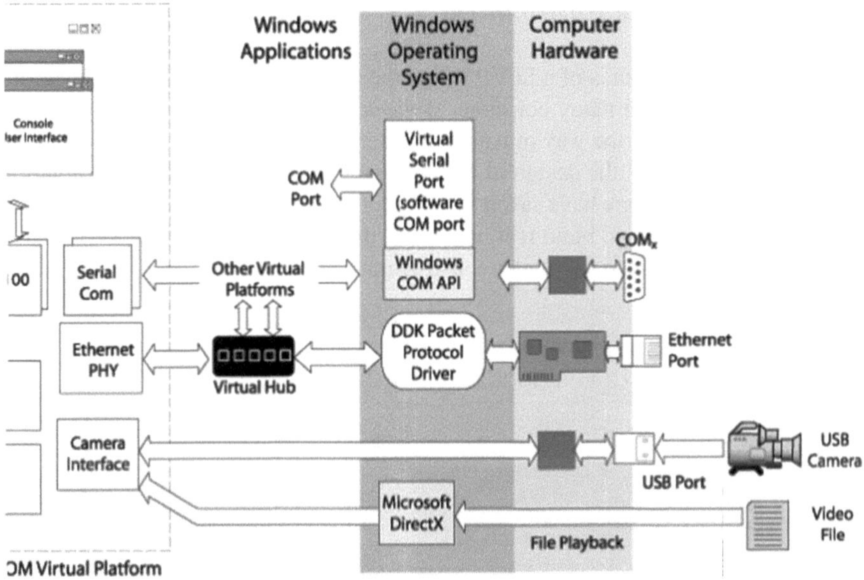

Fig. 5.16 Virtual I/O

model directly utilizes the Ethernet driver in the host system, which in turn uses the actual Ethernet hardware in the host and connects the prototype into the real world. In the case of the host being a Windows machine, this means interfacing via the Microsoft DDK protocol driver into the TCP/IP software stack. As packets come in on the Ethernet port, they are directed either to the host software or to the VHub depending on the MAC address embedded in the packet. The prototype is thus physically connected to the Internet or wherever that Ethernet port is utilized. As with the memory example above, the purpose of this virtualization is not to test that the Ethernet hardware actually works, but to use it to stimulate and drive the prototype with real-data streams.

Other virtual interfaces are supported, such as audio codecs, USB, PCMCIA, ATA drives. This means that almost any real device that can connect using these interfaces can be supported, such as a video camera, disk storage system, or WiFi card.

Similar techniques have often been used in the emulation world. In this case the speed disparity between the emulator, which usually runs at a rate of about 1 MHz, and the real world is quite large – possibly several orders of magnitude. In these cases speed gaskets may be required that can handle the buffering of data between the two parts of the system, such that neither the real world nor the emulation environment are forced to work at the rate of the other environment. Many protocols, such as those used on the Internet, can handle long delay times or response times, as long as the timeout periods are correctly setup.

5.3.6 Modeling the Environment

One of the great attributes of a SLVP is that the system can be tried out long before it has been built. With many consumer devices, the user interface is an important part of the system and the way in which the user interacts with the system can have a significant impact on the design of both the hardware and software. An emulation of the user interface may have saved BMW a lot of grief.

While it is possible to build real interfaces, the first step is probably to construct a virtual one, such as the one under construction in Figure 5.17.

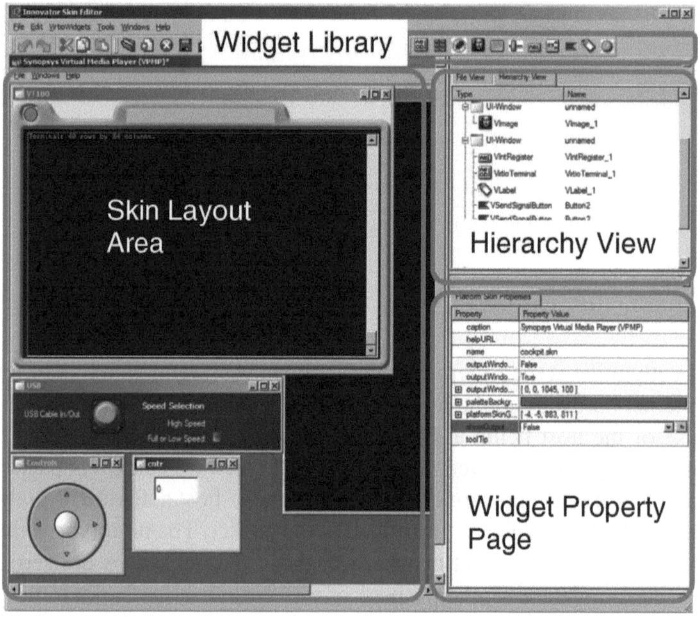

Fig. 5.17 Virtual user interface creator

A virtual user interface is constructed from a number of graphical widgets that can be dropped into an editor. Examples would be buttons, dials, switches, an LCD display, or a terminal. These widgets have several parameters associated with them that can be used to modify the way in which they operate. Alternatively, a user can create any arbitrary bitmap that may have meaning within the environment being constructed. These widgets are then connected into functionality using data types, signal connectivity, and or specialized C++ objects.

Figure 5.18 shows an example of such a user interface prototype for the TI OMAP platform. On the left we can see a virtual device consisting of a display area, plus a number of buttons, and other IO devices spread around the device. To

Fig. 5.18 Virtual user interface

the right are other parts of the virtual prototype including a camera, some configuration switches and other I/O that are related to plugging in or removing the USB device.

5.3.7 Tying It All Together

With all of the pieces of a system having been modeled, it is still necessary to tie them all together into the system model. Innovator fully supports the IP-XACT standard from SPIRIT which was described in Chapter 2. IP-XACT contains a description of the components that are to be interconnected. It is also possible to do this manually in the Innovator editor, by dragging and dropping components and interconnections.

5.3.8 Documentation

Even when a component has been created and completely verified, there are still operations that need to be performed before the component can be reused or placed into an IP library. One of the most important steps is documentation. The creation of the component includes much of the information necessary for those documents, such as the interfaces and the externally accessible registers. An example of the document generation process is shown in Figure 5.19.

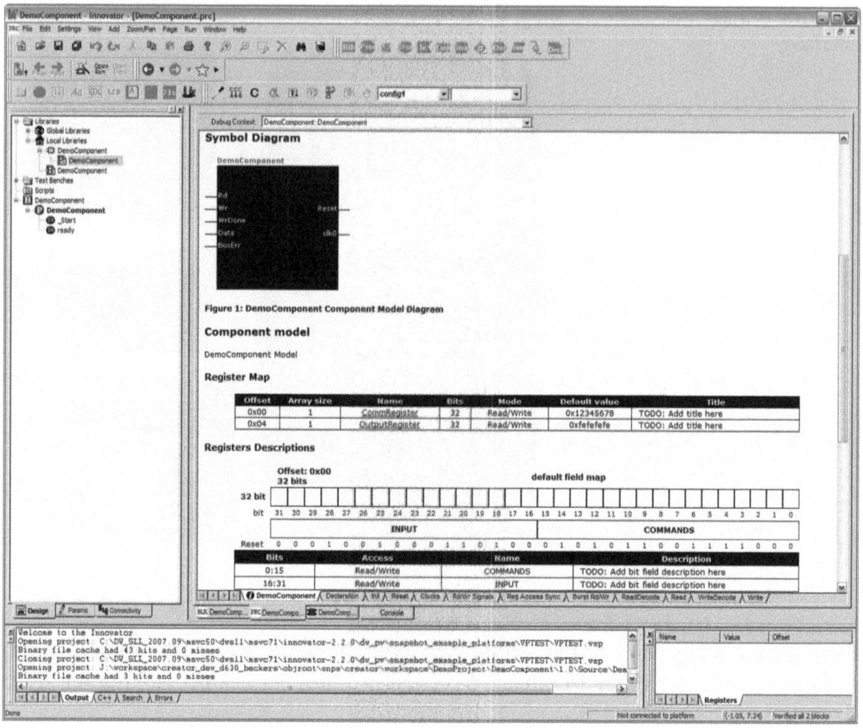

Fig. 5.19 Generating documentation

5.4 Running the Prototype

Once the prototype has been created, and it passes the structural integrity checks, it needs to get compiled for execution. The IDE should make this a quick and reliable process. What gets created is an executable that combines both the model of the platform and an integrated simulation kernel. This executable also contains all of the necessary interfaces to enable the connection back into the IDE such that it can be controlled and debugged within the same environment.

Innovator creates one of two executables based on user preferences. The first is based on the VRE environment as shown in Figure 5.20.

This execution environment utilizes the VRE scheduler as the simulation kernel and models are loaded as dynamic libraries. Magic-C models are run as a group controlled by their own simulation kernel which appears to VRE as a single master. Port binding is specified in an XML file which acts as a netlist.

The second method is based on SystemC where the netlist and port binding are decided at compile time. This creates a monolithic executable including the SystemC scheduler as shown in Figure 5.21.

The code generator has to perform additional work creating wrappers for Magic-C and VRE C++ models that enable them to be integrated into a SystemC

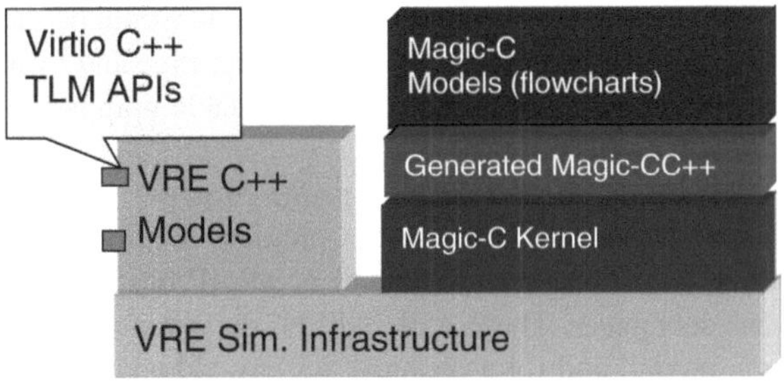

Fig. 5.20 Execution environment based on VRE

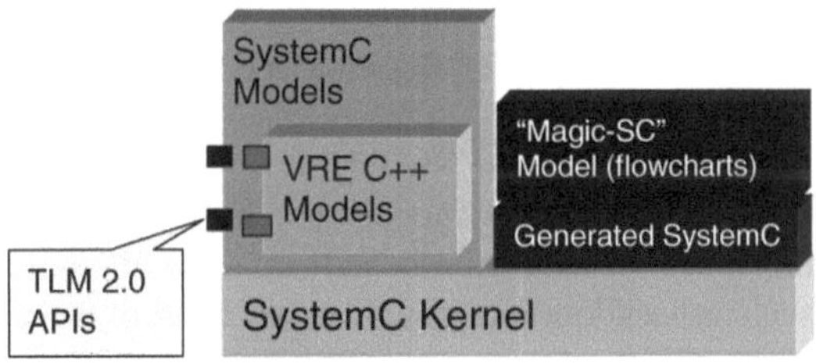

Fig. 5.21 Execution environment based on SystemC

environment. However, this is the environment that must be used if any SystemC models are being integrated into the system.

5.4.1 Debug

There are several types of debugging necessary within a SLVP, each requiring a different kind of support from the tools. The first is the debugging of the models themselves and the others are related to the debugging of the complete system. At the system level, not only do we need to debug the functionality, but we also need to debug some of the non-functional aspects of the system, such as component utilization, or performance. While the verification of functionality at both levels would traditionally be seen as debugging, the second is often the act of understanding the interactions going on in a system and ensuring that synchronization and communications have been properly defined and that the architecture, as much as has been defined, is capable of supporting them according to the specification.

Models within the Synopsys Innovator product line can be written in three languages, SystemC, VRE C++, or Magic-C. With the first two, source-level debugging is the primary requirement and an open framework, such as Eclipse, allows a variety of different debuggers to be used. Specific examples are the Windows MSVC++ or the popular GDB.

While standard text debuggers may be the principle means of interacting with the model, additional debug capabilities can be added that are more attuned to the hardware that they represent. An example would be the display and editing of registers. Being able to see these change over time and by adding additional capabilities, such as color coding, can provide valuable graphical views of data as it progresses through a model.

Processor models are the prime example of a debug environment tailored to a specific piece of hardware. Complete development and debug environments for these are often part and parcel of the IP block, including the models, compilers, and debuggers. These tools will be the same tools that are used both for the system model, and for a physical system, although exact features may differ between these two extremes. It is thus possible to connect the debugger into the SLVP in several ways. One possibility is to use the JTAG port of the processor model and to connect this to the debugger in exactly the same way as would happen in the physical world, except that the interface between the model and the debugger becomes virtualized.

While this may be possible, it is hardly the best way to perform such an integration. Most processor models also have high-level debugger interfaces that can do this much more efficiently and make use of many of the other features of the SLVP, such as backdoor memory accesses. Most of these interfaces are socket based which means that the debugger can be brought up and attached to a running model. In many cases these debuggers are capable of attaching to more than one processor at a time such that even greater control and visibility can be attained.

For the debug of the proprietary Magic-C models, Innovator provides a custom debug environment that supports the same kind of features expected of the language debuggers.

Platform debugging, while requiring many of the same features as a model debugger, requires a lot more visualization capabilities that enable an engineer to be able to quickly and easily comprehend the state of the system and the transactions that pass through the system. Some of these will seem very familiar to hardware engineers, such as waveform displays for events and signals augmented with the ability to display transactions. Given that the quantities of data could be very high, advanced filtering capabilities are required to ensure that only the necessary data is displayed.

Even though a tool may provide many debug aids, users will always be able to think of additional tools that may be design specific. This calls for an open system into which custom analysis tools, performance monitors, and other forms of visualization can be integrated. Innovator again uses the Eclipse framework to do this, allowing user written Java plug-ins to enhance the debug capabilities.

5.4.2 *Analysis*

When performing analysis, there will generally be large quantities of data involved. Statistical analysis only holds true when there is enough data such that the margin of error becomes small enough to not skew the results. This is one of the prime reasons why Synopsys has put execution speed at the top of the priority list, ahead of things like timing accuracy.

Execution speed can be compromised by data collection and logging, so this also requires some compromises, such as separating the collection of data from the analysis. While this means that instantaneous results cannot be seen, it is less likely to be useful when the data is being collected for actions such as performance analysis. It is also important to avoid storing data that can be derived or is redundant data.

An example of this is to consider a trace from a processor. If we execute a sequence of instructions that is part of a basic block of code (there were no branches or other transfer of control), then each address will be an increment of the previous one. Thus the recording of every address is redundant because the address of subsequent instruction execution can be derived from the first instruction.

The logging and analysis of data coming from loosely timed SystemC models creates some challenges. This is due to the temporary decoupling that can happen meaning that the order in which things would be placed into a log file are not necessarily in time incremental order. For example, if we consider the following TLM 2.0 call:

```
void b_transport(T& trans, sc_time& offset)
```

The `sc_time` parameter is used to specify that the `b_transport` call happens logically at time `sc_time_stamp() + offset`. There are two ways to accommodate this, and Synopsys has made both of them possible.

If data is pushed into the file by each of the components, then we have to ensure that they record not only the time that the SystemC kernel is at but also the local time of the model performing the data store. In this way all events are fully timed even if they are not in order. Thus if the following events were logged at SystemC global time 50:

```
log(h1, x, 5);
log(h1, y, 10);
log(h2, z, 7);
```

The analyzer would display them as shown in Figure 5.22.

Alternatively, a pull mechanism can be used where data is only recorded when time advances. At that point, all models that have new data to be logged are requested to provide the data. This ensures that the log file is in time incremental order.

Once the data has been recorded, many functions can be applied to the data. Synopsys generally refers to these as filters and they include logic, event, mathematical, statistical, and so on. While a number of these are pre-built and packaged, such as the cache statistics shown below in Figure 5.23, the user is free to create their own custom analysis screens.

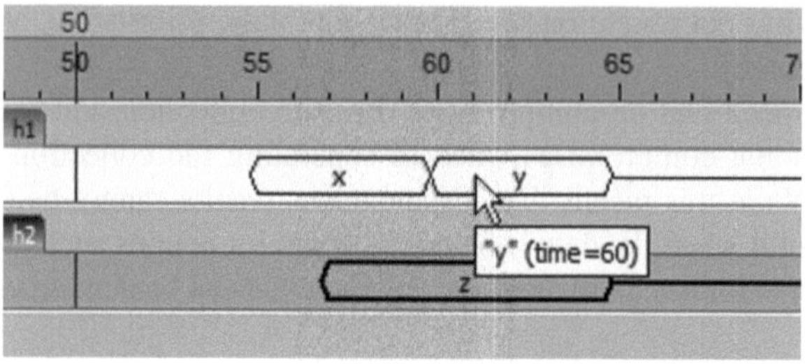

Fig. 5.22 Correctly timed events

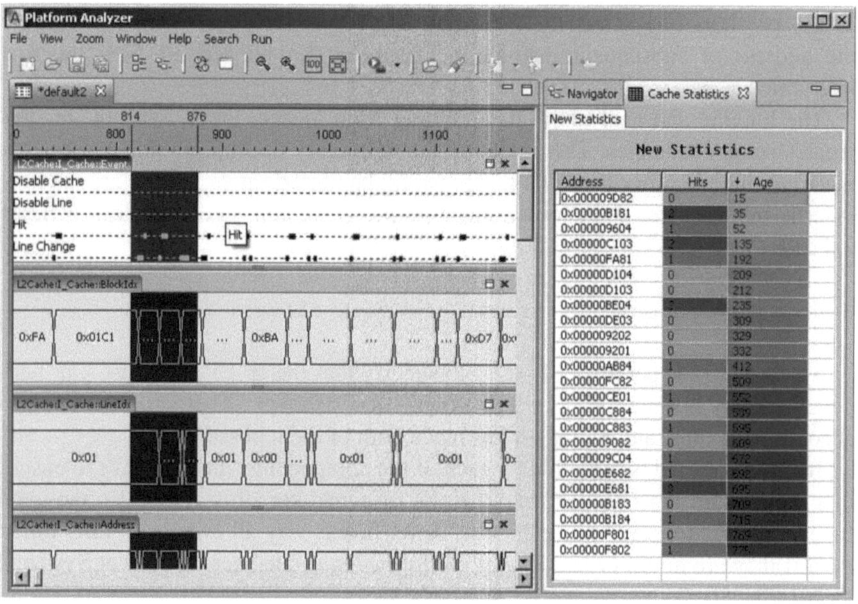

Fig. 5.23 Platform analysis

Many different types of analysis are available. A few of them are shown in Figure 5.23 which is dealing with the analysis of a cache system. At the top left we can see a transaction trace showing when cache hits happen or when line refreshes take place. Below that we see a more traditional waveform trace for several fields related to the cache. On the right we see a memory heat map showing how frequently elements of the cache have been accessed and how long they have been in the cache.

Several different ways to display statistics are possible, such as those shown for bus overhead statistics in the following two figures. In Figure 5.24 we see a pie chart, whereas in Figure 5.25 the same data is displayed as a bar chart.

Creating custom displays can be performed on several levels. The first is to create custom displays given the basic building blocks supplied within the tool

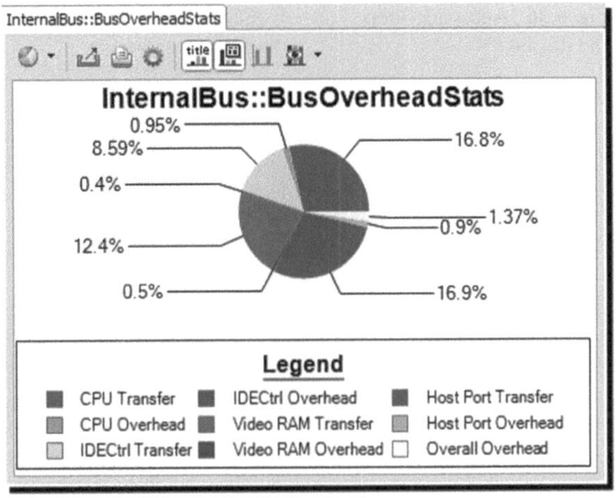

Fig. 5.24 Bus overhead displayed as a pie chart

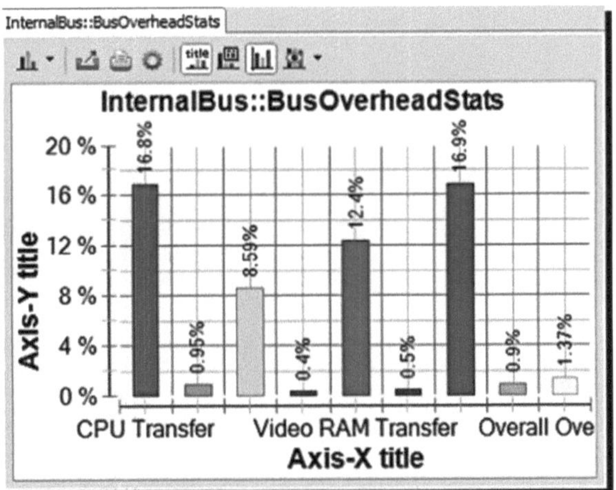

Fig. 5.25 Bus overhead displayed as a bar chart

(Figure 5.26). One or more data sources can be selected and complex functions applied to them from the range of built-in functions.

If by using the inbuilt filters and functions, the tool is not capable of producing the required statistics; complete custom processing can be accomplished by making use of the API supplied (Figure 5.27). This is an Eclipse-based Java plug-in API which means that the user is relieved from tedious data storage manipulations and other bookkeeping functions. It also protects the tool from crashes when the statistics themselves are being debugged.

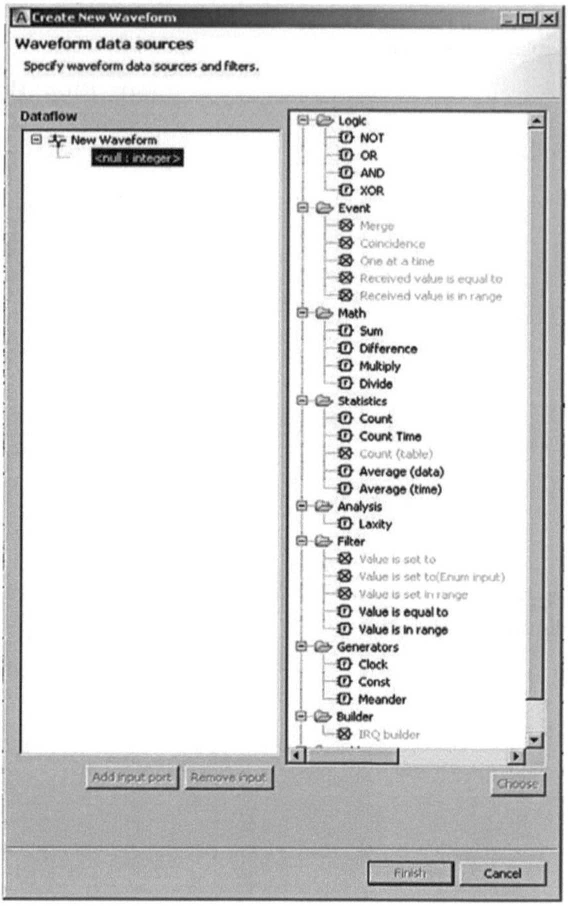

Fig. 5.26 Applying filters and displays

```
}
@Override
public NumericDataset getData(BasicRange timeRange, long intervalCount) {
    return getData(timeRange);
}
@Override
public NumericDataset getData(BasicRange timeRange) {
    checkInitialized();

    ChannelDataset<Number> set = source.getData(timeRange);
    long[][] data = new long[set.size()][2];
    int i=0;
    for (Iterator<ChannelData<Number>> it = set.iterator(); it.hasNext(); i++) {
        ChannelData<Number> point = it.next();
        data[i][0] = point.getKey();
        data[i][1] = range.getEnd() - point.getValue().longValue();
    }
    return (NumericDataset) getChannel().createDataset(data);
```

Fig. 5.27 Custom statistics generation using Java

5.4.2.1 Power Analysis

Recently, power has migrated from a second class design consideration to one that is very much a first-class consideration. Battery technology has not been able to keep up with the ever increasing demands being placed on it in handheld consumer devices, such as smart-phones. This means that companies are now employing many strategies to reduce power consumption. As with many aspects of a design flow, the optimization first starts at very low level by trying to improve the fundamental devices or optimization of the gate or RT levels. While the gains have been considerable, attacking the problem at the system level will lead to much greater gains, but this creates some new demands for modeling and analysis.

Most of the earlier attempts to reduce power consumption came from clock gating. This means turning off the clocks that are fed into components that are not actively being used. Given that large quantities of activity are associated with clocks this is a waste of power if the only thing being accomplished is saving the state of registers that are not going to be read. With leakage power becoming a larger percentage of total power as chip geometries decrease, this is becoming a less effective strategy. Designers are now looking at completely powering down blocks and using variable voltages to trade off between performance and power consumption. Both of these schemes require a lot of additional analysis and are often controlled by software which has to be written, verified, and debugged. Failures of the software in this aspect of functionality tend to be critical because if a block is not powered up in time for when it is expected to be used, a total failure could result.

Performing system-level power analysis early in the development cycle can allow power management policies to be tested and to enable software power optimizations to be assessed in a relative sense. Power figures will not be accurate enough to predict the exact amount of power that will be consumed, but it will allow you to find out if a change would result in a relative savings. This is particularly true of an LT platform since at this level of timing detail, tools can only provide figures for the estimated power consumed per function. When the timing is refined down to the AT level, trade-offs are then possible between performance and power since the SLVP will now be able to provide profiles of power over time.

While clocks do not in general exist when modeling at this level of abstraction, it may be necessary to start modeling the clock controller so that it can be integrated with the other power management features.

Components are characterized by a set of power parameters that are then used in localized power equations to calculate the component power. These are then accumulated by a central logging module for analysis after the simulation run. Different types of components will have parameters. For example, the ones defined for a processor at the LT level are active power, dormant power, power inactive, and power shutdown. For a memory they would be power clock, power, idle, power read, and power write. More accurate models would have additional parameters such as voltage and frequency as well as those representing state.

After simulation, the power can be analyzed in a number of different tools, some of which are shown in Figure 5.28.

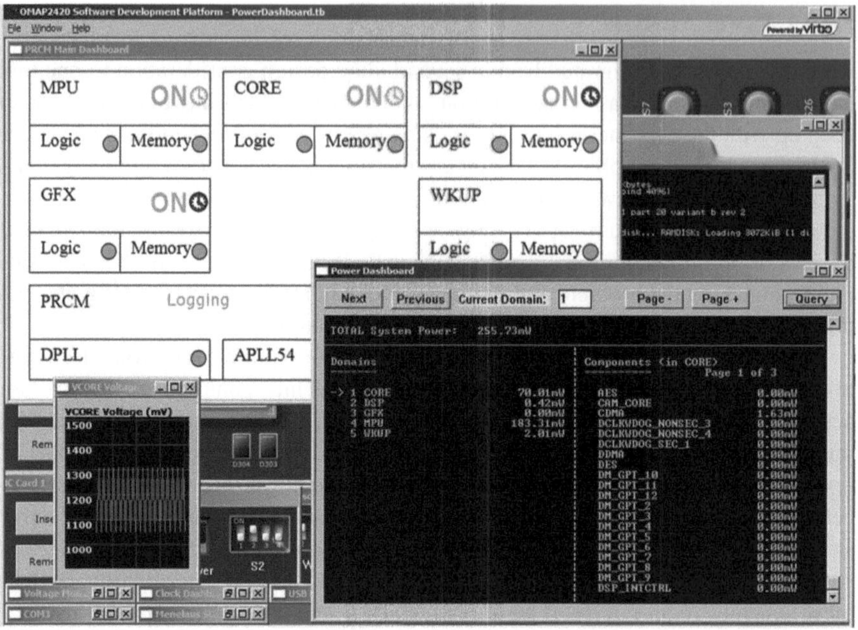

Fig. 5.28 Power analysis

5.5 Verification

So far we have talked about the creation of the SLVP itself, but this is not the only complex piece of software that needs to be developed early in the design. The test-bench is also needed to ensure that the functionality of the SLVP is correct and to become the verification environment for the hardware as the implementation proceeds.

5.5.1 *Platform Deployment*

Once a platform has been constructed and verified, it needs to be distributed either within the company, for example, to the software team or external to the company. Depending on the final recipient the amount of visibility into the model or the ability to change the platform may vary. Even once a platform has been shipped to a customer, a central team in that company may wish to further restrict the model by setting parameters and making certain attributes read only.

Innovator uses XML for platform configuration. A platform configuration contains the model and user settings as well as limitations in what can be modified. The platform model product ships in various forms, some of which have a full interactive environment, while others are run-time only versions that contain the executable model of the platform, but no front-end user interface.

Even after a platform has been shipped, web-based updates can be provided in a similar manner to the way in which Windows or other applications on the desktop are updated.

5.5.2 Verification Methodology Manual

The development of sophisticated testbench creation tools, such as Specman and Vera, turned verification from being a rather boring and tedious task, that consisted of creating long and often complex sets of stimulus that would be fed into a high-level modeling environment into one using modern and very different languages than had been used in the industry before. Users not only need help with the usage of these languages but also need help in how to structure testbenches, how to get the most benefits from the languages and the things to avoid to ensure good execution speeds. This is the kind of knowledge that is generally built up over time, but the EDA companies supplying these new languages had to jump start the process. In the case of Vera [16], Synopsys created or supported the writing of several books that helped in this regard.

However, these early books were more about the usage of the languages rather than the methodologies for creating effective testbenches. With the introduction of SystemVerilog came a new book [17] that was co-written by Synopsys and ARM called the Verification Methodology Manual (VMM). This book concentrated almost exclusively on how to create a scalable, efficient testbench environment by establishing a set of rules that users should follow. It also established a number of other goals, such as testbench model interoperability, and speed of development. Others followed with similar methodologies, such as the Open Verification Methodology (OVM) [18].

While these testbenches are targeted at the verification of RTL models, they are modeled at the transaction level and thus have a lot in common with SLVPs. More importantly, there is a strong desire in the industry to be able to reuse the same testbench to verify both the system-level model and the RTL as depicted in Figure 5.29.

This level of reuse is expanded even further when added to the desire to be able to place an RTL model into a SLVP such that it can be verified in context using either the high-level testbench or with live data feeds coming from things such as the virtual devices described within this chapter. Another scenario that has to be supported is using the SLVP as the reference model against which the RTL is compared. In fact it is likely that the SLVP will be the initial model that is used to develop and debug the testbench.

At its heart, VMM is a class library that implements many of the fundamental things that will be at present in almost every testbench [19]. By providing a reference implementation for these, verification engineers instantly become more productive because they do not need to write and debug these themselves. Over time the VMM has been extended and adapted to new usage scenarios, such as additional concepts

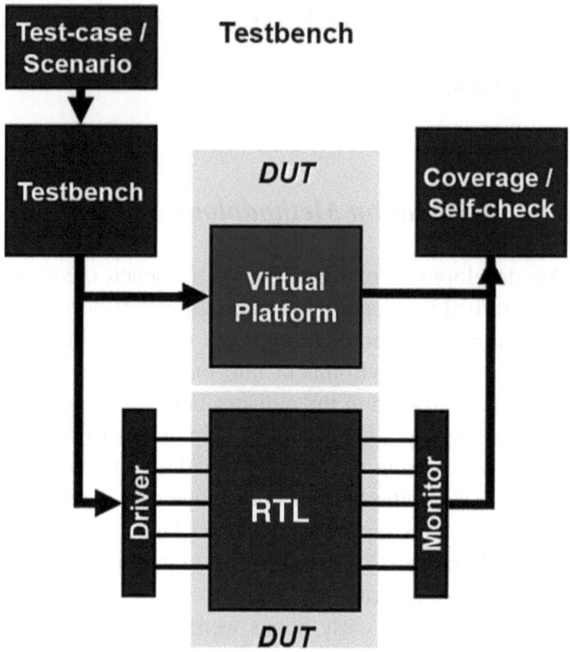

Fig. 5.29 Common testbench

needed for the low-power systems [20] and more importantly in this context, the integration of SystemC such that the VMM can support the ES level.

The SystemC–SystemVerilog connection is made through a special interface that Synopsys calls the transaction-level interface (TLI). While SystemVerilog has an inbuilt interface direct programming interface (DPI) that allows very fast and efficient passing on data between SystemVerilog and any C-based language, it does not support the calling of time-consuming methods on the C side. This makes it difficult to use SystemC as a reference model. Using TLI, SystemC interface methods can be invoked in SystemVerilog and vice versa, overcoming the previous limitations.

5.5.3 Building the RTL Testbench

There will be many cases where the system-level model needs to be constructed so that an RTL model can be inserted into it and the system-level tests re-run. This requires that the structure of the two environments is the same, at least where these insertion boundaries are created. At the insertion point, the abstraction of the two models will not match and thus converters are necessary. These converters are often called transactors and specific instances of them (such as those between the hardware and software) were discussed in Chapter 2 and others (between hardware blocks) will be discussed in more details in Chapter 8.

5.5.4 Regressions

The verification methodologies in use today have been refined over time. One factor that has received a lot of attention is the ability to run regression tests. While this is perhaps a mundane topic, it requires a lot of attention to detail, such as the stability of random number generators. Without this, every run using constrained random stimulus generation could go in a completely different direction after just a small change in the design if it caused a different order of execution.

At the system-level we are just beginning to consider some of these issues – in fact most SLVPs rely entirely on the existing RTL verification methodology, but in many cases these are not appropriate. The primary reason is that for RTL verification we know that a second predictor model is required against which we will compare results. But do we need a predictor model against which to compare the system model?

While in the future the answer to this may become yes, it is not always the case today. If the system model is going to be the model against which the RTL is compared, then this formal checking of the model will eventually happen, even though it may not be at the time it is first developed. This argument breaks down completely if the RTL model is going to be synthesized from the system model. It only holds true if the system model and the RTL are generated independently.

If we fast forward to a time where synthesis is possible from the system model, then a predictor model will be required against which the system model will be compared. When this happens, we will probably not need to verify at the RTL level at all, just as very few gate level simulations are performed today. This is thus not extra work, but a migration of every aspect of verification to a higher level of abstraction.

Verification of the system model is somewhat more ad hoc today. Many runs are created to test architectural possibilities, others based on feeding in captured data streams. However, while ad hoc there is still tremendous value in being able to re-run these tests to detect differences after the model has been updated. For this reason, Synopsys has included regression run capabilities into their tool as depicted in Figure 5.30.

While frowned upon for RTL verification, system testing today often uses a comparison against golden results methodology. This is the best that can be achieved until a predictor model becomes available, but at the same, the dependence on the system model verification also increases when this happens. To make this kind of run possible relies on an extensive scripting system, that enables every variable or parameter that can be modified in the GUI, to also be modifying through the script. Recording of sessions is also provided so that test runs can be captured faithfully.

The solution developed by Synopsys is based on PERL that not only has access to the internal variables via the Innovator command line but also controls the regression runs. Because every company likely has different ways in which they wish to set up the regression environment, Synopsys ships the source code to allow for customization. At the end of the regression runs, several reports can be created that allow classification based on logical "test groups," authors, defect tracking ID, and release numbers.

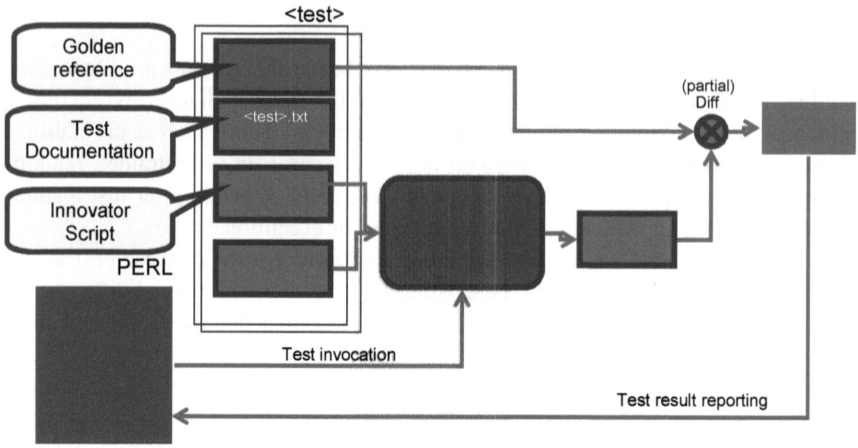

Fig. 5.30 Regression environment

5.6 Example

A common expression is that the proof of the pudding is in the eating, meaning that no matter how well intentioned something is, if it does not work, then it just does not work. Given that Synopsys is a developer of IP, it would seem that Synopsys should be able to gain from the use of their own tools, both in the development of the system-level models, and by enabling them to develop the software drivers for their blocks quicker than they otherwise could.

That was put to the test when Synopsys needed to update their USB on-the-go (OTG) controller, which is part of their existing DesignWare library.

USB enables "outside-the-box" connectivity with hot-swap capability for industrial, consumer, embedded, PC, and PC peripheral products. USB requires a host to manage and control bus traffic to and from USB peripherals such as hard drives, cameras, flash card readers, printers, and scanners. USB peripherals must manage data translation between the peripheral application and the USB Host. OTG products typically must handle both host and peripheral functions and switch between these roles.

Together with the USB IP, Synopsys provides a reference driver enabling developers to design USB host, peripheral, and OTG products. In addition, a transaction-level model is provided as part of their system-level library, which reflects the underlying implementation. While Synopsys may wish that all customers use their implementation IP, they do supply a USB model authoring kit that enables them to modify the transaction-level model to reflect other hardware implementations. Specifically, the enhanced host controller interface (EHCI) and host model authoring library provide generic USB functionality which can be extended by model developers to implement specific model characteristics like the actual register definitions. The libraries support C++ and SystemC model development and contain support for connecting to real-world devices and for test purposes, allowing virtual USB devices to be loaded into and exchange data with the host operating system.

Writing device drivers requires an in-depth understanding of how a given IP and platform functions, both at the hardware and the software level. Because many device drivers execute in kernel mode, software bugs often have much more damaging effects to the system. This is in contrast to most types of user-level software running under modern operating systems, which can be stopped without greatly affecting the rest of the system. Even drivers executing in user mode can crash a system if the device being controlled is erroneously programmed. These factors make it more difficult to diagnose problems and the consequences of problems more impactful.

After Synopsys' initial USB OTG implementation, hardware support for descriptor-based DMA transfers had to be added as a feature. The new feature was added for the device mode only, and this required the Linux software device drivers to be extended to support it. A virtual platform was to be used for developing and testing the required driver updates.

Three primary objectives were defined, namely

- Develop and test Linux device driver for descriptor-based DMA HW enhancement in the OTG IP core
- Prove that the developed device driver can be run "as is" on an FPGA system
- Verify the RTL implementation within the virtual platform.

5.6.1 The Application

There are two basic use models for virtual platforms supporting driver development in the context of a USB core. First, a virtual platform can be used to model a device like a digital still camera, a USB video camera, or a media player. Using virtual I/O capabilities for software emulation of device I/O, the virtualized product can be connected to a Windows or Linux host on which the host controller resides and connects to the USB device. This process is illustrated in Figure 5.31.

In this case the virtual platform simulates the USB device in question. The USB Device Controller model is part of the virtual platform and an ARM processor runs the device driver and application software within the virtual platform. A software debugger can be attached to the processor in the virtual environment and the function of the device driver can be developed and verified on the virtual platform long before RTL or silicon are available. In addition, the USB application software running on the device – a data transfer program for media files in the example of Figure 5.31 can be developed, debugged, and verified.

Pushing the button "USB Cable Input" connects the virtual platform to the USB Host Controller on the host on which the virtual platform executes. In effect, all the steps happening when the real device is connected will happen in this case. The Windows operating system will inform the user that a new hardware device has

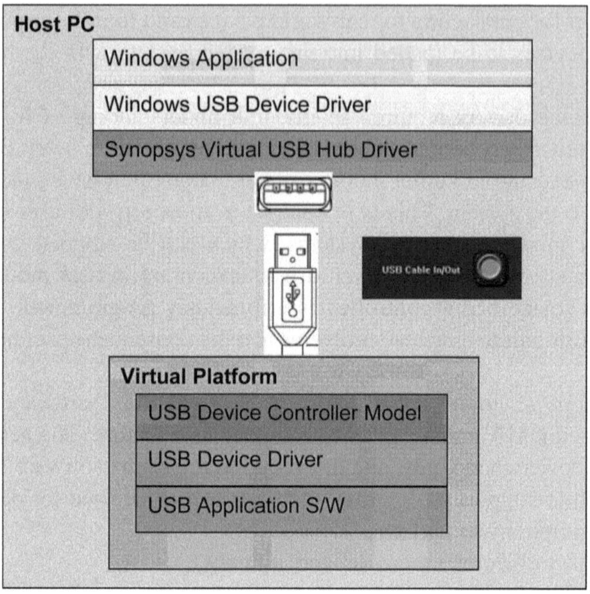

Fig. 5.31 USB device in virtual platform connected to real-world host

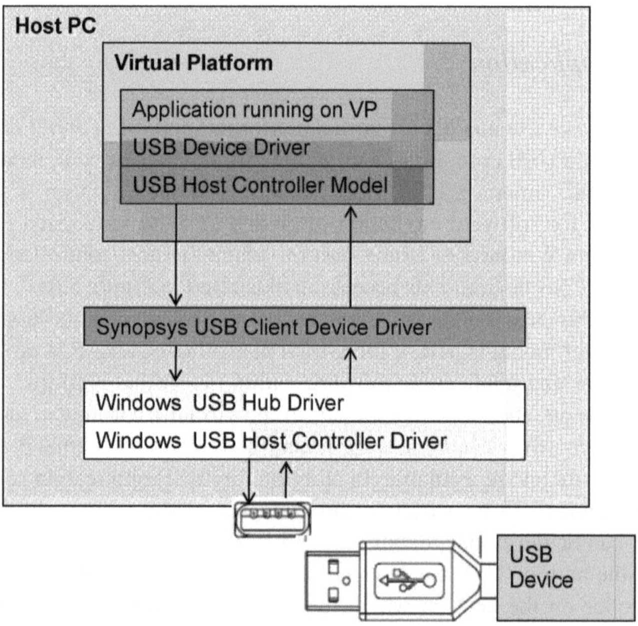

Fig. 5.32 USB host in virtual platform connected to real-world device

been detected. Together with Synopsys' virtual platform offerings, a virtual USB Hub Driver is provided. This Hub Driver interfaces to the Windows USB driver environment connecting the virtual hard disk in the media player to the Windows application in which the virtual hard disk appears as a new disk to interact with.

Figure 5.32 illustrates a second-use model. This time the virtual platform simulates the USB Host environment and the Host PC is connected to a real-world USB device. The real-world device, a memory stick, is connected the Host PC using the Host Controller Driver and Hub Driver provided by the Windows operating system. Synopsys provides a USB Client Device Driver to connect this real-world device to the host executing in the virtual platform. This way it is easy to develop, debug, and verify host controller drivers and applications for different host-operating systems.

While the main use model for the virtual platforms described in this chapter is pre-silicon software development, other uses for the platform become possible as the SoC design cycle progresses. As such a virtual platform can evolve to meet different needs. There are three main use models for software-driven verification which utilize the integration of virtual platforms with signal-level simulation at the register transfer level:

- When an RTL block becomes available, for example, it can be substituted for its transaction-level model in the virtual platform. Software can then be verified on this version of the platform as a way to validate both hardware and software.
- The virtual platform can also provide a head start toward testbench development and post silicon validation tests by acting as a testbench component running actual system software. The virtual platform can be used to generate system stimuli to test RTL, and then verify that the virtual platform and RTL function in the same way.
- As portions of the virtual platform are verified as equivalent to their corresponding RTL, the virtual platform can become a golden or reference executable specification.

The interface from virtual platforms to traditional RTL simulation can be done using transaction-level interfaces, allowing the actual transactor to be written in SystemVerilog and the bus functional model to be synthesizable in order to allow co-execution with hardware-based environments. Alternatively, the transactor can be written in SystemC and the interface to RTL simulation can be at the signal level.

The first-use model – using TLM models to generate stimulus for RTL verification – allows early software validation on the actual hardware implementation, as well as efficient scalable system validation, for which software becomes part of the verification setup. Knowing that real-system scenarios are used increases verification confidence. Furthermore, simulation used in verification is faster, given that as much of the system as possible is simulated at the transaction level. This is shown in Figure 5.33.

The second-use model – early testbench creation and development of post silicon validation tests on a virtual platform – allows development of all testbench

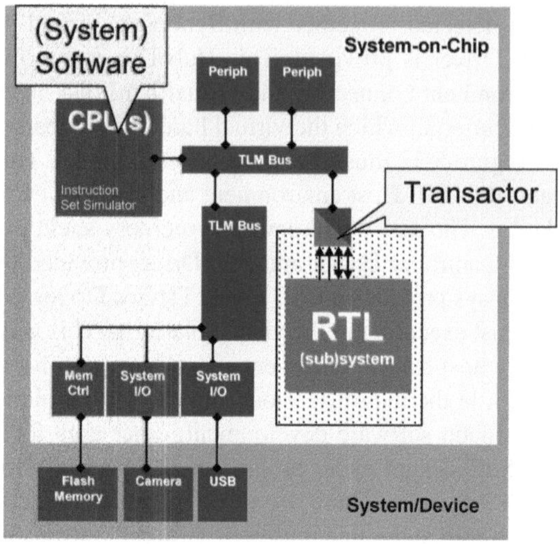

Fig. 5.33 TLM model used to generate RTL stimulus

and validation infrastructure on a virtual platform, as well as early scenario and test-case-creation. Users can efficiently develop embedded directed software tests on the TLM model which can be used for system integration testing. As a result, productivity of verification test-case development increases.

The third-use model – verifying TLM platform and RTL equivalence – helps to validate the correctness of the software development platform against the RTL once it becomes available. As a result users gain a single golden testbench for the transaction level and the RT level.

Another testbench environment that proved to be useful was to perform closed-loop testing using the RTL implementation. This enables a single test to verify aspects of both the host and target functionality. This environment in shown in Figure 5.34.

5.6.2 The Bottom Line

The software engineering team extended the existing drivers based on published IP textual specifications. At the same time hardware support for the descriptor-based DMA enhancement in the DesignWare OTG IP core was added, and the virtual platform modeling team enhanced the transaction-level model to support the new feature.

The software engineering team completely debugged and tested the basic driver functionality early on the virtual platform. As a result the driver became available 4 weeks prior to FPGA prototype availability. In addition to the early availability of the device driver, the actual software bring-up time was reduced by 4 weeks after the hardware prototype became available. The most critical bugs had already been

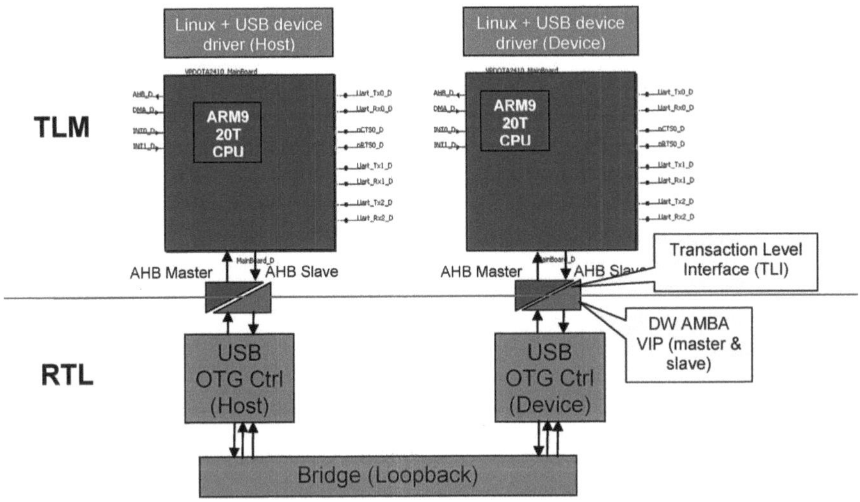

Fig. 5.34 USB OTG RTL verification

removed during the development using the virtual platform, allowing the hardware prototype bring-up to focus on timing critical issues.

Synopsys notes that this case study outlined a relatively simple derivative update of an IP core and that they would have expected even more savings for developments starting from scratch or for more extensive IP updates.

5.7 The Future

We can expect to see many changes and developments in the usage of SLVPs over the next few years. Some of these will be based on changing use cases, others based on the progression of the technology. It is fair to say that we have only seen the first chapter in the development of these tools.

For example, today we may assume that the SLVP is being used to assess architectural decisions. This is tied to a manual decision process. The SLVP may have been executed with several input parameters modified so that a Pareto curve could be generated to locate a local minima (or maxima). If automation is ever to be applied to this process, we need to be able to tell the tool how to locate those optimal points and this starts to indicate the basic capabilities that will be required in the next generation of system-level verification solutions. Some of these may be similar to the techniques provided by analog simulators today that perform parameter sweeps and Monte Carlo analysis. Other aspects will be drawn from the RTL verification world. But many of the languages used will require modifications, such as the ability to reason about concurrency and order, rather than being based on clock edges.

We can also expect to see other modeling languages emerge over time. While SystemC and TLM 2.0 have given us a good starting point, it is not the end of the story. They cannot express all of the domains in which system models may be needed; neither can they express things without architectural bias.

In short, the future for SLVPs is so wide open that predicting the future is like saying that almost everything is likely to change. We are now seeing the first generation of successful SLVPs and these will be used to identify what is working, what is not and establish what the second generation will look like. What we do know is that they will be open frameworks based on standards. That was the acceptance that brought us to this point.

References

1. Apple Press Release: http://www.apple.com/pr/library/2008/03/12iphone.html
2. Apple Press Release: http://www.apple.com/pr/library/2008/06/09iphone_sdk.html
3. John H. Day. Can BMW's iDrive Pass Its Road Test Now? Electronic Design, June 2004.
4. Imperas website. http://www.imperas.com/
5. Verification Vertigo Blog: TLM 2.0 – Good Enough for Now. http://www.chipdesignmag.com/bailey/2009/02/06/tlm-20-good-enough-for-now/
6. Alain Pegatoquet, Filip Thoen, Denis Paterson. "Virtual Reality for Modem Software Development – Enabling Pre-silicon Software Development and Validation for 2.5G Wireless Communication". SNUG 2008.
7. Press Release: Synopsys Defines Next Era of Rapid Prototyping. February 9th 2009 http://synopsys.mediaroom.com/index.php?s=43&item=646
8. Brian Bailey: Verification Vertigo blog entry. "Hybrid prototype - are we there yet?" March 17th 2009. http://www.chipdesignmag.com/bailey/2009/03/17/hybrid-prototype-are-we-there-yet/
9. Claude Helmstetter, Vania Joloboff. "SimSoC: A SystemC TLM Integrated ISS for Full System Simulation". APCCAS08.
10. Moo-Kyoung Chung, Chong-Min Kyung. "Improvement of Compiled Instruction Set Simulator by Increasing Flexibility and Reducing Compile Time". Proceedings of the 15th IEEE International Workshop on Rapid System Prototyping, 2004.
11. Tensilica press release: Tensilica Introduces TurboXim Fast Functional Simulator 40-80X Faster than ISS, Automatic SystemC Model Generation, January 2007.
12. Accellera Standard Co-Emulation Modeling Interface. Version 2.0 dated March 22nd 2007. http://accellera.org
13. Accellera Standards Organization http://accellera.org
14. IEC tutorial on SDL http://www.iec.org/online/tutorials/acrobat/sdl.pdf
15. Synopsys Press Release. http://synopsys.mediaroom.com/index.php?s=43&item=93
16. Faisal I. Haque, Khizar A. Khan, Jonathan Michelson. The Art of Verification with Vera. Verification Central, 2001.
17. Janick Bergeron, Eduard Cerny, Alan Hunter, Andrew Nightingale. Verification Methodology Manual for SystemVerilog, Springer, 2005.
18. Open Verification Methodology website http://www.ovmworld.org/
19. Verification Methodology Manual website http://www.vmmcentral.org/
20. Srikanth Jadcherla, Janick Bergeron, Yoshio Inoue, David Flynn. Verification Methodology Manual for Low Power, Synopsys 2009.

Chapter 6
Processor-Centric Design: Processors, Multi-Processors, and Software

Almost all embedded systems are a combination of software running on embedded processor cores, supporting hardware such as memories and processor buses, and other hardware elements including function accelerators and peripheral interface blocks. As a result, systems design increasingly is taking a processor-centric focus. High-level descriptions of systems functionality must be analyzed and partitioned across multiple elements, including processors, potentially more than one, and hardware components where needed.

After the system is partitioned, various components are implemented and integrated. Implementation may be based on reusable, configurable hardware, and software IP or may require development from scratch. The system may not be fully specified at initial design time, because it may be intended to be a reusable product platform. In this case, in order to accommodate additional functionality unspecified at design time, extra processing resources need to be designed in.

There are many options for designing instruction set processors for such systems, including use of fixed instruction set architecture (ISA) processors, use of multiple processors, either homogeneous or heterogeneous, or the design of application-specific instruction set processors (ASIPs) either from scratch or based on configurable and extensible processor IP. Ensuring a successful product design and implementation after choosing one or more of these options can be a complex juggling act between many different issues.

This chapter looks into the issues involved in processor-centric design using a commercial configurable and extensible processor IP as an example. It will start with a particular, quasi-realistic design application and illustrate the steps taken in a design and implementation flow using a particular set of ESL design tools for processor-centric systems, showing how various options are chosen and issues are resolved.

6.1 Choices and Trade-Offs in Processor-Centric Design

Processor-centric design describes a family of design approaches that includes several alternative methodologies. What is common to all of them is a bias toward implementing product functionality as software running on embedded processor(s),

B. Bailey, G. Martin, *ESL Models and their Application*, Embedded Systems,
DOI 10.1007/978-1-4419-0965-7_6, © Springer Science+Business Media, LLC 2010

as opposed to dedicated hardware blocks. This does not mean that there are no dedicated hardware blocks in a processor-centric design; rather, these blocks are present as a matter of necessity rather than choice. In other words, dedicated hardware blocks will be present in the design where they *must* be, rather than where they *could* be. This could be necessary to achieve the required level of performance, to achieve the desired product cost target, or to minimize energy consumption.

The traditional fixed ISA processors offer very stark trade-offs for embedded product designers. In general, they are generic for a class of processing and have few configurability options that allow them to be tailored more closely to the end application. The rise of ASIPs meant that designers could no longer just consider the use of fixed embedded processors for an increasing number of the end-product applications. ASIPs can now offer enough performance and sufficiently low energy consumption, at a reasonable cost, to take over much of the processing load that would have heretofore relied on dedicated hardware blocks. Thus ASIPs have really been a key development enabling a much more processor-centric design style.

Traditional fixed ISA processors can be simply divided into control and dataplane processors. Control processors, such as ARM and MIPS cores, are often used for non-data-intensive applications or parts of an application, such as user interfaces, general task processing, high-level user applications, and protocol stack processing. Dataplane processors are often fixed ISA digital signal processors (DSPs) that have special instructions and computational and communications resources that make them more suitable for data-intensive computation, especially for real-time signal and image processing.

ASIPs have grown in variety, number, and importance in recent years [1]. Because an ASIP can be configured and extended to optimize its performance for a specific application, ASIPs may offer much greater performance (say, 10–100×) and much lower energy consumption (perhaps half to one-quarter) than the same algorithm compiled for a fixed-ISA standard embedded processor – even a DSP. There are a few simple reasons to account for this advantage:

- ASIPs allow coarse-grained configuration of their basic structure to better match the particular applications. If an application is mainly control processing, an ASIP may offer a fairly basic instruction set, but if an application is mainly intensive data processing (from the "dataplane") – for example, audio, video, or other image processing – it may offer special additional instructions tuned to media or DSP kinds of applications. Zero-overhead looping is particularly appropriate to optimize execution of many DSP-type algorithms. Multiply and multiply accumulate hardware is especially appropriate for signal and image processing algorithms.
- The size and widths of registers can be tuned to be appropriate for the particular application domain. If an ASIP is intended for a single task, it may need a limited register set, and the register widths may be specialized for that task.
- Interfaces, such as memory interfaces, and caches can be configured or left out of the design dependent on data and instruction locality and the nature of the underlying algorithmic data access patterns. Sometimes caches may be more

effective than local instruction and data (scratchpad) memories; sometimes the opposite may be the case. Caches may be configured to match the underlying computational requirements for access to memory and data.

- Memory or bus interfaces may also be configured as to width and protocol – e.g., bridges between proprietary bus interface protocols and standards such as AMBA AHB or AXI may be automatically added.
- Diagnosis and debug features such as trace ports or JTAG interfaces may be added or left out.
- Interrupts and exception handling may be configured according to design needs. Often the elaborate exception recovery mechanisms used in general-purpose processors may be unnecessary in an ASIP tuned to run a very specific algorithm deeply embedded in a system. Similarly, only a few specialized interrupts may be necessary, as the processor may have a very limited set of interfaces and peripherals when used in its intended context.
- Very long instruction word (VLIW)-style multi-operation instructions may be added to processors to support applications with a large amount of irregular instruction-level parallelism that can take advantage of such features. This is usually limited to a few operations (e.g., 3 or 4).
- SIMD (Single Instruction, Multiple Data)-type instructions may be added to processors to support vector-style simultaneous instructions acting on large chunks of data at a time. 2-, 4-, 8-, and sometimes 16-way or larger SIMD hardware may be supported.
- Instructions may be tuned to specific algorithmic requirements. For example, if two 13-bit quantities need to be multiplied in an inner loop that dominates an algorithm, use of a 32-bit multiplier is wasteful in area, energy, and possibly performance. A specifically designed 13-bit multiplier (or for more generality, still with significant savings, a 16-bit multiplier) may be much more appropriate.
- Fine-grained instruction extensions including instruction fusions drawn from very specific algorithmic code can lead to significant increases in performance and savings in power. For example, a sequence of arithmetic operations in a tightly nested loop might account for 90% of the cycles when executing the algorithm on a data sample and may be replaced with a single fused instruction that carries out the sequence in one or a few clock cycles. In addition, the operations using general-purpose ALU instructions may end up using specialized hardware tuned to the specific algorithmic requirements, thus saving energy. If combined with SIMD-style hardware and appropriate compiler support, even greater savings in cycle time and power are possible. However, instruction extensions are added to a base ISA and thus consume extra area. In addition, automatically inferring the use of instruction extensions is often difficult to achieve in compilers, especially if the new instructions are irregular. In this case, manual modification of the source code to insert pragmas or macro-calls may be necessary.

Figure 6.1 shows the range of configurability and extensibility possible with the Xtensa processor IP as an example.

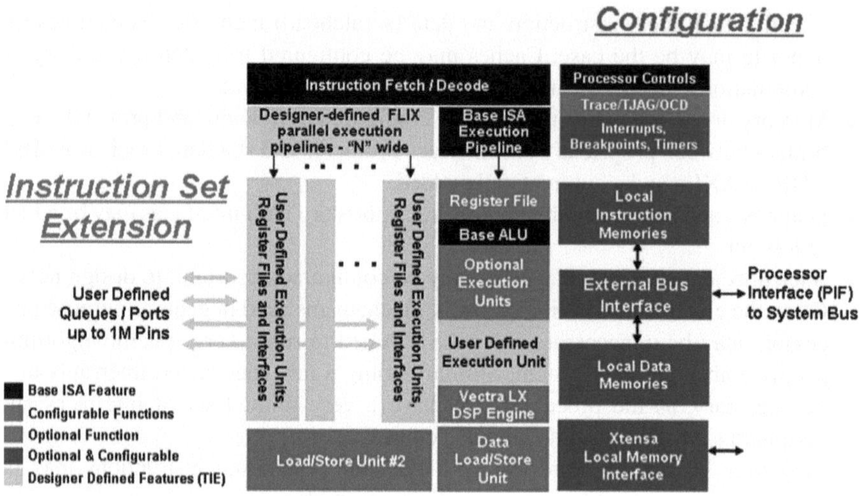

Fig. 6.1 Example of configurable, extensible ASIP

We can see that use of ASIPs instead of general-purpose processors can thus lead, for known algorithms, to a radical improvement in performance and power consumption. This is true whether an ASIP is totally designed to support one specific algorithm or if it is designed to support a class of applications drawn from a single domain. A specialized audio-processing ASIP could be designed just to support MP3 decoding or could be slightly generalized so that it will support many different audio codecs – possibly optimizing one codec such as MP3 that is very widely used, but with general audio instructions added so that new codecs can still take advantage of the specific instructions and hardware incorporated in the ASIP. Similar results can be achieved in video, baseband, network processing, and a wide variety of other computationally intensive dataplane applications.

Sometimes the complexity of a specific application domain may lead to a heterogeneous multi-processor, multi-ASIP design being optimal for a certain target range of process technologies. Video decoding and encoding, for example, may require at least two processors along with specialized additional hardware units and DMA style memory subsystems in order to achieve reasonable results depending on the rates, resolutions, and target frequencies of the resulting systems.

A processor-centric design methodology needs to support design space exploration when deciding whether particular functional requirements for a design can be mapped to a single fixed ISA processor running at a suitable rate, a multi-processor implementation (such as a cache-coherent symmetric multiprocessing "multicore" cluster), a special fixed ISA processor such as a DSP, a single ASIP, a set of ASIPs configured to work together as a heterogeneous multi-processor subsystem, a combination of fixed ISA processor(s) and ASIP(s), and finally, mapping any part of the function into dedicated hardware blocks, almost certainly working in conjunction with the processors. The processor and multi-processor architectures may be organized into a wide range of communications architectures, from shared memory accessed via buses through dedicated local memories, use of DMA blocks to permit concurrent data and instruction movement, use of direct communications such as FIFO queues between processors and from processors to hardware, and almost any combination including new emerging network-on-chip approaches.

6.2 An ASIP Integrated Development Environment (IDE)

In this chapter we will illustrate a real ASIP-oriented design flow using the Tensilica Xtensa ASIP [2, 3] as an example and using its Xtensa Xplorer integrated development environment (IDE) as the design framework within which the flow will be demonstrated. Xtensa Xplorer, hereafter abbreviated "XX," is based on the Eclipse open-source IDE project [4], as are a number of other development environments and system-level design tools.

Figure 6.2 shows the XX environment.

Along the top of the graphical user interface (GUI) is a normal row of command menus, with another row below it of IDE menu items providing navigational aids and control commands. Below it are several panels. We will describe four of them,

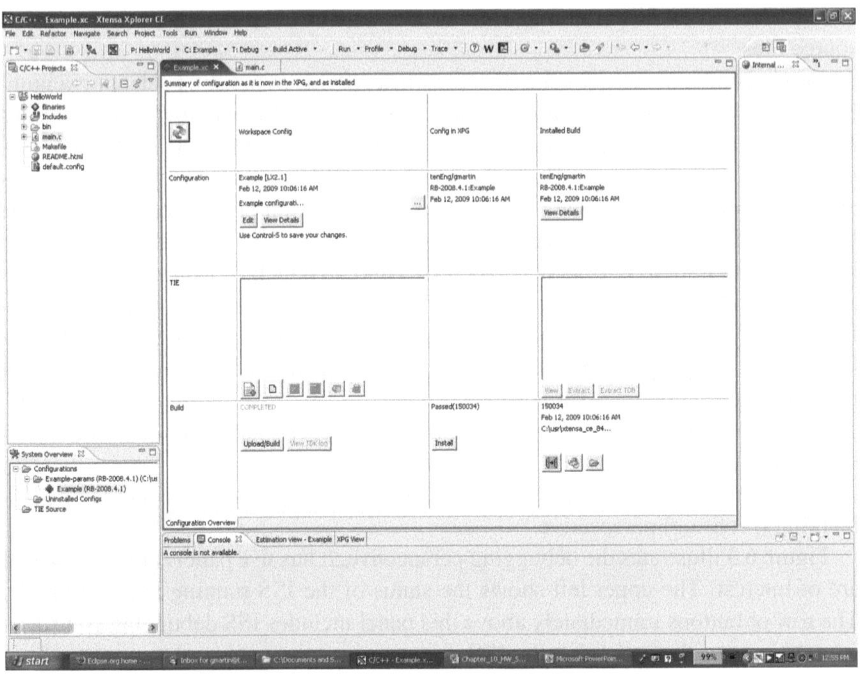

Fig. 6.2 The Xtensa Xplorer IDE

in this perspective (the C/C++ development view – there are alternative perspectives for debugging and profiling (benchmark perspective) described later). Two are familiar views to anyone acquainted with Eclipse: the C/C++ projects view on the upper left and the console window on the bottom. The C/C++ projects view shows the file and folder structure of projects including source and header files, binaries for various targets, makefiles, and other associated project files. The console window at the bottom shows the results of running software on the target instruction set simulator (ISS), with console messages from both the software and the ISS. The ISS for target configurations made available within XX can run in two modes: cycle-accurate and a fast-functional mode called TurboXim. The turbo mode can be up to 100 × the speed of the cycle-accurate mode, for a single processor ISS model, and is useful for verification of software functionality where exact and precise timing is not necessary. Turbo mode is more or less similar to the OSCI TLM 2 loosely timed mode (LT), incorporating the direct memory interface (DMI) and out-of-order execution. There is no OSCI equivalent currently in TLM 2 to the cycle-accurate mode, although this is similar to OCP/IP TL1. The cycle-accurate mode is essential for detailed profiling and comparison of different target configurations and for the development and verification of timing-dependent software (such as device drivers and other hardware-dependent software).

The most interesting addition to this perspective are the window in the middle, which is the Xtensa Processor Generator (XPG) view for a particular example configuration; and the one on the lower left, which is a navigation window for all user-defined processor configurations and Tensilica Instruction Extension (TIE) source files.

Configuring a processor using the Tensilica XPG is a matter of selecting a number of coarse-grained configuration options using the XPG window and its various sub-panels and also adding instruction extensions that are defined using the TIE language. A TIE file can be attached to a number of different processor configurations and can be evolved over time as the processor generation process converges on the final desired version. The configuration and TIE files can be uploaded via the XPG to build a configuration on Tensilica's servers. This can then be downloaded to the customer's own local workstation after one or two hours when the configuration is built. In addition, the TIE files can be iterated and compiled locally using a local compilation loop, allowing design space exploration of processor configurations without re-building each one from scratch.

The C/C++ perspective allows the editing of target source files which are also opened up in the central panel of the overall window.

Two other perspectives are important: the debugging perspective and the benchmarking (profiling) perspective.

Figure 6.3 illustrates the debugging perspective. It has five panels in this view that are of interest. The upper left shows the status of the ISS running the target code. The row of buttons immediately above this panel includes ISS debugging controls – for starting, pausing, stopping the ISS, for stepping into, over, and step-return from instructions. The panel on the upper right is a variable display; the user can also view breakpoints, registers, expressions, and other debug information. The middle

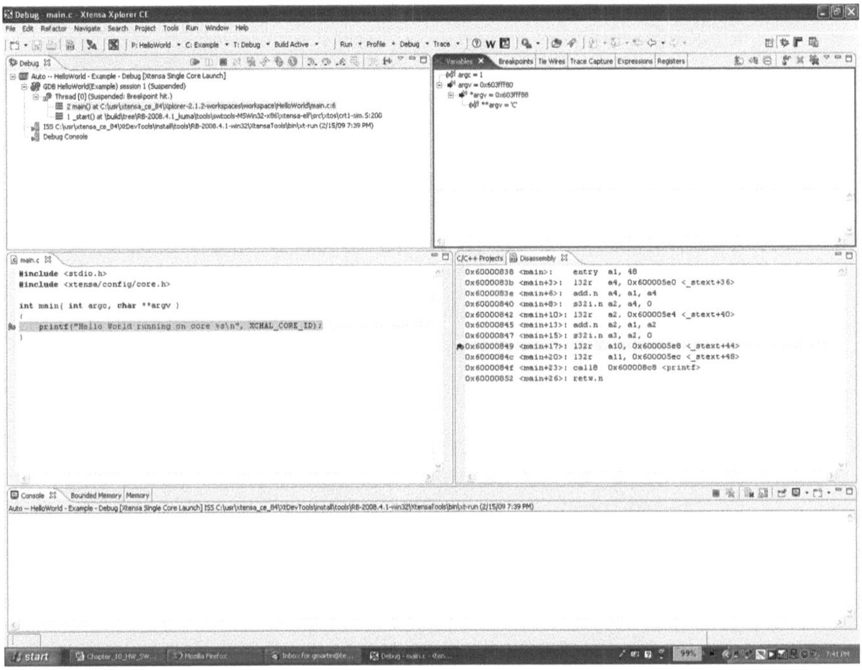

Fig. 6.3 Debugging perspective

left panel shows the source code (in this case "Hello World") and the user can set and control breakpoints using the source code. The middle right panel is a disassembly view. Finally, the bottom panel shows the console view and also can be toggled to show various memory views.

The benchmark perspective, as shown in Figure 6.4, allows a variety of profiling experiments to be run and compared. These profiling experiments are carried out with the cycle-accurate ISS accessed via XX in order to obtain enough accuracy for decision making. Although they can in theory be run on the TurboXim fast-functional models, these only report total numbers of instructions executed and may be misleading if the user equated instruction counts with cycle counts. In the figure, we can see source code with cycle annotations in the middle top; a disassembly view of the currently examined function on the upper right, and below, one of many profile views showing functions, their total cycles, size, and other relevant parameters. There are many possible profile views including comparisons between one run and another. This will be illustrated later in this chapter.

Tensilica currently offers two product lines – the Xtensa configurable processor IP and families of predefined processors for various application areas, ranging from general-purpose controllers through audio and video processors and subsystems through to a variety of DSPs. To support customers who use only a preconfigured core, there is a special version of XX that does not have the regular XPG view nor the ability to use the TIE compiler. In this chapter we will illustrate the processor-centric

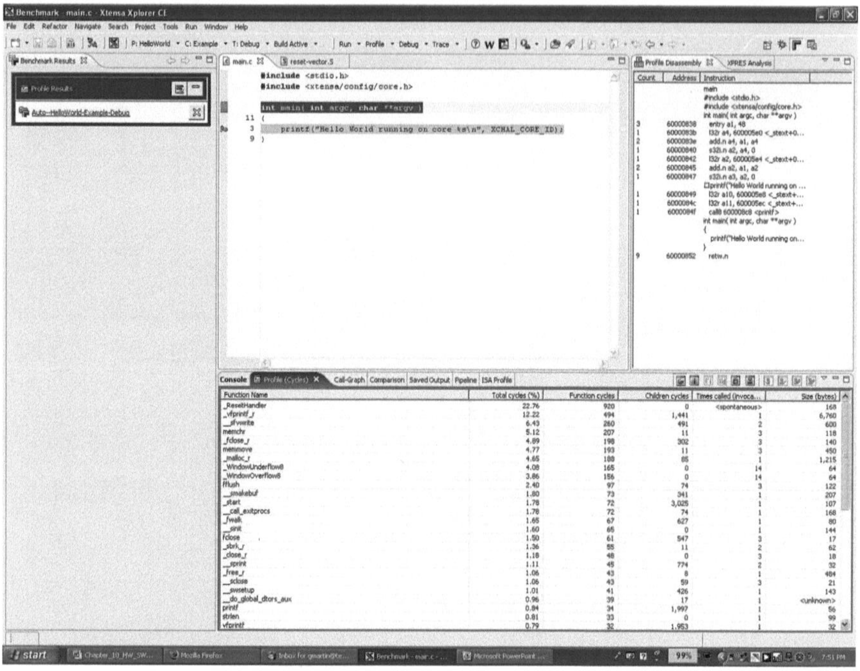

Fig. 6.4 Benchmarking perspective

design methodology using both predefined cores and configurable Xtensa cores to show the advantages of configuring a processor and adding instructions to improve the execution of specific applications.

6.3 Introduction to Flow and Example

We will illustrate a processor-centric design flow using, as an example, JPEG decoding. This will be shown on a variety of Tensilica processor configurations. The djpeg application – decode jpeg – relies on a library libjpeg and is provided with XX as a sample application. Thus it is possible for readers who have access to an evaluation of Tensilica technology to reproduce many of the results shown here.

Readers should note that use of Tensilica technology is done for convenience, but it is merely one example of many different alternatives that may be available for ASIP generation, optimization, and analysis.

In general, the processor-centric design flow has the following steps:

- Start with an algorithmic description. This is very often reference C or C++ code obtained from a standards organization, especially in the communications and multimedia areas where reference implementations are common. Alternatively, it may be reference code generated from an algorithmic description captured in a modeling notation such as the Mathworks MATLAB or Simulink, or in UML or one of its profiles, and using code generation to obtain executable C or C++. As described in [5], such code may not be ideal to use as a starting point because it may contain many structures and code that would be heavily optimized in any real target implementation, but it is still a better starting point than either artificial code that must be written to somehow represent the computational and communications patterns of the real application or having to write the application from scratch. In the case of our djpeg example, the reference implementation comes from the Independent JPEG Group (IJG) (1991–1997).

- Characterize the algorithm by running it on a very generic target processor. This will give you some idea of the general computational and communications requirements of the algorithm (communications being defined as both data access and control access communicating into and out of the algorithm). A target processor (e.g., a 32-bit RISC processor with no special instructions, and normal 32-bit interfaces) is a reasonable basis for processor characterization as it will allow the advantages of particular configuration options and instruction extensions to be seen during comparative profiling. Such a processor may have no cache, using only system memory; or small instruction and data caches; or local instruction, data, or unified memories. Data and instruction access patterns found during characterization will be useful in optimizing the memory subsystem.
- Identify "hot spots" in the target application. These will very often be loop nests in which multiple instructions are executed over large data samples. Techniques such as instruction fusion (combining multiple instructions into one); vectorization (SIMD – Single Instruction Multiple Data) methods, where the same instruction is applied to many data items; and multi-operation instructions – where several operations without dependencies could be executed simultaneously on a VLIW-style architecture – are very commonly identified in these hot spots. In addition, there may be bottlenecks in memory access in these hot spots which are stalling the processing for many cycles. A variety of techniques are available for improving these interface bottlenecks, including local memories, multiple load-store units, wider memory access, various cache configurations, direct FIFO interfaces, and improvements to shared system memory via better buses or other possible interconnect schemes.
- Configure the processor and add instruction extensions to accelerate the execution of the algorithm. Re-characterize the code running on the new modified target. It may be necessary to restructure the code or insert pragmas into it in order that the compiler can take full advantage of vectorization (SIMD) or fused instructions.
- If the performance targets for the algorithm are met and the estimates of power consumption and cost (area in terms of gates) are satisfactory, stop: this processor is now a reasonable choice for the function and it may now be integrated into the overall design. Otherwise, further code restructuring and further configuration exploration and additional instruction extensions may be important. In this case, repeat the last few steps until either a satisfactory result is achieved or it is necessary to add specialized hardware blocks as coprocessors in order to achieve the desired results.

- If hardware blocks are necessary, they may be created using ESL synthesis tools based on the algorithmic description for that part of the algorithm which must migrate to hardware. The design team may explore a variety of mechanisms for tying such accelerating blocks to the main processor – via hardware FIFOs, via coprocessor interfaces, or loosely coupled via systems buses, possibly using communications mechanisms such as DMA in order to speed up data transfer between processor and hardware blocks. This will be discussed in detail in Chapter 9.

6.4 Starting with Algorithms

In the jpeg decoder (djpeg case), the algorithm is divided into a general-purpose library, libjpeg, with 12 C header and 47 C source files, used as a library by both the jpeg decoder and encoder sample code, and the djpeg application, consisting of 2 C header files and 8 C source files. As mentioned, this code came from the International JPEG Group. It is most definitely reference code, and not really target-optimized code, but suffices to illustrate design space exploration. The decision as to whether to modify the source code on a real design project before beginning to explore the processor design space is an important one and will depend on a design team's judgment as to the bias introduced into the evaluation stage by using the unmodified original source. It should be noted that all software embeds some notion of an implementation and that as changes are made to it, the software will reflect the ultimate implementation architecture.

6.5 Processor Definition

6.5.1 Designing the Design Space Exploration

We are going to define several experiments in order to find a good processor tuned to this application.

First, we are going to do a coarse-grained design space exploration across a range of predefined Xtensa cores available as preconfigured products. This is a quick and easy way to get a preliminary assessment of the impact of a number of different configuration options for an application and may indicate the kind of starting point which is most suitable for creating an optimized ASIP. We will look at four preconfigured cores:

1. DC_B_106micro – a basic 32-bit microcontroller.
2. DC_B_108mini – also a basic 32-bit microcontroller; its most significant difference to the 106micro is the presence of fast 32-bit integer multiplier. The DC_B_106micro has a 16x8 integer multiplier that is used iteratively in combination with other instructions to provide 32x32 integer multiplies, and thus takes several cycles. The DC_B_108mini has a 2-cycle fully pipelined 32x32 integer multiplier that can produce a new result every cycle. This has an impact on the JPEG decoding results as we will see later.

3. DC_B_545CK – an 8-MAC VLIW DSP with an extensive set of additional vector-oriented DSP-style instructions.
4. DC_570T – a VLIW-style machine allowing up to three operation slots per instruction for simultaneous operation execution when independent.

Much more information about these preconfigured cores is found in [3].

These four cores are of quite different sizes and capabilities. The estimated areas for each of them, in square mm in a TSMC 90G process, using SAGE-X libraries, when optimized for speed (as opposed to area), and post-layout, is as follows:

1. DC_B_106micro – 0.21
2. DC_B_108mini – 0.36
3. DC_B_545CK – 1.76
4. DC_B_570T – 0.70

We compiled the "release" version of the target library and application for each of these cores using aggressive (level 3) optimization, inter-procedural analysis, and SIMD optimization where possible (on the 545CK only, as it has support for SIMD-style execution units).

We then utilize a Tensilica tool for automated design space exploration. This is called XPRES, which has been described in detail in [6]. Further details can be found in [7].

XPRES is a tool that analyzes C/C++ source code and explores many different possible processor configurations using three dimensions of concurrency: pipeline parallelism – that is, creating fused instructions that combine several sequential operators into one instruction; data parallelism, using vectorization and SIMD, that duplicates functional units so that the same operation can be applied to multiple data items in parallel; and instruction parallelism – the use of VLIW-style hardware so that multiple independent operations can be scheduled at once. XPRES will either produce a single "best" result or a Pareto-style curve showing the best results along a performance–cost (in estimated gates) trade-off curve. The results of XPRES are sets of TIE files specific for the application whose design space is explored.

We will start the XPRES experiments using a very basic configuration and then exploring what the additional TIE-defining configurations and instruction extensions provide.

XPRES offers the distinct advantage in developing an ASIP that it shares the same front-end as the compiler. Thus, any instruction extensions explored by XPRES will be automatically inferred and utilized in compiling the source code of the application for the target processor(s) defined using XPRES. This allows much quicker design space exploration.

XPRES results come out in source TIE format and thus can be used as the basis for further manual optimization and fine-tuning the code to optimize results. For example, XPRES analysis will give feedback on loops where vectorization is difficult due to the data access patterns in the loops, giving the designers an opportunity to do some fine-tuning of the application and the ASIP. We will conclude our exploration by discussing fine-tuning.

6.5.2 Exploring the Processor Design Space: Preconfigured Cores

Let us start by compiling the djpeg reference code along with the jpeglib library for each of the four target cores and then run a profile execution using the cycle-accurate ISS for each target (Figure 6.5). The first chart shows the results of these four runs (targeting release rather than debug, with aggressive optimization):

The chart shows total cycles for the djpeg application running on each of the four target processors. The actual numbers of cycles are as follows:

1. DC_B_106micro: 18,490,019
2. DC_B_108mini: 17,731,352
3. DC_B_545CK: 19,164,659
4. DC_B_570T: 11,609,358

We can see that the 570T is the best, by about 1/3, in comparison to the 108mini. Perhaps surprisingly, we see that the 545CK, which has many DSP instruction extensions and SIMD execution units, does the worst. A little analysis will explain why.

Let us look at the detailed cycle counts reported for the top several functions in each profile and zero in on one function. First, for 106micro:

Function name	Total cycles (%)	Function cycles	Children cycles	Times called (invocations)	Size (bytes)
pass2_fs_dither..BJ	20.34	3,761,029	3,619,681	162	596
jpeg_idct_islow	19.68	3,639,270	0	1,764	976
fill_inverse_cmap..PI	19.57	3,619,681	0	14	1,120
decode_mcu..KK	9.11	1,685,237	663,051	588	1,080
put_pixel_rows..KB	7.69	1,423,238	393,675	162	252
ycc_rgb_convert..BK	6.16	1,139,508	0	162	200

Then, for 108mini:

Function name	Total cycles (%)	Function cycles	Children cycles	Times called (invocations)	Size (bytes)
pass2_fs_dither..BJ	20.99	3,723,562	3,603,864	162	532
fill_inverse_cmap..PI	20.32	3,603,809	55	14	1,076
jpeg_idct_islow	17.56	3,113,686	0	1,764	940
decode_mcu..KK	9.45	1,675,630	557,795	588	884
put_pixel_rows..KB	7.82	1,386,902	388,178	162	224
ycc_rgb_convert..BK	6.62	1,174,176	1,786	162	184

Then, for 545CK:

Function name	Total cycles (%)	Function cycles	Children cycles	Times called (invocations)	Size (bytes)
jpeg_idct_islow	22.89	4,388,032	714,098	1,764	1,300
pass2_fs_dither..BJ	19.8	3,794,722	3,485,208	162	548
fill_inverse_cmap..PI	17.29	3,315,274	168,214	14	1,268
decode_mcu..KK	8.82	1,690,374	544,626	588	884
put_pixel_rows..KB	7.94	1,522,262	265,929	162	236
ycc_rgb_convert..BK	5.39	1,033,398	0	162	188
mulsi3	4.93	944,859	0	88,962	40

Then, finally, for 570T:

Function name	Total cycles (%)	Function cycles	Children cycles	Times called (invocations)	Size (bytes)
pass2_fs_dither..BJ	20.77	2,412,214	2,107,386	162	664
fill_inverse_cmap..PI	18.15	2,107,331	55	14	1,256
jpeg_idct_islow	15.51	1,801,110	0	1,764	1,124
decode_mcu..KK	11.02	1,279,360	416,984	588	1,152
put_pixel_rows..KB	9.95	1,155,607	247,422	162	296
ycc_rgb_convert..BK	7.35	854,064	1,789	162	208

Now, in particular, let us look at the function jpeg_idct_islow (we would actually explore all the "hot spots" in all the top functions in the profile, but let us explore this one). Here is a table of the function cycles vs. processor type:

Processor	Number of function cycles for jpeg_idct_slow
106	3,639,270
108	3,113,686
545	4,388,032
570	1,801,110

We note a significant reduction in cycles (about 500,000, or 13%) for this function between the 106 and the 108. To explore why, we look at the profile disassembly view. Here we note that both processors, in their compiled code, use the "mull" function for many multiplies. The disassembled code is actually very similar for the 106 and 108 (and note that the size in bytes is very similar as well – 976 bytes for 106 and 940 bytes for 108). The "mull" instruction in the 106, which is for 32 × 32 bit integer multiplies, is mapped, as discussed above, into a multi-cycle use of more primitive instructions, whereas the "mull" on the 108 uses the inbuilt 2 cycle multiply unit. This will no doubt account for some of the differences.

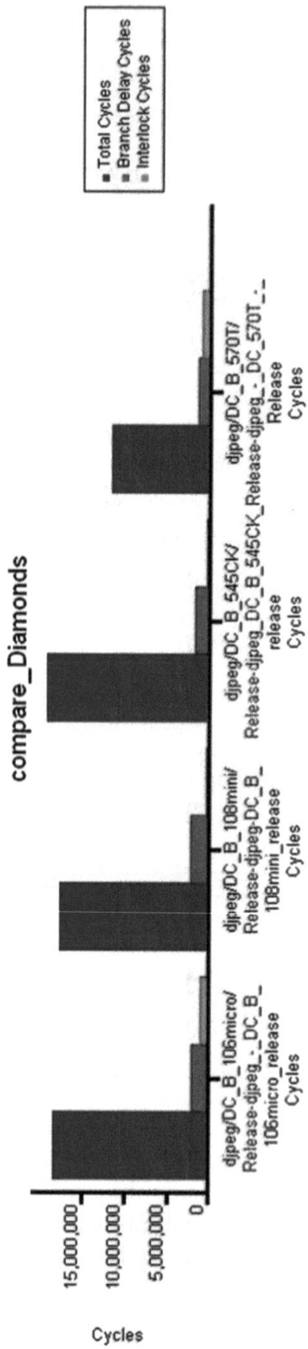

Fig. 6.5 Comparison of several preconfigured core profiles for djpeg

When we look at the 545CK and the 570T we notice something at first counterintuitive – the 545CK jpeg_idct_slow takes about 700,000 more cycles than the 106. Here again we look at the disassembly code and zero in on the multiply in particular. In the 545CK disassembly for this function, we notice some use of SIMD instructions, such as addx8, addx4, addx2, subx8. But we also notice a number of calls to a function called mulsi3. In fact, looking at the bottom of the 545CK profile table above, we note that about 89,000 times, for total cycles of over 944,000, or on average more than 10 cycles a call. Looking at the 545CK description in the data books, we see that the 545CK is optimized as a DSP for 16-bit operations (not general 32-bit operations) as is very common in signal processing DSPs. As a result, its eight parallel MACs (multiply accumulators) are intended for 16-bit operations, and when 32-bit multiplies are required it uses alternative means such as general 32-bit multiplier emulation. Looking in the Tensilica C application programmer's guide, we find that __mulsi3 is an integer multiplication emulation routine.

What then can we conclude from the 545CK results? It's a simple lesson, but worth noting: using the wrong instruction extensions or using a preconfigured core intended for a different application space may be worse than using a general-purpose RISC processor (such as the 106 microcontroller). For 16-bit signal processing applications, the 545CK offers many advantages. For JPEG decoding using unoptimized code and general 32-bit arithmetic, it may function correctly (as of course it does), but not necessarily advantageously.

Finally, the 570T does significantly better than any of the other cores on this function. In general, it has wider bus widths and a 3-operation VLIW capability. When we look at the disassembly of this function, we notice, in the tight loops (in between loads and stores), considerable use of multi-operation instructions – with at least two of the three slots filled. For example,

```
7,589    60003f18 { l32i a10, a3, 0; mull a5, a5, a10; nop }
7,589    60003f20 { l32r a8, 60003314 <_stext+0x64>; mull a6, a6, a8; nop }
```

where a mull (32×32 integer multiplier) operation is executed in parallel with loads and a "nop" (no-op) instruction.

It does not require too many executions of two instructions in parallel in loops to lead to a significant cycle reduction – and we note that the cycle count for the jpect_idct_slow function in the 570T is about half the number of cycles on the 106 (1.8 million cycles as compared to 3.6 million on the 106).

We can thus draw some conclusions from this very simple coarse-grained ASIP design space exploration:

- Sometimes an implementation of a single instruction such as a multiplier can have a reasonably large impact on overall execution time.
- If an ASIP is designed for one domain, its instruction extensions may not be very useful for a new application drawn from a different domain. Careful attention must be paid to matching ASIP to application.

– If there is a large amount of independent, non-data-dependent concurrency in an application, using VLIW-style multi-operations (in Tensilica terms, FLIX, for flexible instruction extensions) may be quite advantageous.

In the next section, we will do a more comprehensive exploration of the fully configurable XTENSA architecture using the XPRES tool.

6.5.3 Exploring the Processor Design Space: Automatically

We are going to illustrate design space exploration for an ASIP using the Tensilica tool XPRES. As mentioned earlier, XPRES is an interesting way to automatically examine thousands or millions of configurations drawn from the range of its search space; but XPRES does not cover the complete design space possible with Tensilica processor configurations and instruction extensions and the Tensilica TIE language. Primarily, it explores the three dimensions of instruction fusion, SIMD, and VLIW configuration as discussed earlier.

After covering an example of using XPRES on our djpeg application, we will look at some other tools for ASIP automation including the cache and memory exploration tool. Although it is possible for a designer to carry out the entire process of ASIP configuration, extension, and design space exploration manually using the ASIP tools from Tensilica (or other providers), the use of these automated tools can be very useful in both teaching design teams the use of TIE and helping them quickly explore design sensitivities.

For the XPRES experiments we will start with a simple "xpres_test" configuration built on a template provided by Tensilica as a standard starting point, "xpres." As you can see from the following summary, it is a basic 32-bit RISC with instruction and data caches, wide interfaces to these caches, and a standard 32-bit processor system memory interface.

The first step is to profile the djpeg application on the xpres_test configuration using the cycle-accurate ISS. Figure 6.7 shows the basic profile chart. Again, we use a "Release" target rather than "Debug," and apply the same aggressive optimizations as in the earlier experiments (level 3, inter-process analysis, and SIMD where possible).

The number of cycles for this application is 16,701,310. Note that this is a little better than the 106micro or 108mini results shown earlier. Branch delay cycles are due to stalls caused by program branching; interlock cycles are stalls due to pipeline hazards.

We now apply XPRES to automatically optimize a configuration for this application. We will enable fusions, SIMD, and VLIW. Figure 6.8 shows the XPRES output, compared to the default xpres_test configuration:

The XPRES-generated configuration runs djpeg in 12,704,117 cycles, which is a saving of about 4 million cycles or approximately 24%. Note that this cycle count

Instruction options		
16-bit MAC with 40 bit Accumulator	no	
MUL16	yes	
MUL32	Fully Pipelined	
32 bit integer divider	yes	
Floating Point (coprocessor id 0)	no	
CLAMPS	yes	
NSA/NSAU	yes	
MIN/MAX and MINU/MAXU	yes	
SEXT	yes	
Boolean Registers	yes	
Number of Coprocessors (NCP)		0
Enable Density Instructions	yes	
Enable Processor ID	no	
Zero-overhead loop instructions	yes	
Synchronize instruction	no	
Conditional store synchronize instruction	no	
Count of Load/Store units		1
Max instruction width (bytes)		8
Pipeline length		5
Vectra LX DSP coprocessor instruction family	no	
HiFi2 Audio Engine DSP coprocessor instruction family	no	
Thread Pointer	no	
Interrupts enabled?	yes	
Interrupt count		1
Int 0 type/priority level	Timer/1	
High Priority Interrupts	yes	
Interrupt Level count		2
Medium Level Interrupts	no	
Timer count	yes	
Timer count		1
Timer 0		0
Byte ordering (endianness)	Little Endian	
Number of physical AR registers. 16 forces Call0 ABI		32
Miscellaneous Special Register count		0
Generate exception on unaligned load/store address	align	
Enable Processor Interface (PIF)	yes	
Write buffer entries		4
Enable PIF Write Responses	no	
Prioritize Load Before Store	no	
Widths of Cache and Memory Interfaces		
Width of Instruction Fetch Interface		64
Width of Data Memory/Cache interface		128
Width of PIF interface		32
Width of Interface to instruction cache		64
Instruction Cache size (Bytes)/Line size (Bytes)	1024/16	
Associativity	Direct	
Line Locking	no	
Instruction Cache memory error	none	
Data Cache (Bytes)/Line size (Bytes)	1024/16	

Fig. 6.6 xpres_test configuration

Associativity	Direct
Write Back	no
Line Locking	no
Data Cache memory error	none
Debug	yes
Data address breakpoint registers	0
Instruction address breakpoint registers	0
Debug interrupt level	2
Trace port (address trace and pipeline status)	no
Add data trace	no
On Chip Debug (OCD)	yes
Use array of 4 Debug Instruction Registers (DIRs)	no
External Debug Interrupt	no
Full scan	no
Make latches transparent	no
MemoryProtection/MMU	Region Protection
System RAM start address/size	0x60000000 / 512M
System ROM start address/size	0x40000000 / 128K
Vector configuration	
Reset Vector start address/size	0x40000000 / 0x2e0
Kernel (Stacked) Exception Vector start address/size	0x6000019c / 0x1c
User (Program) Exception Vector start address/size	0x600001bc / 0x1c
Double Exception Vector start address/size	0x600001dc / 0x1c
Window Register Overflow Vector start address/size	0x60000000 / 0x178
Level 2 Interrupt Vector start address/size	0x6000017c / 0x1c
Relocatable Vectors	no
Target and CAD options	
Functional Unit Clock Gating	yes
Global Clock Gating	yes
Register file implementation block	Flip-flops
Reset All Flops	yes
Software Target Options	
Xtensa Tools should use Extended L32R	no
Software ABI	windowed
C Libraries	newlib

Fig. 6.6 (continued)

is still greater than that required on the DC_B_570T discussed earlier. This is a result of using a non-VLIW configuration in the XPRES experiments, which ends up leaving some potential non-SIMD independent concurrency unexploited in this configuration. Combining the XPRES TIE with VLIW would achieve additional performance.

XPRES generates TIE in source form and it is thus possible both to examine the TIE and to use it as a basis for further manual optimization. One example of an automatically produced fusion instruction is fusion.add.ne.movtx8 which operates on eight 32-bit and eight 17-bit quantities and produces eight 32-bit outputs.

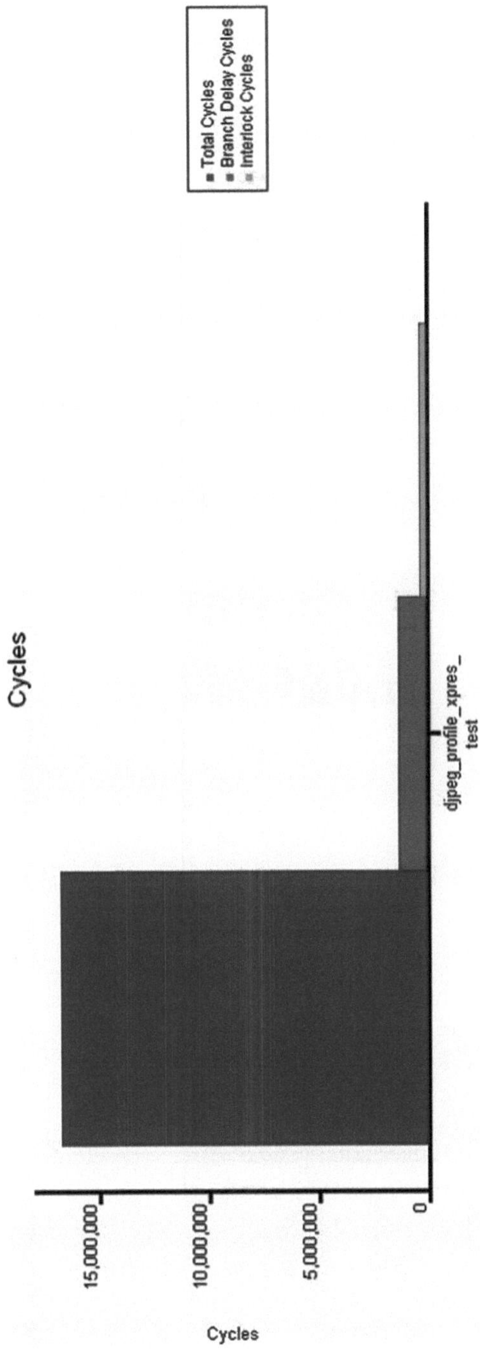

Fig. 6.7 djpeg profile on xpres_test configuration

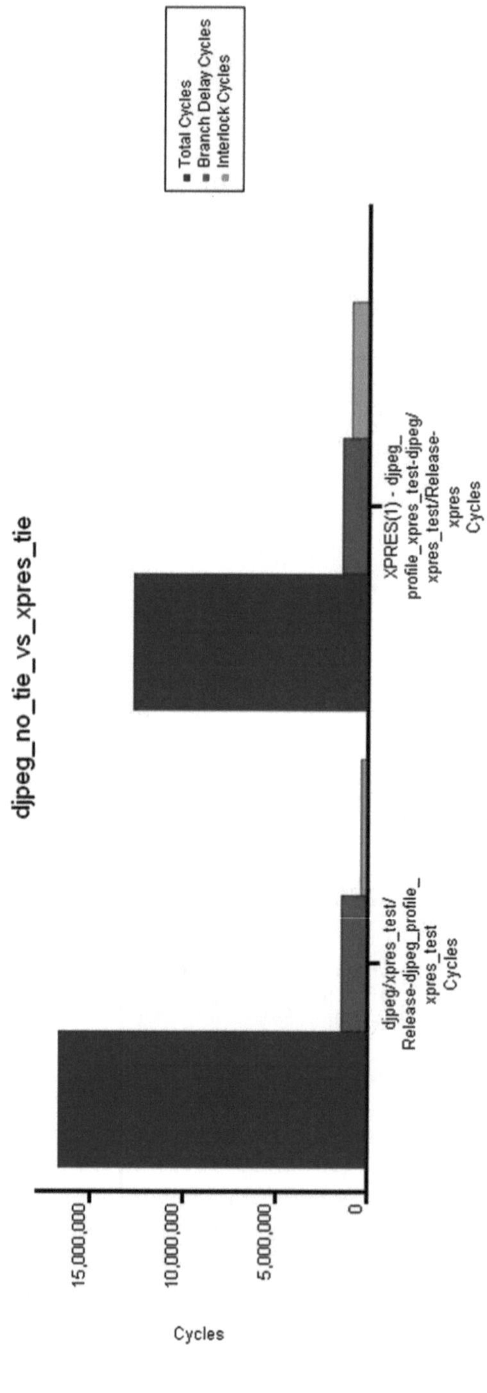

Fig. 6.8 djpeg on xpres_test configuration – without and with XPRES-generated TIE

```
// Fused Vector Operations:
//
// fusion.add.ne.movtx8
//  A fused operation composed of the following operations:
//  add32x8, movt32x8, ne17x8u.
//
operation fusion.add.ne.movtx8 { out regfile32x8 a,
in regfile32x8 b, in regfile17x8 c } { }
{
 wire t0 = |c[135:119];
 wire [31:0] t1 = 32'h1 + b[255:224];
 wire [31:0] t2 = t0 ? t1 : b[255:224];
 wire t3 = |c[118:102];
 wire [31:0] t4 = 32'h1 + b[223:192];
 wire [31:0] t5 = t3 ? t4 : b[223:192];
 wire t6 = |c[101:85];
 wire [31:0] t7 = 32'h1 + b[191:160];
 wire [31:0] t8 = t6 ? t7 : b[191:160];
 wire t9 = |c[84:68];
 wire [31:0] t10 = 32'h1 + b[159:128];
 wire [31:0] t11 = t9 ? t10 : b[159:128];
 wire t12 = |c[67:51];
 wire [31:0] t13 = 32'h1 + b[127:96];
 wire [31:0] t14 = t12 ? t13 : b[127:96];
 wire t15 = |c[50:34];
 wire [31:0] t16 = 32'h1 + b[95:64];
 wire [31:0] t17 = t15 ? t16 : b[95:64];
 wire t18 = |c[33:17];
 wire [31:0] t19 = 32'h1 + b[63:32];
 wire [31:0] t20 = t18 ? t19 : b[63:32];
 wire t21 = |c[16:0];

 wire [31:0] t22 = 32'h1 + b[31:0];

 wire [31:0] t23 = t21 ? t22 : b[31:0];

 assign a = { t2, t5, t8, t11, t14, t17, t20, t23 };
}
```

The other interesting thing to consider is where this XPRES-generated configuration comes from. It is possible to generate the whole Pareto surface and display it as a performance vs. cost chart, using XPRES in a multiple-output TIE mode (Figure 6.9)

This is quite an interesting output chart. First, the "Base" xpres_test processor configuration is shown as a dot on the vertical bar about 40% of the way along the *X* axis. The *X* axis shows an estimate of additional gates, added to the Base processor

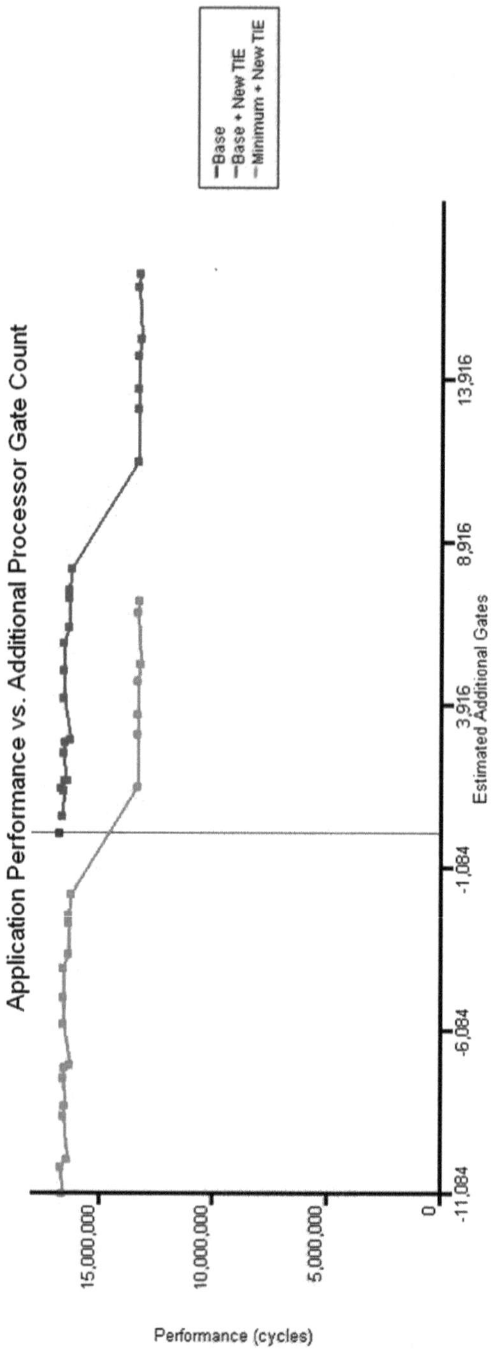

Fig. 6.9 XPRES results from multiple TIE output mode for djpeg

configuration, for the various new TIE options that are on the Pareto curve from the XPRES design space exploration. The right-hand curve shows Base plus new TIE. Note that the curve consists of two portions: performance more or less equivalent to Base, then a distinct descent in number of cycles about half way along, and then a flat curve. (The *Y* axis shows performance in terms of cycle count, so better is lower on the *Y* axis.) This occurs somewhere about 11,000 additional gates. To interpret this curve, no additional performance gain is achieved by XPFRES until hitting about 11,000 gates of TIE, and the result achieved at 11,000 gates is about the same as that achieved on the right-hand most point which is about 16,000–17,000 gates. Twenty-two different TIE configurations are shown on this chart.

Another interesting curve is the left most one, which is actually the same curve shifted to the left. This represents a "Minimum" starting configuration, because it turns out that not all the configuration parameters of the xpres_test starting configuration were used in this design space exploration. Notes on the aspects of xpres_test that were actually useful can be found in the output TIE file we looked at before. Near the beginning of the TIE we see

```
//
// This TIE requires the following configuration settings:
//
//   Required Endian: Little
//   Required Instruction Width: 64 bits
//   Minimum Data-Memory Width: 128 bits
//   Required Load/Store Units: 1
//   Requires Byte Enables: No
//   Requires Booleans: Yes
//   Pipeline Length: 5 stages
//
// This TIE was generated on a processor configuration
with the
// following ISA instruction options enabled:
//
//   CLAMPS
//   MUL32
//   MUL16
//   NSA/NSAU
//   MIN/MAX and MINU/MAXU
//   Sign Extend to 32 Bits
//   Enable Density Instructions
//   Enable Boolean Registers
//   Zero Overhead Loop Instructions
```

However, there are other features of the configuration not used in the particular djpeg TIE. If instead of starting with xpres_test, we had started with a more minimal configuration (denoted "Minimum" in the chart), we could have achieved the same performance but saved about 11,000 gates.

The optimal XPRES-generated TIE based on the xpres_test configuration turns out to be what is called xpres_18 TIE, adding 11,383 gates to the processor. It is then possible to "bake" this TIE file into the processor configuration. After we have explored cache configurations in the next section, we will combine this TIE together with the optimal cache choice into a new configuration. For the case where one wants to optimize a configuration to handle multiple codecs, it is possible to run XPRES using profiling based on all the codecs considered together.

6.5.4 Exploring the Processor Design Space: Cache and Memory

There are additional tools provided in the Tensilica ASIP toolset for design space exploration, including an energy estimation tool and a cache/memory exploration tool. Here we will examine the cache and memory exploration tool, which is a CAD (computer-aided design) tool rather than an EDA (electronic design automation) tool. XPRES with its automated exploration is more of an EDA tool; the cache/memory exploration tool assists the user in defining the design space to be explored and automatically runs all the profiles for each point in the space, allowing the various design points to be compared (Figure 6.10).

The experiments are based on defining data points using a definition panel shown here:

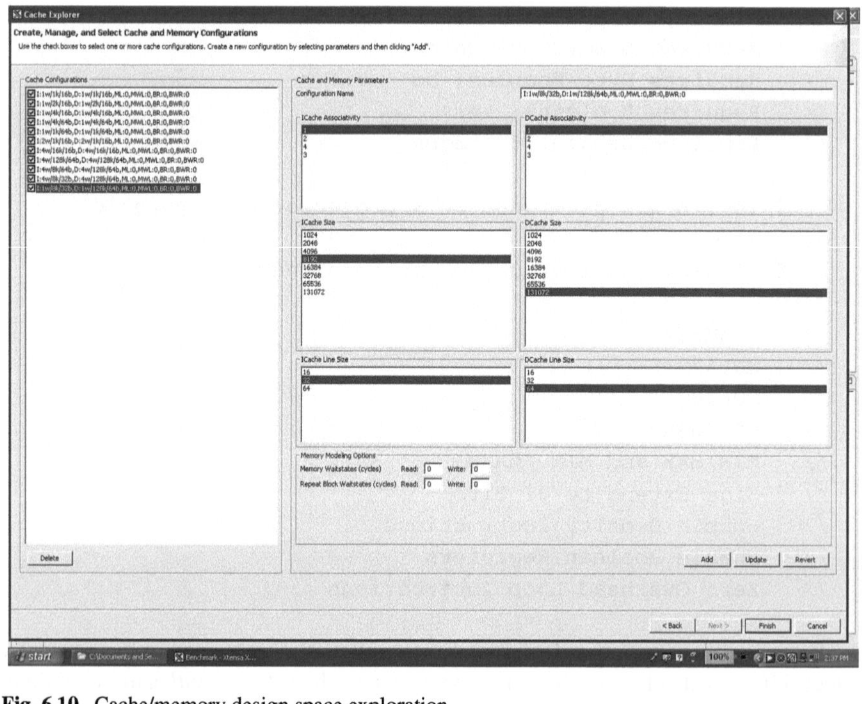

Fig. 6.10 Cache/memory design space exploration

There are six basic axes of variability for Instruction and Data cache (Icache and Dcache): three for each type: Associativity (1,2,3,4-ways); size (8 choices, from 1,024 through to 128 K bytes) and line size (16, 32, 64 bytes). This defines $4\times8\times3 = 96$ choices for each cache and thus $96^2 = 9{,}216$ experiments. It is also possible to define various memory delay parameters to place the cache fetches and writes into an overall system context.

For our xpres_test configuration (without added TIE) and djpeg application, we defined 21 different experiments to run and profiled each one. The comparison chart shows the results from all of these (Figure 6.11).

Cache configurations on the left-hand side have small data caches (the default 1 K cache, for example), or small caches with large line sizes (which take longer to fill but still miss quite frequently). Configurations on the right tend to have large data caches and vary the line sizes, associativity, and cache sizes. Clearly, too large an instruction cache will not be optimal in hardware cost terms if the working sets of the application do not require that size of Icache to achieve good hit rates. The two configurations on the far right achieve near-optimal total cycles with 4-way associativity, 8 K Icache, 128 K Dcache, 32 or 64 byte line size for Icache, and 64 byte line size for Dcache. Let us pick the 32 byte line size for Icache and the other parameters and explore combining these with the optimal XPRES-generated TIE from the previous section.

Of course, physical design considerations will be important in optimizing cache memory sizes as well, but for these experiments, let us assume that there will be no significant impact in choosing such a large data cache.

After configuring the resulting processor, which we called xpres_18_cache, we reran the djpeg application on it and compared its execution with our original xpres_test configuration (Figure 6.12).

This is very interesting. The performance of the cache-optimized configuration combined with the xpres_18_tie is almost identical to that of the configuration with small caches and the xpres_18_tie: 13,276,799 cycles vs. 13,270,215 cycles. Combining the TIE and larger cache configurations does not seem to be a good combination. One axis of optimization plays off against the other in such a way that it does not seem to be a reasonable combination. Almost all the gain in performance is the result of the TIE, rather than the cache, even if in isolation the cache optimization looked beneficial.

In fact, very large caches may be, in general, the wrong way in which to optimize a processor, as they come with a considerable memory area cost due to the large data cache. In addition, such a memory size tends to slow down the processor configuration due to its size. Later, we will compare some of these experiments against each other for estimated speed, power, area, and overall performance.

6.5.5 Exploring the Processor Design Space: Fine-Tuning

We are not finished with what can be done to configure an ASIP with these tools. One of the things that XPRES gives feedback on is vectorization of loop nests.

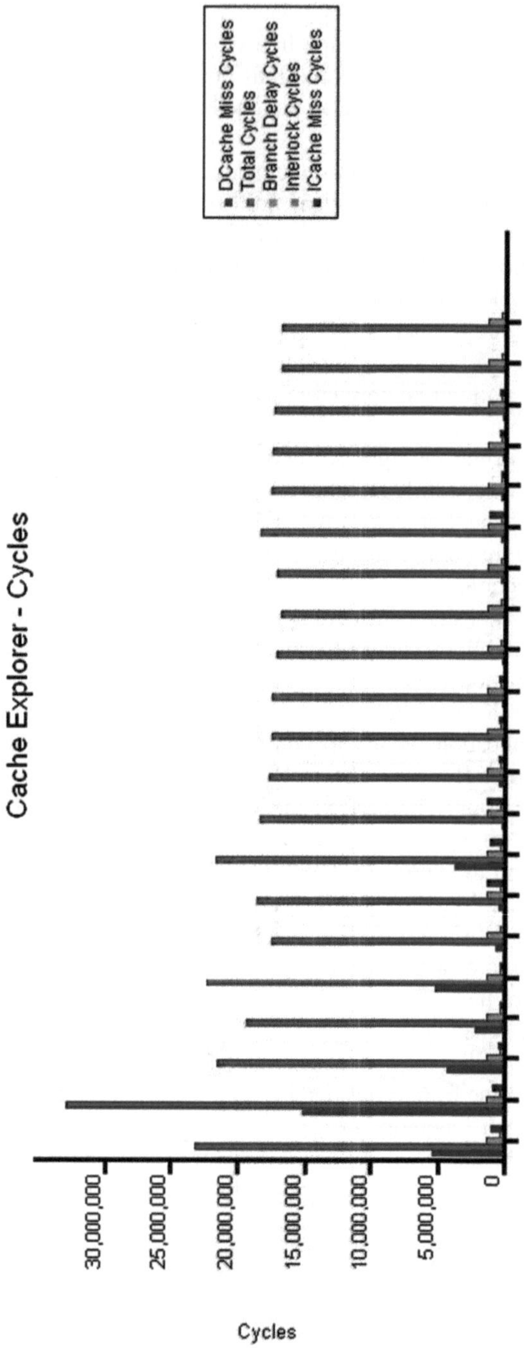

Fig. 6.11 Cache design space exploration

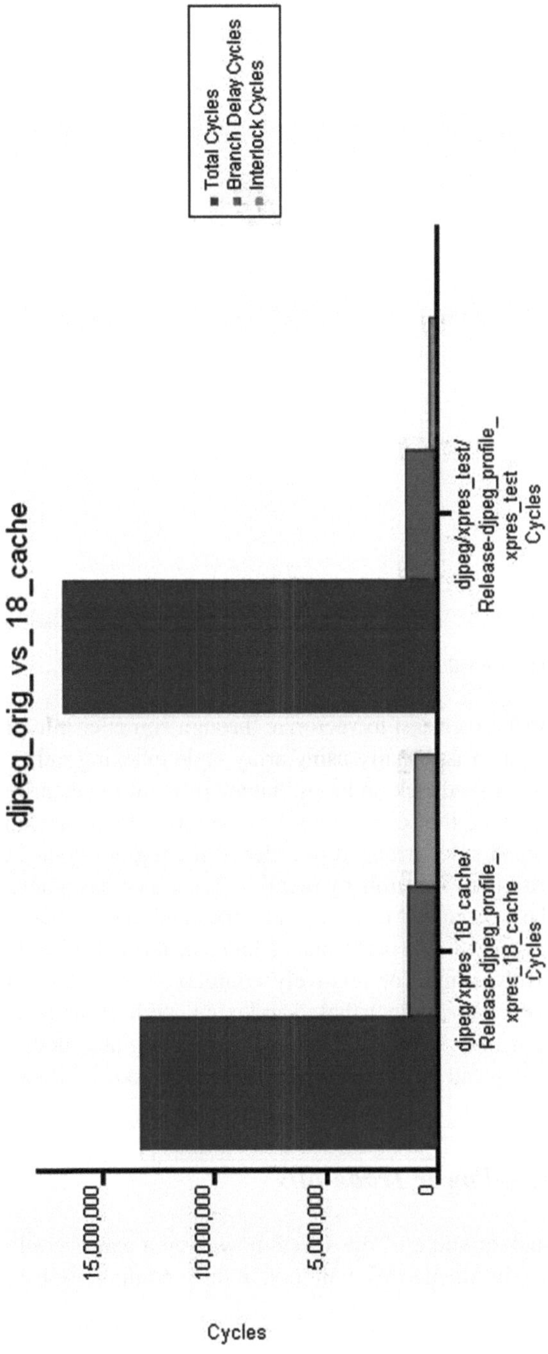

Fig. 6.12 Profiling combined TIE 18 with cache optimization vs. original

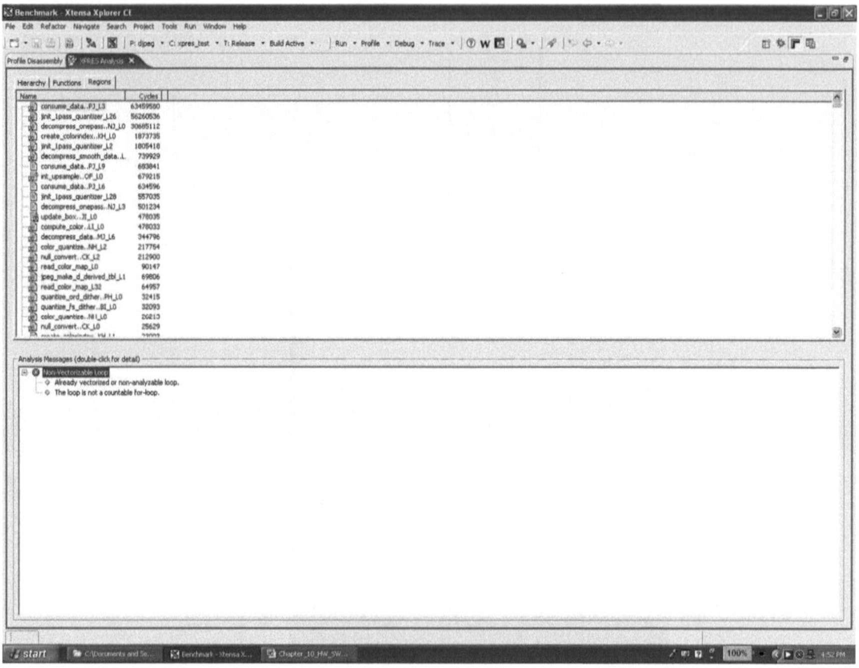

Fig. 6.13 XPRES analysis results

The type of loops that are easiest to vectorize through compiler inference are those with regular access patterns, usually using array-style indexing rather than pointer indexing. XPRES gives feedback on loops that are difficult to vectorize.

In Figure 6.13 regions that contain hard to vectorize loops are highlighted by a mark on the left-hand side, arranged in order of the regions contribution to total cycles. It is also possible to view this by function. Each loop has pointers to the specific reasons why the loop cannot be easily vectorized. All these areas are candidates for code restructuring to allow a performance increase through SIMD mechanisms. This may be relatively simple or relatively complex – often the reference code for an application may use pointer-based indexing, with copious variable aliasing, and improving it may require considerable rewriting and analysis. The code restructuring phase may end up overlapping with the full code productization phase.

6.5.6 Speed–Area–Power Trade-offs

Here we want to look at some of the speed–power–area and overall performance trade-offs from our experiments. We will look at three configurations:

1. A simple "example" processor, less complex than our xpres_test configurations (smaller instruction size, no FLIX, small caches)

2. The xpres test with the more optimal TIE added to it (earlier referred to as xpres_18_tie)
3. The combination of the cache parameters and the xpres_18 TIE
4. The xpres_test with just the cache parameters.

We utilize the estimators of speed, area, and power consumption provided with the processor generator, XPG, as guidance to the designer exploring various configurations. These are estimators, using statistical modeling and extrapolation, that try to predict the resulting characteristics. Of course, real post-synthesis, post-layout results may vary significantly. The process technology used for the estimates is a 130 nm CMOS process, LV (low-voltage) variation.

Here is a spreadsheet with these exploration results:

Configuration	Speed MHz	Core size mm^2	Core power mW	Memory area mm^2	Total area mm^2	djpeg cycles	Run time sec
Example	332	0.49	32.62	0.27	0.76	20845874	0.0628
xpres_test_18	312	0.85	38.75	0.49	1.34	13270215	0.0425
xpres_18_cache	257	0.95	33.5	6.51	7.46	13271799	0.0516
xpres_cache	257	0.95	33.5	6.51	7.46	16702894	0.0650

Let us look at these plotted along the various axes of comparison:

The first chart shows the four configurations vs. the maximum estimated speed (in the 130 nm LV process) (Figure 6.14). We see a minor drop-off in maximum speed when we add TIE and other changes between "Example" and "xpres_test_18" – changes that we hope will have a greater positive impact on performance than the drop-off in maximum speed. There is a much bigger drop-off when we increase caches (especially the very large Data Cache) in the latter two configurations on the right.

In Figure 6.15 we can see the estimated sizes for the four different configurations. Notice that the right hand two configurations with the large data caches are significantly larger than the two configurations on the left due to the very large memory sizes required for caches. Depending on the relative performance of the resulting configurations, this (plus the slowdown in maximum speed) would tend to bias us away from the large caches.

In this final figure we have computed and compared the application run time in seconds (parts thereof) for our four configurations, by dividing the number of cycles by the processor configuration speeds (Figure 6.16). Here we see clearly that xpres_test_18, with "normal" cache sizes, does significantly better than any of the other configurations. When combined with the area for the configuration (total area, including memory) as shown in the previous graph, xpres_test_18 looks like the optimal trade-off for performance, power, and speed.

Of course, it may well be that some very different trade-off of cache parameters, when combined with a different set of TIE, may have a significantly lower impact

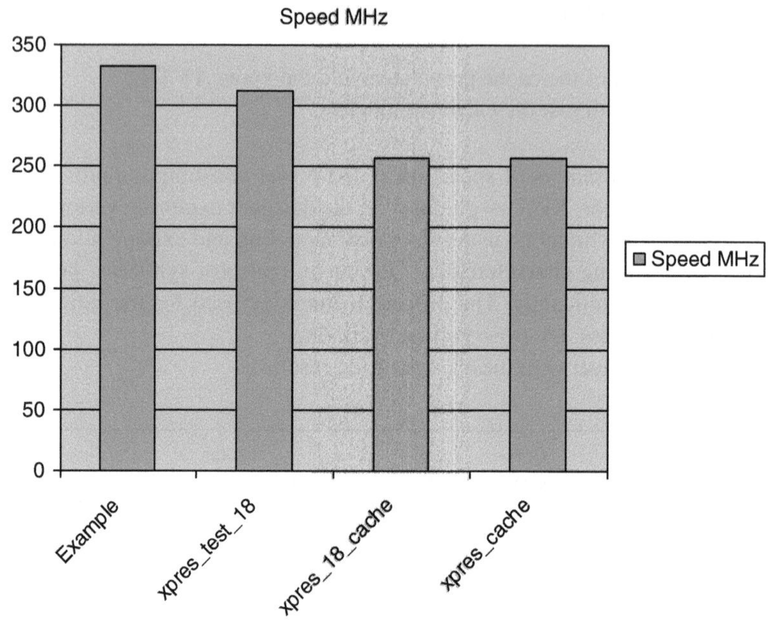

Fig. 6.14 Configuration vs. maximum estimated speed

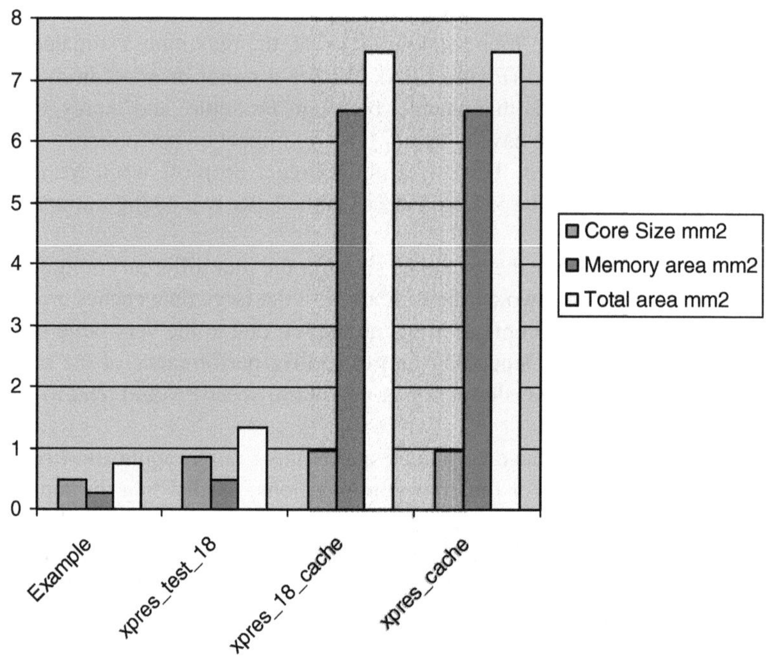

Fig. 6.15 Configurations with core, memory, and total areas

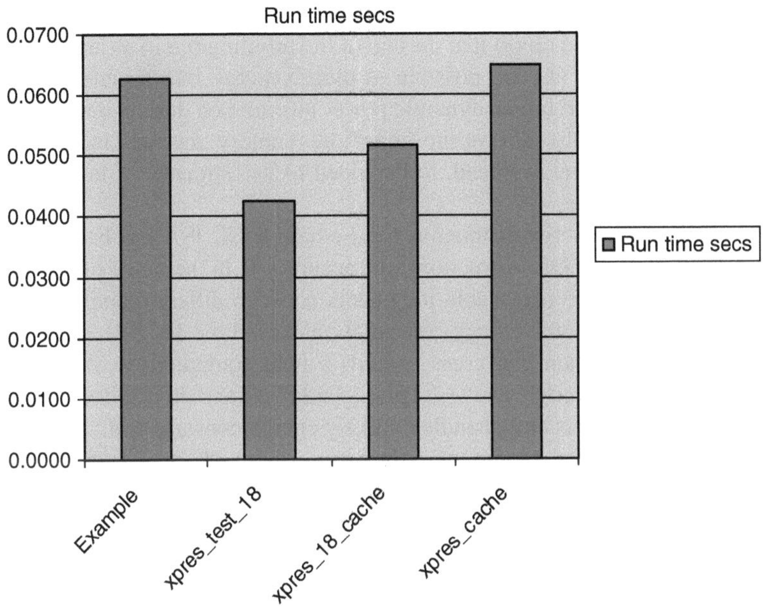

Fig. 6.16 Configurations vs. application run time in seconds

on overall speed and still produce enough of a performance gain that such a different configuration may be superior to the one that is best here. This is one reason why design space exploration is an art, not a science, and the toolset illustrated here allows many experiments to be devised and run quickly in order to get a good handle on the best ASIP paths to take.

6.5.7 Detailed Energy Space Exploration

Although the basic processor configuration generator provides an energy/power estimator to explore the energy space as one explores the configuration design space for an application or applications, it is essential to have a more precise energy estimator. At the same time, it is also important that the energy estimator runs at system level, as more accurate RTL, gate, or post-layout energy analysis tools have very long runtimes and are often unable to be driven (for a processor) with real software.

System-level energy estimators are still very much in their infancy, and system-level power/energy analysis frameworks are not common. Since energy estimation is extremely important in the configurable, extensible ASIP design space Tensilica developed a system-level energy estimation tool, Xenergy, that runs in conjunction with the ISS. While the base ISA for the Tensilica processor family can be estimated for energy consumption with a high degree of precision, the job of estimating energy consumption for arbitrary TIE is much more difficult. Xenergy relies on information generated during the TIE compilation process, to give it information about the resources consumed by each new instruction added to the ISA via TIE.

This then allows the energy estimator to compute on a cycle-by-cycle basis the energy consumption and report it at the end of the profiling run in a variety of ways. Theoretically it would also be possible to plug Xenergy into an energy analysis framework which would utilize dynamic power information during simulation.

The Xenergy tool also allows the impact of memory accesses in energy consumption terms, and bus overhead, to be added to its estimates of instruction-by-instruction consumption.

The accuracy for energy estimation for the base RISC ISA has been measured to be ± 3–5%. For the TIE instructions, the accuracy is in the range of ± 15–20%. This is enough to make reasonable judgments between different configurations if the differences in predicted energy consumption exceed the 15–20% error margin. Clearly, as a design team converges towards a final configuration, it can carry a candidate processor through synthesis, place, and route and do detailed post-layout analysis to get a more accurate handle on likely energy consumption.

To support the development of TIE functions which may improve energy consumption (as well as performance), the profiler allows a detailed energy consumption function profile to be generated. Figure 6.17 shows the profile for our xpres_test_18 configuration. Rather than show the complete function list, we will show it in order down to the cumulative 99.99% energy point.

Specialized TIE may be developed for the function energy "hot spots" (literally!). Such TIE instructions may both improve performance and reduce energy consumption, since there is often a reasonable correlation between these two objectives.

We can compare the total energy consumption, divided into key components, for the four configurations discussed earlier. Figure 6.18 shows the profile data for all four running jpeg decode.

The various candles show the different contributions to overall energy consumption. The two configurations on the left show the two configurations with large cache – the medium instruction cache and large data cache. (The one on the left also has the XPRES-generated instructions.) There is a reasonably large difference between total energy consumption, although falling within the bounds of possible error. Note that the large caches contribute to a high proportion of leakage energy for these configurations.

However, the two configurations on the right – the simple "Example" and the xpres_test_18 configuration – have greatly reduced energy consumption – well outside the 15–20% error margin. From an energy basis there is not much to choose between them, but xpres_test_18 shows up as the best of the four, and when combined with the earlier superior performance on the djpeg application clearly shines as the right choice.

6.6 Software Implementation

Production-worthy software implementation for the final product is often done in the same IDE used for processor configuration and design space exploration. The Eclipse-based CDT (C/C++ Development Toolkit) gives the developers a way to

	Total energy (%)	Function energy (pJ)	Cumulative energy (%)
pass2_fs_dither..BJ	21.74	450,982,453	21.74
jpeg_idct_islow	19.38	402,082,550	41.13
fill_inverse_cmap..PI	17.81	369,464,185	58.94
decode_mcu..KK	10.35	214,729,642	69.29
put_pixel_rows..KB	8.79	182,451,347	78.09
ycc_rgb_convert..BK	6.44	133,696,705	84.54
prescan_quantize..GI	3.9	81,035,138	88.44
jpeg_fill_bit_buffer	2.93	60,831,270	91.38
update_box..JI	2.18	45,352,950	93.56
memset	1.19	24,732,307	94.75
decompress_onepass..NJ	1	20,822,353	95.76
memchr	0.99	20,672,787	96.76
memmove	0.74	15,392,723	97.5
finish_pass1..DJ	0.64	13,388,287	98.14
sep_upsample..LF	0.13	2,765,109	98.28
jpeg_huff_decode	0.13	2,762,251	98.41
__sfvwrite	0.12	2,664,111	98.54
start_output_gif..JB	0.11	2,415,456	98.65
memcpy	0.11	2,392,590	98.77
compute_color..LI	0.1	2,203,169	98.88
access_virt_sarray..CH	0.1	2,167,650	98.98
jpeg_make_d_derived_tbl	0.1	2,124,890	99.08
_WindowOverflow8	0.1	2,105,001	99.18
_WindowUnderflow8	0.09	1,989,921	99.28
post_process_prepass..IF	0.08	1,665,160	99.36
post_process_2pass..JF	0.06	1,333,488	99.42
main	0.04	893,475	99.47
put_3bytes..HB	0.04	862,916	99.51
jpeg_start_decompress	0.03	761,999	99.55
_malloc_r	0.03	752,501	99.58
process_data_simple_ main..BL	0.03	693,855	99.62
fwrite	0.02	620,751	99.65
_free_r	0.02	578,386	99.67
fflush	0.02	503,441	99.7
flush_packet..BB	0.02	488,205	99.72
_fwalk	0.02	422,665	99.74
jinit_color_deconverter	0.01	390,753	99.76
process_data_crank_post.. CL	0.01	330,816	99.78
fullsize_upsample..MF	0.01	329,250	99.79
get_dht..ID	0.01	292,808	99.81
__swrite	0.01	271,877	99.82
get_dqt..JD	0.01	247,650	99.83
keymatch	0.01	212,888	99.84
__srefill	0.01	210,138	99.85
alloc_small..JG	0	192,362	99.86
lflush	0	182,769	99.87
_write_r	0	176,877	99.88
alloc_large..KG	0	155,634	99.89
alloc_sarray..LG	0	139,200	99.89

Fig. 6.17 Energy profile for configuration xpres_test_18

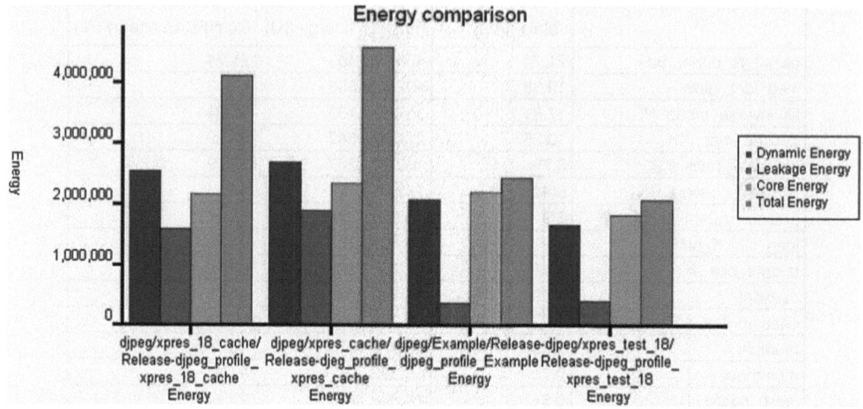

Fig. 6.18 Energy comparisons of four configurations running jpeg decode

collaborate within the XX-Eclipse environment, to partition the application into separate libraries and projects, to assign standardized sets of build properties, to integrate their various efforts into the full application, and to manage externally built projects and libraries. Because the IDE allows the software to be continually simulated and re-profiled using the ISS (in both cycle-accurate and turbo modes), the developers are able to continually revalidate that their software will be both functionally correct on the target processor and also measure its performance on a regular basis.

Of course, half the battle in developing software is debugging it when it goes wrong; we will discuss the debugging capabilities offered by the processor IDE and modeling environments later.

XX contains certain links to externally created models (using either SystemC or a proprietary internal C/C++ modeling environment) that allow simulations involving Tensilica cores to be controlled and debugged from within the XX environment. This allows the validation of software in a more elaborate system model context, which may bring in more realistic system timing effects to interact with the target software. This can be very useful in developing working target software whether for a single-core or multicore system. We will discuss this further in the next section.

6.7 Predicting Software Performance via Sampling

The simulation technology in the Tensilica ISS has some interesting capabilities that expand its range of use. One example is a statistical sampling technique that combines the TurboXim fast-functional simulation and the cycle-accurate modes of the ISS. This can allow significant amounts of real software to be run on a target configuration ISS in the turbo mode, but by using sampling, an accurate prediction of its performance in cycle-accurate mode can be made. We will illustrate this by running audio decode software (an mp3 decoder) on the audio processor DC_B_330HiFi configuration mentioned earlier. The test stream used was an approximately 5 min song ("Phantom Limb" by The Shins).

Statistical sampling to speed up analysis and simulation is not new (one recent reference is [8]). However, it is not available in many commercial software development environments. The Xtensa sampling capability has been used in a number of development projects with good accuracy.

Sampling runs the target ISS in both fast-functional and cycle-accurate modes for sequences that are biased to the fast-functional mode. For example, the default for the sample option is to run in cycle-accurate mode for 1,000,000 instructions and then to switch to turbo mode for 100,000,000 instructions. At the end of the simulation, the ratio between the number of instructions and the number of cycles for the cycle-accurate part can be used to predict the overall application software number of cycles. This can end up being surprisingly accurate as we will see.

Options are provided to control the sampling behavior. The ratio between the numbers of instructions in turbo vs. cycle-accurate mode can be varied. In addition, the length of each cycle-accurate simulation sequence can be modified from the default 1,000,000 instructions. In this example, we will modify the ratio but leave the cycle-accurate sequence length alone.

The table illustrates the simulation wall clock time, the number of simulated instructions for turbo and cycle-accurate modes, the number of cycles reported for the cycle-accurate portion, the predictive ratio of cycles/instruction based on cycle-accurate mode, and the predicted overall number of cycles. This is reported for two baseline runs – 100% turbo, 100% cycle accurate, and three hybrid sampling runs, with the ratio of turbo to cycle-accurate sequence length being 50, 100, and 200. We will also show the prediction error (where 100% cycle accurate is the comparison point). A simple memory model was used for the cycle-accurate portions, which models cache misses and system and local memory accesses. TurboXim does not model cache and system memories – it essentially treats all memory as an infinite cache that always hits. This is one reason for speed, but one reason why it is not a good predictor of real cycles.

Although there is some variance in the experiments, a few observations can be made. First, the turbo simulation is about 25× faster than cycle accurate. The hybrid sampling simulations add some time as one would expect, but still are much faster than 100% cycle accurate. The three sampling runs have predictions for number of total cycles that are all less than 1% error (it is unclear why the error increases slightly as the amount of cycle-accurate simulation increases but some amount of variation is to be expected). The default ratio of 100 to 1 for turbo sequence to cycle-accurate sequence is a good compromise on speed and accuracy. Although this is just one simple example, it is clear that good performance predictions are possible with high-simulation speed through careful use of well-calibrated statistical sampling.

This can be a very valuable tool for production software development, where overall performance can be predicted for large samples and scenarios run quickly in an accelerated simulation environment.

Type of run	Simulation time (sec)	Number of instructions in turbo	Number of instructions in cycle accurate	Cycles for cycle accurate	Total instructions	Predictive ratio	Predicted cycles	Error
Turbo	408	2,308,087,489			2,308,087,489			
Sample 200	519	2,302,672,559	11,000,000	15,622,406	2,313,672,559	1.4202	3,285,877,768	0.14%
Sample 100	534	2,292,541,235	22,000,000	31,246.496	2.314,541,235	1.4203	3,287,342,916	0.19%
Sample 50	708	2,270.095.655	45,000,000	63,951,086	2.315.095,655	1.4211	3,289,982,435	0.27%
Cycle accurate	11,021		2,309,744,260	3,281,161,346			3,281,161,346	O %

6.8 Multicore Issues

We have concentrated so far on optimizing a single ASIP for an application and optimizing the application to make it more ASIP suitable, as well as turning reference code into a final product using all the capabilities of the IDE.

However, there are some times when a single ASIP is not enough. Or, a single ASIP may need to run at too high a frequency or using a higher supply voltage than is optimal for minimizing the energy consumption of an embedded product. With this in mind, it is time to think about using multiple processors for a product. In particular, we will think about using multiple ASIPs for an application – what is called heterogeneous asymmetric multiprocessing (AMP). However, there are other possible multi-processor architectures that could be considered.

In particular, the last few years have seen a growing interest in multicore and manycore architectures [9-11], These are only loosely defined terms. When many people say "multicore," they mean symmetric multiprocessing (SMP) architectures with a cluster of homogeneous processors that are cache coherent, usually with hardware assistance. SMP multicore is the kind of multicore that most people are talking about. In particular, processor vendors such as IBM, Intel, AMD, and Sun have been advocating the use of 2-, 4-, 8-, and upcoming 16-way SMP multicore devices as the basis for laptops, desktops, and larger servers (the large servers scaling potentially to hundreds or thousands of cores in a large box). "Manycore" usually refers to devices with several tens of processors scaling to several hundreds – and some people are talking of thousands of processing cores on a single device. Sometimes manycore devices are homogeneous, although not necessarily SMP; sometimes they use a master control processor and have an array of homogeneous processing engines. Rather than use hardware cache coherency and totally regular or bus-based interconnect, they might use specialized interconnect such as network-on-chip meshes. However, currently most manycore devices are either research devices, emerging new devices from new IP companies, or emerging from graphics devices and being applied more widely.

In the embedded SoC domain, however, most multi-processor approaches are heterogeneous or a combination of simple SMP together with specialized cores such as ASIPs. The classic cell phone baseband control processor (e.g., an ARM) and DSP is now almost a cliché of heterogeneous multi-processor embedded products. These days, mobile wireless devices may include specialized ASIPs for audio and video processing as well as control processors and DSPs. In the generalized control domain for mobile embedded devices, small SMP clusters (2-way and 4-way) are beginning to emerge: for example, the ARM MPCore architecture. These multi-way SMP cores are potentially a good way to offer control processing for general applications (as opposed to dataplane-intensive applications such as baseband, audio, and video processing), because cores may be shut down when not needed; tasks can easily migrate from core to core if paused (e.g., for memory access) and restarted; tasks or threads can be migrated for load sharing and rebalancing; and because real-time performance is not usually an issue with these applications. The focus on SMP multicores in recent years has encouraged developers to develop multi-threaded applications that can take advantage of these parallel processors.

However, general multicore processors are not efficient in energy, area, and performance for the highly specialized dataplane tasks. Furthermore, although there have been many efforts in recent years to develop general programming models for parallelizing applications – often provided as APIs and libraries – there has been little progress in automating the creation of concurrent programs – despite many decades of research. Multi-threading models with simple concurrency mechanisms for thread creation and synchronization currently represent much of the state of the art, although researchers continue to work on this problem.

In the dataplane, creating concurrent architectures to exploit multiple ASIPs and asymmetric multiprocessing is a more ad hoc art. Nevertheless, there are practical ways to proceed. In particular, in the development of a multicore, heterogeneous AMP video subsystem, a very practical methodology was developed to partition the processing into multiple ASIPs, to optimize the ASIP architectures, and to allow programming of new video codecs onto the now fixed multi-ASIP architecture after it was specified. We will use this as an example of a practical methodology for multicore.

6.8.1 A Practical Methodology for Multi-processor ASIP Definition and Programming

Figure 6.19 shows a two core video subsystem. This subsystem is meant to operate with a third processor – a system controller, along with a system memory linked to the system controller, DMA memory port, and video system memory port. Of course, the complete embedded product may include many other subsystems as well, such as an audio subsystem.

The video subsystem itself consists of two ASIPs specialized for different parts of the video decoding and encoding task. Each of these is a configured, extended Xtensa processor core – a stream processor and a pixel processor. The video subsystem also includes a multichannel DMA controller and other associated hardware.

The decision made to partition the video processing requirements into two cores (as opposed to one, three, four, or more) was made based on video reference code and processing requirements for standard image formats up to standard definition (D1 – from 704×480 to 720×576) resolution.

The methodology for defining the subsystem into two ASIPs, and configuring the ASIPs with the right instruction extensions, consisted of taking video reference code and running it on a generic processor, using extensive profiling. This identified computational and communications kernels ("hot spots") that were candidates for acceleration. It also identified clusters of computation that could be considered to be mapped onto a single core based on the sequence of computations being carried out in the video reference code and the type and amount of data being passed (or referenced) between various video functions.

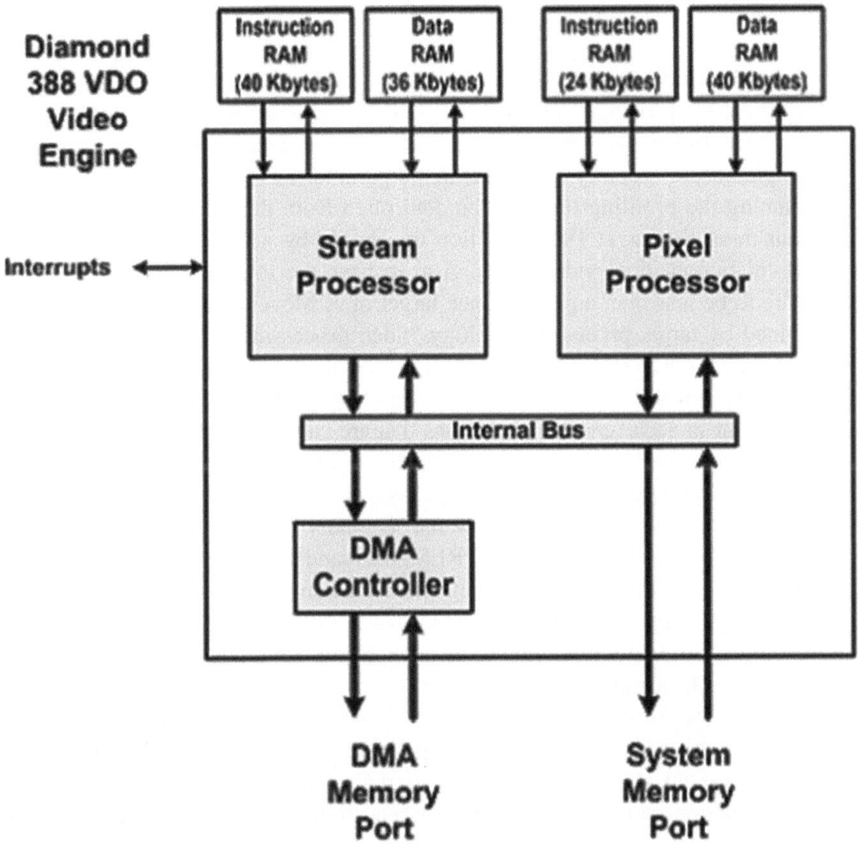

Fig. 6.19 Tensilica 388 VDO video subsystem

Note that the structure of the reference code can have a very large influence on this analysis, and design teams need to take care that they consider code refactoring and restructuring in order not to be overly biased by the reference code in designing a system. Moving some functions or computations around may have a large impact on data locality and avoid the need to move or copy data. Similarly duplicating some functional code may actually lead to a speedup in performance especially if variations in the function are required at different points in the video processing.

After identifying hot spots and cut lines, the design of a system of ASIPs will begin to fall out from the analysis. (In general, despite many years of research, very few automated algorithms exist to do this effectively.) Designers can then explore speeding up the hot spots with instruction extensions and other configuration options such as FLIX and SIMD, using tools like XPRES, sets of predefined instructions, and manual analysis and TIE creation and optimization. Then the team would complete the virtual partitioning across the multiple ASIPs, taking care to load balance the result. A multicore simulation model can be created to test the system structure and to be used as the basis for final code porting and optimization.

In detail, the steps are as follows:

1. **Compile and profile** the reference video codec(s) on a single starting core configuration (perhaps a very simple one or a predefined core), using the cycle-accurate ISS model with realistic input data for detailed profiling. Assume a single access memory and set realistic parameters for memory latency when running the profiling (this can be controlled from the XX environment when launching the target ISS). Partition the profile by assigning functions to one of 'n' hypothetical video cores. Aim to have the total amount of cycles per core to be less than a performance target of X Mcycles/sec, where X is determined by target process technology, video processing rate, and video stream characteristics. This emphasizes the need for using realistic input test streams.
2. **Identify hot spots:** The execution profile reports how many processor cycles are spent in each function. Functions that are called frequently or require high processing bandwidth are considered hot spots and are good candidates for acceleration by using the specific TIE instructions extensions.
3. **Create candidate TIE instructions,** manually or using technology such as XPRES, or a combination of XPRES and manual modifications. Compile the TIE using the TIE compiler, which will create an include file of TIE intrinsics. Modify the application to call TIE intrinsic functions in place of the software code that originally represented the hot spots of the application.
4. **Verify TIE usage:** after modifying the source code using TIE intrinsics, compile and simulate the application. Verify the results of the TIE-enabled application against the results of the original source code. Profile the application again to check if optimization goals are reached for hot spots. Once basic optimization goals are reached, it is time to partition the code for multiple ASIPs in order to meet overall system performance needs.
5. **Identify cut points** where functions and code can be refactored to map into separate tasks running on separate ASIPs. At these cut points, it is very likely that data will need to be moved between ASIPs. There are several mechanisms that can be used for data movement, including DMA between local memories, shared local memories, use of system memory, and direct FIFO queues between processors. In the case of Xtensa, direct HW FIFO connections can be created with TIE. This can be modeled in the code with calls to push and pop routines which model FIFO channels. For the other implementations, DMA, shared memory, system memory can all be modeled in the application using memcpy functions to copy chunks of data from one structure (representing one processor) to a different structure (representing DMA into another ASIPs local memory or writing into shared memory). Use memcpy, perhaps in a standard API library, to represent data movement. Verify TIE and data movement C code.
6. **Compile and profile** the codec on the single-core ISS with the data movement functionality now implemented with C functions in an API library. Assume a single access memory.
7. **Partition functions for multicore.** Aim to balance the computational requirements of each core, including time required for data movement. Refactor the code: completely separate the code destined for each core by placing them in

different functions. This will also define the TIE instructions required for each ASIP. It is possible that some TIE functions will need to be duplicated on more than one ASIP. Aim to have the total amount of cycles per core to be less than the computed performance requirement of X Mcycles/sec. If they meet the performance, go to the following step; otherwise return to Step 2 (identifying hot spots) for further optimizations.

8. **Build** the separate ASIP configurations as defined above. Once each configuration has been generated, build a multicore system-level model using SystemC or other modeling approach (this will be described later). Tensilica offers two modeling methods – a proprietary C-based modeling library called XTMP and a SystemC-based library called XTSC.
9. **Modify** the target code for each ASIP to incorporate real data movement between cores. As described earlier, data movement has been modeled first with memcpy operations, wait delays, and library-based implementations of memcpy and HW FIFO queues. Now the real mechanisms must be substituted for the modeling approaches. This target code can be compiled and run on each ASIP and within the context of the system model built in the previous step.
10. **Profile** the multicore partitioned application on the system model. This will use real models of the data movement implementation and the TIE extensions to speed up each task. The results of this profiling should be quite accurate and at this point, further optimization of the task code on each ASIP is possible.
11. At this point, we have a reference implementation partitioned among the number of ASIPs desired for the target system and each ASIP has a basic configuration defined, as well as a system model incorporating the individual ASIPs and models of the system communications mechanisms. The reference code used to define this system can be further developed to turn it into real production code. In addition, new codecs can be ported to this system using a derivative methodology, which will be discussed later.

6.8.2 Developing Multicore System-Level Models

As mentioned, Tensilica has two system-level modeling environments, XTMP and XTSC. Both can be used to develop a system model, and XTSC has been used to interface the Tensilica ISS to external ESL system modeling tools. Because these modeling libraries use the Tensilica ISS as a basic element, they allow both cycle-accurate and fast-functional (Turbo-style) execution. Later we will show how to invoke the system model from the XX environment and use its multi-processor debugging capabilities for control.

More information on these system modeling approaches can be found in [12].

6.8.3 Porting Methodology for New Video Codecs to the Multicore system

Once a multi-ASIP system has been defined for a particular application, the development methodology defined above can be used to port new applications to this system. We developed a simple derivative methodology for the video system to support porting.

This methodology started with the creation of a single-core, artificial configuration model that combines most of the functionality of the multicore ASIP-based system. For example, this single processor core includes all of the application-specific TIE instructions of the multi-ASIP system. It does not model any DMA controller or memory subsystem of the multicore system. However, it can include any HW FIFO interfaces from the multi-ASIP system, looping back into the core.

This single-core configuration model is the starting point for porting an application to the real multi-processor system. Again, we assume we are starting with reference C-code for a new video codec, that we want to port to the multi-processor system. Since the methodology is a derivative one based on the original system definition methodology, we will summarize it briefly here:

1. **Compile and profile the codec** on the single-core ISS model with realistic input data. Assume a single access memory. Partition the profile by assigning functions to one of the video cores. Since in 130 nm technology, a reasonable performance target for video codecs is 200 MHz, aim to have the total amount of cycles per core to be less than 180 Mcycles/sec, in order to keep a reasonable margin.
2. **Identify hot spots:** The execution profile reports how many processor cycles are spent in each function. Functions that are called frequently or require high processing bandwidth are considered as hot spots and are good candidates for acceleration by using the VDO-specific TIE instructions.
3. **Instantiate VDO TIE intrinsics:** VDO TIE instructions are available for use by C/C++ programs through intrinsic functions. Modify the application to call VDO TIE intrinsic functions in place of the software code that originally represented the hot spots of the application. The VDO TIE instructions from different cores should not be mixed in the same function because these will be executing on separate processors after multicore partitioning.
4. **Verify TIE usage:** After modifying the source code for TIE usage, compile and simulate the application. Verify the results of the TIE-enabled application against the results of the original source code. Profile the application again to check if optimization goals are reached for hot spots.
5. **Implement DMA functionality** with C functions. Verify TIE and DMA C code.
6. **Compile and profile** the codec on the single-core model with the DMA functionality now implemented with C functions. Assume a single access memory.
7. **Partition functions for multicore.** Completely separate the code destined for each core by placing them in different functions. Ensure that the functions destined for each core use only TIE instructions belonging to that core (i.e., Stream Processor TIE instructions or Pixel Processor TIE instructions in the 388VDO). Aim to have the total amount of cycles per core to be less than 180 Mcycles/s, to target a 200 MHz frequency for the system with reasonable margin. If it meets the performance, go to the following step; otherwise return to Step 2.

8. **Data sharing for multicore:** Each processor in the dual-core 388VDO Engine has separate local instruction and data memories. The 388VDO uses shared external system memory and a hardware DMA engine for data movement between cores. Because the single-core virtual model does not contain a DMA engine, data movement between the two cores can be simulated using memcopy functions. Those functions can then be replaced by DMA control functions when the code is ported to the VDO Engine.
9. **Profile the multicore partition:** Profile the application to make sure the cycle counts for the functions that will be executing on different cores is balanced. Note that the memcopy cycle counts may be removed since these functions will be performed by the DMA controller of the VDO Engine. These profile results will still be an approximation of the final performance since some overhead will be introduced for synchronization and communication between the two cores when the application is ported to the VDO Engine.

It is important when using this porting methodology to verify on an ongoing basis that the functions are properly mapped to the right core. Designers need to verify that only TIE intrinsics from a particular core are being used within a single source code file. A simple way to check the use of the correct TIE instructions is to use a C #define to indicate the appropriate TIE file to use and set the value of this #define as a compile-time parameter set in the compile instruction. By toggling the value of such a define over a set of source code files, it will be relatively easy to determine which of them uses instructions from both VDO cores vs. which of them uses instructions from only one core. Those source files containing functions that use instructions from both cores may gradually be split up into new source files that are core specific, and migrated from the single-core software directory or project into core-specific directories or projects.

6.8.4 Using the IDE for Multicore Simulation and Validation

Figure 6.20 shows the XX IDE running a multicore system simulation. In this case, it is a three core simulation using the 388 VDO subsystem and a system controller, built using the XTMP modeling environment, and able to be run in both cycle-accurate and turbo modes. The MP launch option is used in XX, and the console window shows the output from running a video codec on the multicore model. In this case, it is an H264 decoder running on a small sample stream. The figure shows the simulation after two frames have been decoded. In addition, this is being run in turbo mode which runs about 20–40 times faster than the cycle-accurate mode. The turbo model still incorporates a model of the HW DMA engine which is part of this system, as well as the external controller and multi-port system memory and controller.

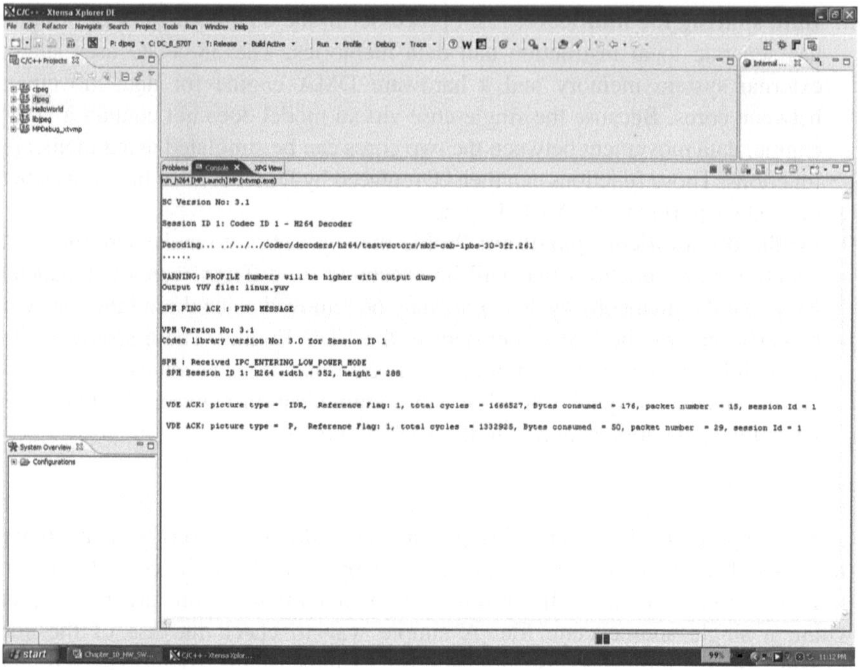

Fig. 6.20 Multicore model simulation

6.9 Debug

6.9.1 Single-Core Debug in the IDE

Figure 6.21 shows XX for a single-core debug session using the djpeg application compiled as a debug target, for the 570T predefined core. The upper left window shows the process stack; currently the execution is suspended in the prescan_quantize function in the jquant2.c source file. The middle left window shows the exact line in the current source file where the execution is suspended. The right-hand middle window is the disassembly view, and the top right window shows the current set of variables and their values. The console window is at the bottom. Various controls are offered to single step, step over, into, and step return from each instruction to set breakpoints, capture executions traces, examine registers, etc.

6.9.2 Multi-processor Debug in the IDE

Figure 6.22 shows XX running the multicore XTMP model of the video system containing three processors – a control processor (the system controller) – in this case, a DC_B_330HiFi_be, a big endian version of a preconfigured audio processor that

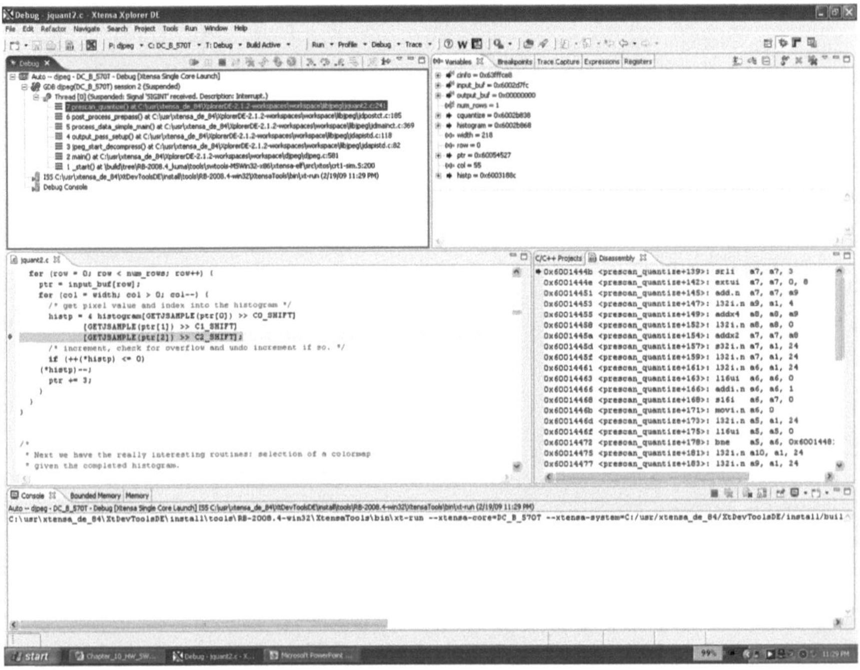

Fig. 6.21 IDE debug view of single core

can also run generalized control functions; and the two cores of the video subsystem. The XTMP model also models a system memory controller with three ports, the system memory, the local memories in the video subsystem, a DMA engine in the video subsystem, the video interconnect network, and special hardware assistance for certain video instructions that is part of the video subsystem. This model can be run in either cycle-accurate or turbo mode. It is possible to dynamically switch between turbo and cycle-accurate modes through target source code commands to the ISS that are trapped and executed in just a few cycles.

This mode switching allows the model to boot up in turbo mode and run in fast simulation to a point where a problem occurs in the target code, at which point, the switch to cycle-accurate simulation may be needed to debug the problem in detail.

The simulation is synchronized – all the cores advance one cycle at a time and when any core is stopped, the whole simulation stops. It is also possible to write a model where cores run asynchronously (as they might on an emulation board or in real hardware, where a core may be put into a suspended mode while the rest of the system continues to clock), but in general, for debugging multi-processor systems with tight interactions, this is not very useful and can be confusing at times. The process of debugging video codecs on the multicore model relied on synchronized execution.

The simulation control panel in the upper left of the XX IDE display shows all three cores and their status (running, suspended due to interrupt, suspended at

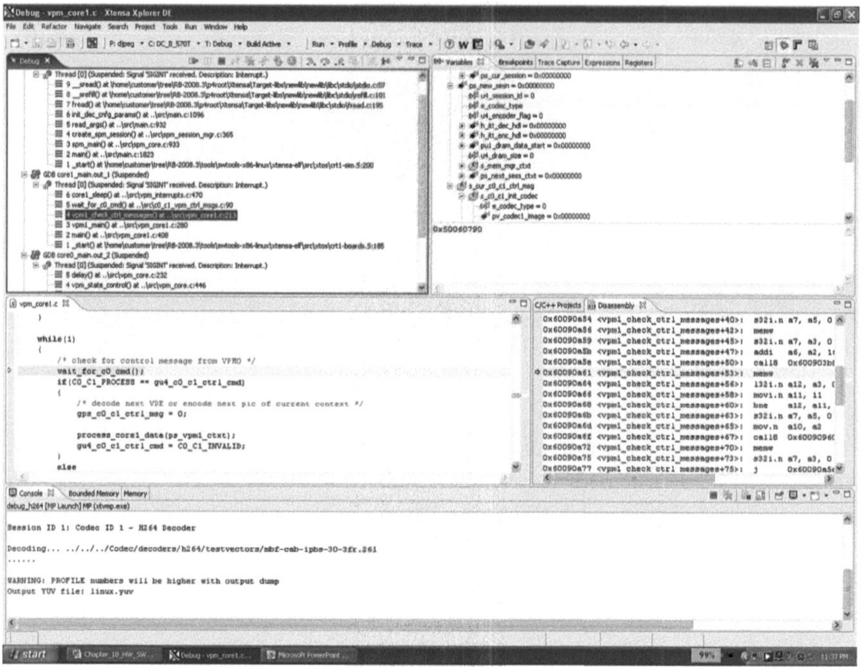

Fig. 6.22 Multi-processor debug of Video subsystem

breakpoint, etc.). The locus of control can be switched from one core to another by selecting it. When this happens, the current execution context is switched to the core that is being focused on. Thus, the source code display in the middle left panel, the disassembly view in the middle right, the variable display in the upper right, etc., will all switch to the core that has been selected. To get a view of each core's state and what they are executing is as simple as selecting each core in turn in the simulation control view. The simulation control also shows the calling stack of procedures for each core and the execution context can be switched to any process on the stack. When this is done the source code, disassembly, variable, and other views will switch as well.

Use of the XX IDE for debugging can be a powerful way to solve problems with the code and complex multicore synchronization issues and has indeed been used as such with the video system and its codecs.

There are also commercial ESL tools and IDEs with multicore simulation and debug capabilities that go beyond the processor offerings of a single vendor (and, for example, the Tensilica ISS is available in ESL tools from CoWare [13], Carbon Design Systems [14], VaST [15], and Imperas [16]), but that is beyond the scope of this chapter. We illustrate below a snapshot from an example system in the CoWare Platform Architect tool and below that from the Carbon Design SoC designer tool to illustrate what is possible (Figures 6.23 and 6.24).

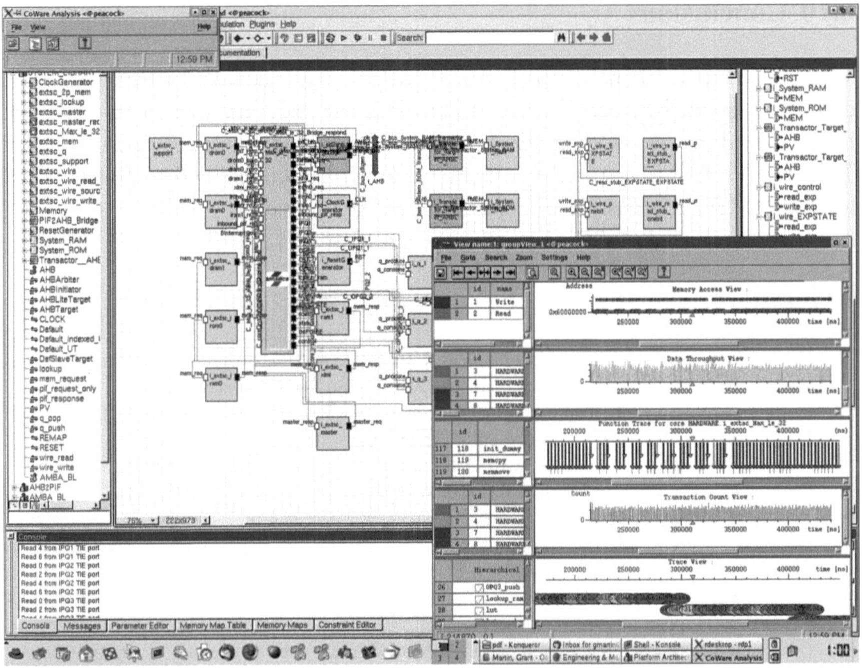

Fig. 6.23 Example of processor-based system modeled and analyzed in CoWare Platform Architect tool

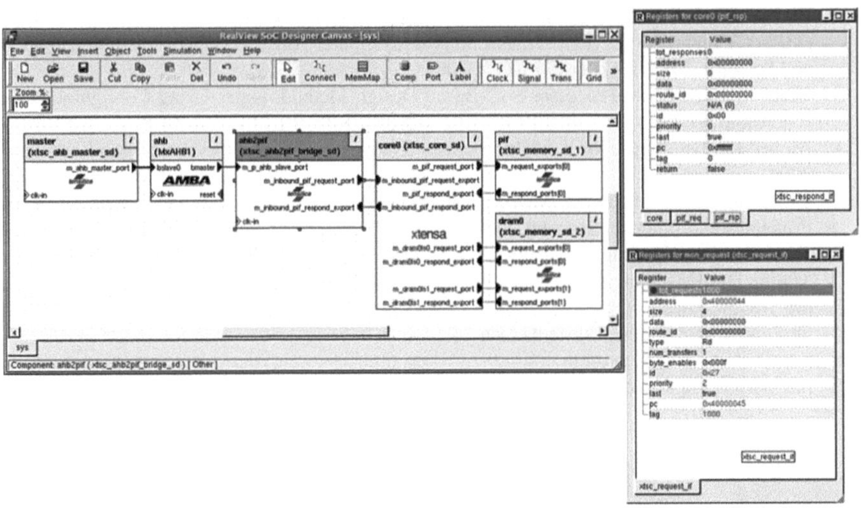

Fig. 6.24 Example of processor-based system modeled and simulated in Carbon Design Systems SoC Designer

6.10 Conclusions

We have illustrated in this chapter a variety of practical mechanisms for a processor-centric design methodology as part of an overall ESL methodology. In particular, we have focused on the use of application-specific instruction set processors using the Tensilica ASIP IP as an example. Many of the methods illustrated could be applied to other ASIPs, although details will of course vary.

References

1. Customizable Embedded Processors: Design Technologies and Applications, edited by Paolo Ienne and Rainer Leupers, Elsevier Morgan Kaufmann, San Francisco, 2006
2. Chris Rowen, Steve Leibson. Engineering the Complex SOC: Fast, Flexible Design with Configurable Processors, Prentice-Hall PTR, 2004.
3. Steve Leibson. Designing SOCs with Configured Cores: Unleashing the Tensilica Xtensa and Diamond Cores, Elsevier Morgan Kaufmann, 2006.
4. http://www.eclipse.org/
5. Brian Bailey, Grant Martin, Andrew Piziali. ESL Design and Verification: A Prescription for Electronic System Level Methodology, Elsevier Morgan Kaufmann, 2007.
6. David Goodwin, Darin Petkov. Automatic Generation of Application Specific Processors, CASES 2003, pp. 137–147.
7. David Goodwin, Steve Leibson, Grant Martin. Automated Processor Configuration and Instruction Extension, chapter 6 in [1]
8. Roland Wunderlich, Thomas Wenisch, Babak Falsafi, James Hoe, "Statistical Sampling of Microarchitectural Simulation", ACM Transactions on Modeling and Computer Simulation, 16(3): 197–224, July 2006.
9. Wayne Wolf, Ahmed Amine Jerraya, Grant Martin. "Multiprocessor System-on-Chip (MPSoC) Technology", IEEE Transactions on Computer-Aided Design of Integrated Circuits and Systems, 27(10): 1701–1713, Oct. 2008.
10. Grant Martin. "Multi-Processor SoC-Based Design Methodologies Using Configurable and Extensible Processors," Signal Processing Systems, 53(1–2): 113–127, 2008.
11. Ahmed Jerraya, Wayne Wolf. Multiprocessor Systems-on-Chips, Elsevier Morgan Kaufmann, 2004.
12. Wolfgang Ecker, Wolfgang Müller, Rainer Dömer (ed.). Hardware-dependent Software: Principles and Practice, Springer 2009. Chapter 7: Generation and Use of an ASIP Software Tool Chain by Sterling Augustine, Marc Gauthier, Steve Leibson, Peter Macliesh, Grant Martin, Dror Maydan, Nenad Nedeljkovic and Bob Wilson.
13. http://www.coware.com
14. http://www.carbondesignsystems.com
15. http://www.vastsystems.com
16. http://www.imperas.com

Chapter 7
Codesign Experiences Based on a Virtual Platform

7.1 Introduction

One of the key elements of an electronic system level (ESL) methodology [1] is the concept of platform-based design. Platform-based design (PBD) allows extensive reuse of components, which reduces the time-to-market for the first release of a product, maintenance, and subsequent releases. When the platform is modeled at a high level of abstraction, we often call the model a "virtual platform" (see Chapters 5 and 8). By recreating – in a simulation model – the full architectural environment of a system as well as the algorithms that it implements, virtual platform-based design enables quick design space exploration through experimentation with different design possibilities, without having to first invest time and effort in the design of a physical prototype. The final, optimal, solution might be one that was not usually explored using past methods.

Design teams usually perform a functional verification of their specifications by creating executable C/C++ or Java models of their specifications instead of directly working with the final target platform. Working at this higher level of abstraction both enables and speeds up the verification and validation of the specifications (see Chapter 4, Section 4.1.2), as developers do not need to worry about micro-architectural details such as clock rates, exact pin-outs, line signals, or bus protocol designs. ESL promotes the use of virtual platforms, by mixing the executable application model with a model of the architecture, resulting in a full simulation of the architecture executing the application. There is a trade-off between the simulation speed of these applications and architecture models and their timing accuracy. Therefore, several levels of abstraction are used to first perform quick, but less accurate, simulation as typified by Chapter 5 and later in the design process as typified by Chapter 8, to perform slower but more accurate simulations.

This chapter presents a hardware/software codesign approach based on a virtual platform, named Space Codesign [2], applied to a joint photographic experts group (JPEG) decoder application. First, we define the concept of a virtual platform and describe, from the experience of the developers, what the requirements are for a virtual platform that will enable hardware/software codesign. We also propose a classification of virtual platforms that allows one to compare this platform

B. Bailey, G. Martin, *ESL Models and their Application*, Embedded Systems,
DOI 10.1007/978-1-4419-0965-7_7, © Springer Science+Business Media, LLC 2010

with others. Then, we present the methodology based on four main steps: system specification, exploration, analysis, and integration. All these features are illustrated through the JPEG decoder application by refining its specification from reference code to an implementation.

7.2 Virtual Platforms

7.2.1 Introduction

A virtual platform is a fully functional software representation of a hardware/software SoC design based on a mix of processors, software, peripherals communication links, memories, and other intellectual property (IP) cores. In virtual platforms, these components are often modeled at the level of processor instructions, function calls, memory accesses and data packet transfers, as opposed to the bit-accurate, nanosecond-accurate logic transitions of register transfer level (RTL) models [3, 4]. Common use cases for virtual platforms include system architecture development, analysis, optimization, verification, and validation [4]. This is what is shown in this chapter. By adding a design methodology to the virtual platform, two results are expected:

- A system architecture that meets performance and cost goals with optimized hardware/software partitioning. It establishes an optimal hardware architecture before moving on to the physical implementation. The optimized hardware architecture will also act as a "golden" reference for RTL implementation and verification.
- Hardware/software co-development that produces application and system software before a real hardware prototype becomes available.

In the rest of this section, we present a quick overview of the evolution of platform complexity. Then, we discuss a methodology associated with virtual platforms by exploring concepts within the Space Codesign solution. We complete the section with an updated summary of recent surveys of existing virtual platforms [1, 5].

7.2.2 Evolution of Platform Complexity

Before the 1990s, designs were based on a sea of digital logic (e.g., custom RTL). In the early 1990s, designs typically contained a few islands of IP in a sea of digital logic. From the mid-1990s on, the opposite situation was observed: many IP islands were stitched together with less and less RTL. In the late 1990s, (configurable) processor IP was integrated while in the early 2000s solutions were based on multiple (configurable) processors with IPs as hardware acceleration engines or as software code. Finally, during the last few years, considerable efforts have been spent on

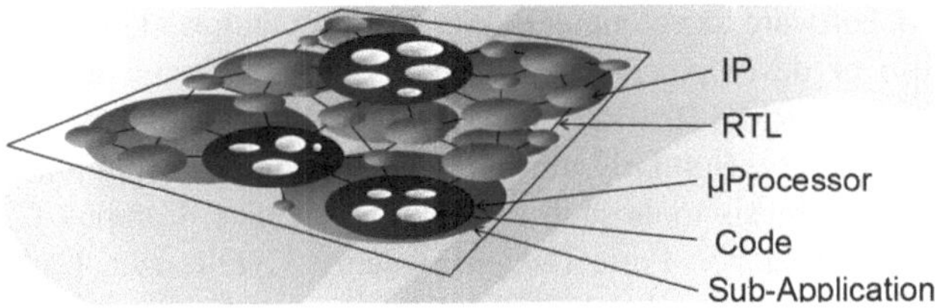

Fig. 7.1 Platforms today

the development of complex sub-system applications that need to communicate and therefore require hardware–software partitioning. Figure 7.1 illustrates the result of this evolution. A smartphone with features such as video telephony, Bluetooth, a 10 megapixel camera, JPEG, and VGA displays represents a good example of a complex system with many applications.

Market forces are driving electronic system developers to design ever more complex products whose architectures are based on a large and growing mix of processors, software, complex peripherals, and intellectual property cores. With the added pressure to integrate these sub-systems as much as possible into a single piece of silicon (e.g., an ASIC or a large FPGA) managing the development and analysis cycles of these systems has become an overwhelming challenge. This challenge has given rise to new design methods that require a different system-level approach. This electronic system-level approach, based on a virtual platform, provides the ability to address both software and hardware development simultaneously, allowing optimization in each domain and between them.

To cope with this challenge, five features have been implemented in the system-level approach:

- An Eclipse-based flexible framework for verification and validation through simulation from a set of specifications using C/C++/SystemC
- The rapid creation of multi-processor architectures with IP, I/O devices, and hardware accelerators
- Easy exploration and mapping of complex applications to multi-processor architectures, allowing users to save various mappings and partitioning choices for comparison
- Creation of virtual platforms, simulation, SW debugging and non-intrusive monitoring, data collection and analysis and
- Exporting the system that has been specified to an HDL flow for implementation.

7.2.3 Methodologies

Virtual platforms should support existing best-practice design, debug, and verification processes for both hardware and software developers. The platform interfaces with a range of hardware and software tools, including standard languages for hardware and software development. For software engineers, virtual platforms enable the use of development environments such as Texas Instruments' Code Composer Studio, ARM's RealView, the GNU tool chain, and debug environments from other vendors [6]. Similarly, for hardware engineers virtual platforms enable the use of ESL synthesis tools such as Mentor Graphics' Catapult-C, which will be discussed in Chapter 9, Forte Design Systems' Cynthesizer, Cadence Design Systems' C-to-Silicon Compiler, and others.

As shown in Figure 7.2, rather than waiting for an RTL description that can be mapped into a rapid prototype (such as an FPGA-based system) to start the software development, virtual platforms allow hardware and software development to progress in lockstep as shown in Figure 7.3. Indeed, as the hardware development process progresses and the hardware is further refined, updated models can be generated and distributed to software development teams [7].

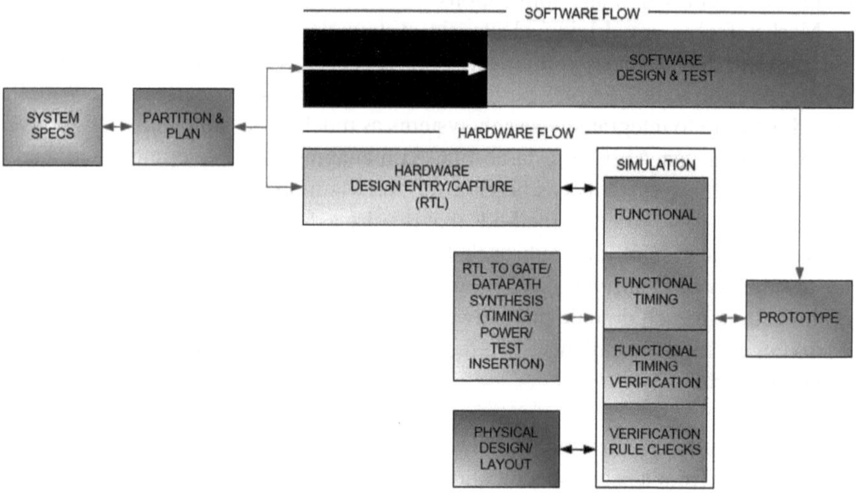

Fig. 7.2 Conventional methodology

In addition, the methodology is a combination of the refinement-based methods [1] [8] going through several levels of transactional-level models (TLM) and the explicit mapping-based methods of platform-based design [1, 9, 10].

The approach comprises four steps: system (functional) specification, exploration, analysis, and integration. System specification, exploration, and integration steps introduce additional information in the form of refinements of the model abstraction, by adding or making the specification more precise. Throughout these steps, we further refine communication and computation, two orthogonally independent components of the model taxonomy (Chapter 1), based on the levels of abstractions proposed in [8]. In general, the refinement follows the path A, B, C, D, and F of these taxonomies as indicated in Figure 7.4. The dark states of this figure are considered to be TLM levels.

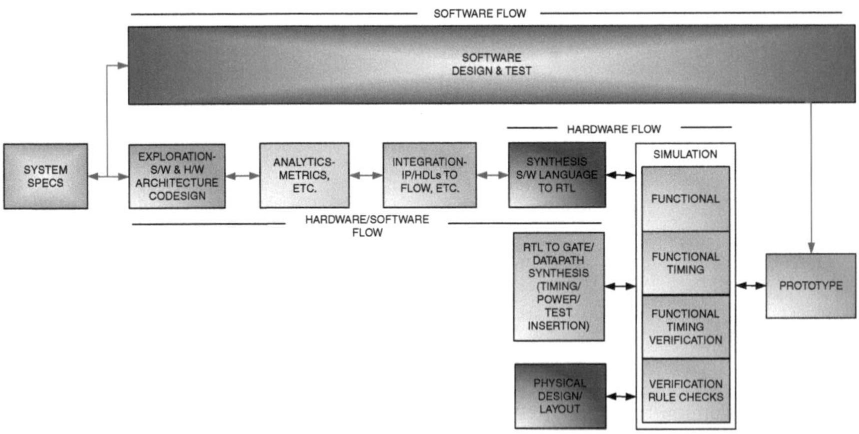

Fig. 7.3 Hardware and software development in lockstep

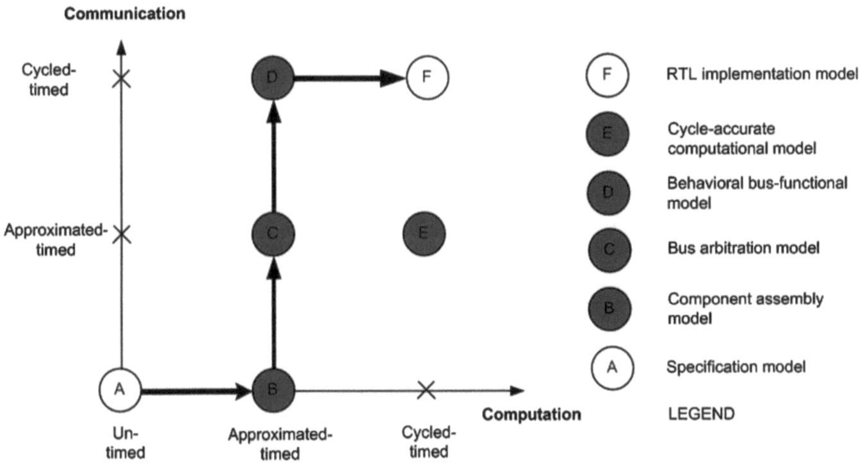

Fig. 7.4 Model refinement path

Also, during the exploration process, a set of mappings or partitions combine the functional specification with a given architecture (Figure 7.5). The resulting mapping can be simulated and analyzed based on a set of metrics. Then, the designer can modify the mapping, change the specification to better suit the architecture, or alter the architecture to better suit the application [11].

7.2.4 Commercial Technologies for Virtual Platform Development

Inspired by [6] (Figure 7.6), it is possible to classify existing virtual platforms into two main groups: (1) functionally accurate and (2) cycle approximate.

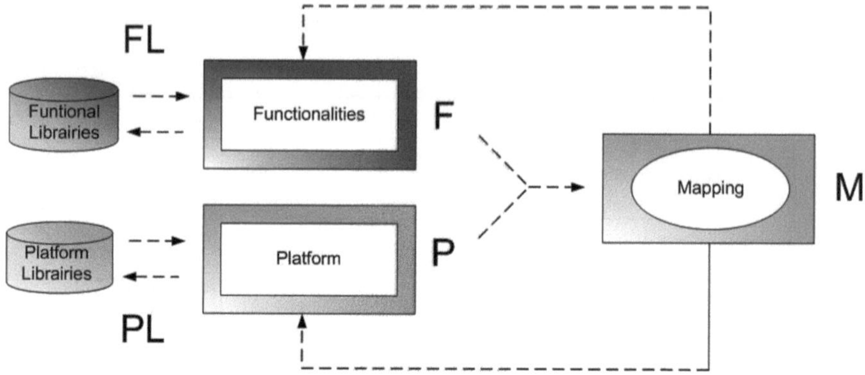

Fig. 7.5 The platform-based design (PBD) approach [10]

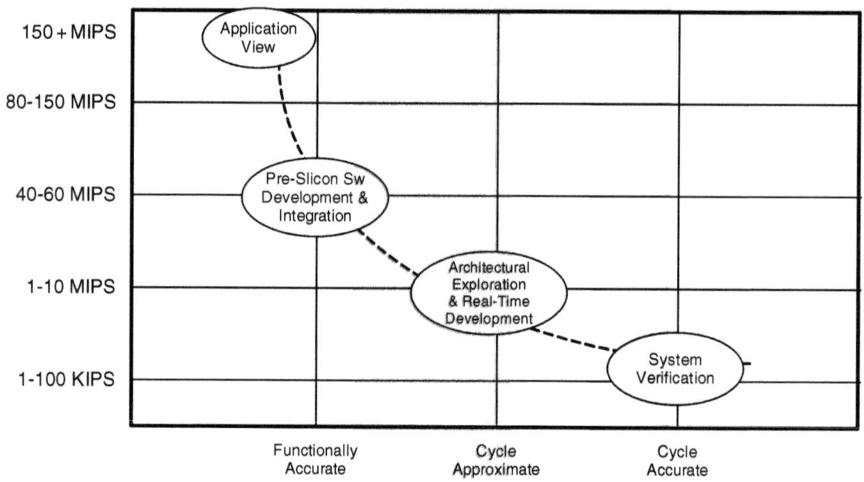

Fig. 7.6 Simulation speeds vs. abstraction levels

Functionally accurate platforms are mainly used for software application development (point A in Figure 7.4). They can be divided into two subgroups: (1) programmer's view (PV) [12] and (2) application view. Programmer's view is generally based on non-pipelined processor models that offer instruction and register accuracy at higher simulation speeds. It may also offer statistics on cache hits and misses. When less detail about the hardware model is required, the application view allows users to increase the simulation speed (150 MIPS and higher) by executing the application as a native process on the host workstation without using any target platform feature modeling.

Cycle approximate platforms are mainly used for architectural exploration. They offer more timing details than functionally accurate platforms. For example, these platforms might include:

- Cycle-approximate bus models (e.g., arbitration, pipelining, and concurrency details)
- Cycle-approximate processor models which take into account the instruction pipeline and the memory hierarchy
- Starvation and deadlock detection when using multi-bus topologies
- Latencies for accesses to on- and off-chip peripherals
- Cycle-approximate models of hardware accelerators and other IP cores
- Execution of the target-embedded software including its real-time operating system (RTOS), hardware abstraction layer (HAL), and interrupt service routines (ISR).

Profiling and analysis tools can exploit this additional timing information to give metrics about bus usage, communication times, task execution times, memory hotspots, processor usage, context switches, and interrupt latencies. This allows the platform to better support allocation, partitioning, and mapping decisions.

Under this classification, Table 7.1 enumerates some well-known commercial platforms.

Table 7.1 Some commercial virtual platform technologies/tools

Vendors	Functionally accurate	Cycle approximate
Coware	Virtual platform	Platform architect
Synopsys	Innovator	
Mentor Graphics		Vista
Virtutech	Simics	
Vast	Comet	Meteor
ARM	System generator	
Imperas	OVPsim	
Carbon	SoC designer	SoC designer

The virtual platform used in this chapter's experiments (Space Codesign) supports some features of the functionally accurate group, but corresponds more closely to the cycle approximate category. Indeed, all the features enumerated above are supported, as well as additional important features less frequently found in Table 7.1 tools:

- system-level non-intrusive monitoring,
- extensive automation when moving blocks from hardware to software or vice versa, and
- down-to-the-chip system prototype generation (bridges with ESL synthesis and logic synthesis tools).

7.2.5 Models of Computation

The model of computation used by Space Codesign is based on Kahn Process Networks (KPN) [13]. For a given input stream, a process in a KPN will always produce exactly the same output stream: it is thus a function from input streams to output streams. Also, all communication channels in a KPN are unbounded and lossless FIFO channels. This means that all write operations are non-blocking and that all read operations on an empty channel will block (stall) until a value is written to the channel. KPNs have the desirable property of having a uniquely defined behavior which depends neither on the timing characteristics of its processes and communication channels nor on their scheduling. For example, a KPN will produce identical results whether its processes are executed sequentially on a single processor or in parallel on multiple processors and/or hardware accelerators. This model of computation is therefore well suited for the specification, hardware/software codesign, and design space exploration of embedded systems.

However, KPNs have practical limitations that are addressed in this virtual platform. First, they assume infinite-length (unbounded) FIFO channels, whereas clearly only finite-length channels are available in a real implementation. This is of course true whether the FIFO channels are implemented as a dedicated hardware FIFO or through a shared memory. The virtual platform addresses this problem by providing application designers with communication functions with flow control. When flow control is enabled for a communication, the sender first sends a given number of bytes to the receiver over the channel and then waits for the receiver to reply with an acknowledge before it sends further data. Conversely, after having read the given number of bytes, the receiver replies with an acknowledge to the sender over the channel. This flow control ensures that the number of bytes that needs to be stored in a channel is always bounded.

A second limitation of KPNs is that they cannot model several important concepts in real-time processing such as timeouts, polling, race conditions, or real-time constraints due to the absence of non-blocking read operations [14]. Therefore, as with previous implementations of KPNs for hardware/software codesign such as YAPI [15], TTL [16], and WFIFO [17], this virtual platform extends the KPN model of computation by allowing non-blocking reads on a channel (polling). These non-blocking reads can be used to model non-deterministic input events and accesses to time-sensitive input/output peripherals.

However, this flexibility comes with a price: an embedded system that uses polling loses the formal properties of KPNs and the functional behavior of the system may depend on the timing characteristics of its processes and channels or on their scheduling. In any case, the virtual platform enforces a minimum delay of one clock cycle between two non-blocking reads on the same channel by the same process. This ensures a full execution of the system by preventing processes from polling a channel an infinite number of times in the same clock cycle. In the terminology of timed process networks [18, 19], the channels are delta-causal. This ensures the system has a uniquely defined behavior for each given schedule and set of timing characteristics for its processes, channels, and input events.

7.3 Platform and Application Description

As mentioned above, the virtual platform used in this experiment allows architecture exploration using a refinement design flow for the development of hardware–software embedded products. SystemC user modules (blocks) are created or integrated according to coding guidelines which ensure that these modules can be transparently swapped between the hardware and software domains. This important feature creates a large number of possible mappings, which are called "configurations." (A definition of configuration is provided in Section 7.3.2.1.). The designer may simulate each configuration in order to find the optimum.

As an example application, we model a picture processor application able to decompress a JPEG image [20]. This application consists of five user modules and two RAM blocks. Table 7.2 outlines the behavior of each module, while Figure 7.7 illustrates the components and communication paths. As a requirement, we will consider a frame size of 128 × 128, a targeted speed of 10 frames per second and an implementation on a Xilinx Virtex II-Pro FPGA [21] based on the Microblaze soft-core processor (100 MHz clock speed).

Table 7.2 JPEG modules and their behavior

User modules	Behaviors
Extractor (mExtractor)	The extractor module first reads the JPEG image header (markers) from jpegRAM. Then it sends commands to mHuffman and the dequantization matrix to mInvQuant. Finally, it stops the simulation when decoding is complete.
Huffman (mHuffman)	The Huffman module first reads data and the symbol table from jpegRAM. Then, it performs the Huffman decoding operation based on the symbol table and sends the result by groups of 8*8 blocks to mInvQuant.
InvQuant (mInvQuant)	The Inverse Quantization module reads the 8*8 blocks sent by the Huffman module, performs an inverse quantization operation based on the dequantization matrix, and sends the resulting blocks to mIDCT.
IDCT (mIDCT)	The IDCT (Inverse Discrete Cosine Transform) module receives 8x8 pixel blocks, performs the IDCT operation for 6 blocks of 8*8, writes the result in bitmapRAM, and sends a signal to mY2R.
Y2R (mY2R)	The Y2R module performs the color space conversion from YCbCr to RGB. The input is an array containing an YCbCr 4:2:0 formatted image which is read from bitmapRAM and converted to an RGB image. The result is sent back to bitmapRAM

The goal is not to optimize the JPEG algorithm and processor resources used in the implementation, but rather to demonstrate, through a known application, the potential of a virtual platform to explore different configurations.

The starting point for the platform model is a transaction-level model (TLM) in which each user module is a behavioral module communicating with other modules

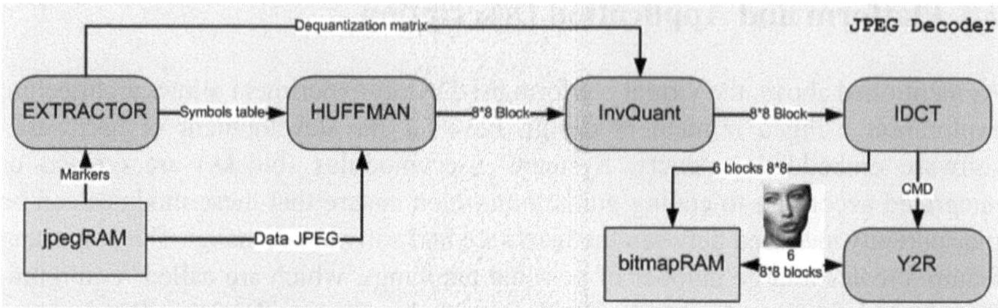

Fig. 7.7 Flow of JPEG application

or with peripheral slaves only through TLM read() and write() functions. Primitive C/C++ data types or pointer-free user-defined data structures can be transferred with these communication functions. Communications between modules take place through abstract FIFO channels and blocking or non-blocking semantics can be used for both read and write requests. Requests to peripheral slaves are handled synchronously. All modules and devices (e.g., memory) are addressed by communication functions through unique identifiers. At lower levels of abstraction, these identifiers are used to address both hardware and software modules. A TLM bus routes communications between modules and peripherals.

7.3.1 System Specification and Functional Verification

Based on the C/C++ application code of JPEG [20], one can, in a straightforward but specialized SystemC environment, create first an untimed functional model (point A in Figure 7.4) of the JPEG application. In the Space Codesign platform this level of abstraction is named Elix and it can be classified as functionally accurate (Figure 7.6). The coding style of the user modules is close to the SystemC behavioral style [22]. Also, designers do not have to deal with SystemC signals (sc_in, sc_out and sc_signal) or interfaces, since these are hidden in the platform. Therefore, encapsulation of the C/C++ application code in SystemC user modules is easy. The most difficult task is to decompose the sequential application in order to exhibit enough parallelism to exploit the concurrency offered by the virtual platform. For the JPEG application, it means decomposing the sequential code into a set of concurrent threads (modules) (Table 7.2 and Figure 7.7). Also, communication between user modules must be provided. Here, communications between SystemC user modules are message-passing based on rendezvous principles (Figure 7.8), but the use of a shared memory to exchange data between user modules is also possible (Figure 7.9).

At this stage, users may wish to include behavioral timing annotations to better simulate the execution latency of the hardware user modules. Approximately timed simulation (points B or C in Figure 7.4) is used to verify specifications, making sure

```
void HUFF::thread_huffman(void){

...

ModuleWrite(ID_IQUANT,SPACE_NORMAL_PRIORITY,
SPACE_NON_BLOCKING, &outjpegmessage);

....

}
```

a) Called from mHuffman

```
void IQTZ::thread_inverse_quantize(void){

...

ModuleRead(ID_HUFFMAN,SPACE_NORMAL_PRIORITY,
SPACE_BLOCKING,&injpegmessage);

...

}
```

b) Called from mInvQuant

Fig. 7.8 Modeling of communication between mHuffman and mInvQuant, based on the *rendezvous* principle

```
void HUFF::thread_huffman(void) {

...

    // write image header into memory
    // WORD 1 is X width
    // WORD 2 is Y height
    DeviceWrite(ID_JPEGRAM, SPACE_NORMAL_PRIORITY,
                m_infoImageHeaderAdd, &m_jpeg_decoder_structure.number_of_samples_per_line);
    DeviceWrite(ID_JPEGRAM, SPACE_NORMAL_PRIORITY,
                m_infoImageHeaderAdd+0x4ul, &m_jpeg_decoder_structure.number_of_lines);

....

}
```

Fig. 7.9 Modeling of shared memory (jpegRAM) communication for Huffman

the design works, checking whether appropriate responses are obtained for given inputs, and obtaining a sketchy estimate of system timing.

Figure 7.10 illustrates a mapping of the JPEG application in Elix to achieve functional verification. More precisely, it represents the PBD paradigm of Figure 7.5 applied to the JPEG application at the functional level. Figure 7.10a illustrates SystemC user blocks (F) and platform components (P) selected from their respective libraries (FL and PL), where FL contains available user blocks and PL contains architecture components for the current level of abstraction (e.g., timed/untimed channels and memories). Figure 7.10b illustrates a possible mapping M between F and P where user blocks and RAM blocks are connected to the timed functional channel named *crossbar*. The mapping result (the configuration) is named "validation."

Finally, this abstraction produces code that can be debugged, results that can be analyzed, and performance that can be profiled (see details later in Section 7.3.3). These operations are performed inside SpaceStudio, an Eclipse-based development environment [23] that simplifies third-party tool integration.

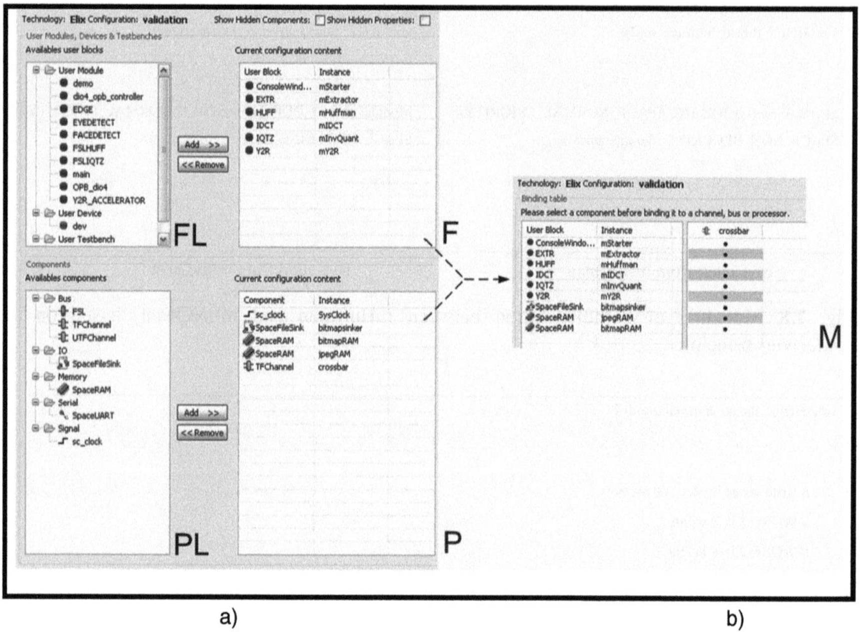

Fig. 7.10 PBD paradigm for the JPEG application to achieve functional validation

7.3.2 Architectural Exploration

During architectural exploration, previously verified SystemC user modules (blocks) are reused without modification and placed in a more detailed architecture. Although no modifications are required a priori, user modules can be refined if desired in terms of latency, using SystemC wait() statements, to give more accuracy to the simulations. The computation abstraction level is thus approximately timed (AT). On the communication side, the abstraction level is cycle timed, more precisely transaction-based bus cycle accurate (T-BCA) [24]. The main idea of the T-BCA modeling abstraction is to verify the timing of a possible future implementation without modeling all the pins at the communication interface of user modules and signals that make up the buses, and instead use the notion of channel and transactions on the channel [24]. The combination of the computation and the communication levels of abstraction (point D in Figure 7.4) provides a cycle approximate platform (Figure 7.6). In the Space Codesign platform this level is named Simtek. The designer allocates a multi-bus architecture and assigns user modules and peripherals to these buses. A T-BCA functional model is associated with each allocated bus.

Transaction-level models can transfer arbitrary-length data structures between user modules in a single transaction. However, refined communications at a cycle-accurate level must divide such data structures into bus-width slices to be streamed on the bus. Therefore, for each user module, a TLM-BCA transactor is provided. These transactors communicate with BCA bus adapters through a generic cycle-accurate communication protocol. Bus adapters translate this protocol into a bus-specific protocol. Adapters thus handle bus requests, process acknowledgments, and associate a bus-based address to each user module and peripheral. This communication refinement can therefore be used to target several bus protocols. Functional models and adapters for the IBM CoreConnect On-chip Peripheral Bus (OPB), and Processor Local Bus (PLB) [25] have been integrated into the virtual platform.

At this level, user modules can be mapped as embedded software running on an instruction set simulator (ISS) connected to a bus [26]. This enables functional co-simulation between refined software user modules and unrefined user modules. Since architecture-specific communication details are handled by platform components such as bus adapters, the mapping of a given user module to hardware or software is transparent to other modules and messages are routed to the right destination, either in software or hardware.

In summary, the platform includes transactions over a specific bus protocol, hardware wrappers to encapsulate user blocks, a collection of user accessible platform-specific IP devices, such as memories, timers, controllers, and I/Os, as well as efficient ISS models for software execution.

7.3.2.1 Definition and Configuration

Architectural exploration may be decomposed into three tasks:

1. *Hardware and software partitioning:* determine which SystemC user modules must be implemented in hardware (e.g., as accelerators) and which modules must be implemented in software (e.g., as tasks).
2. *Processor allocation and task binding:* determine the number of processors in the system and bind each software task defined in the first step to one of the allocated processors.
3. *Bus (channel) allocation and hardware component binding:* determine the number of buses in the system and bind each hardware component (including accelerators from step one, processors from step two and slave peripherals) to one of the allocated buses. This task can be generalized for a network-on-chip.

Figure 7.11 illustrates one possible solution to the first task: a hardware/software solution with one channel, where mExtractor and mHuffman are software tasks executed on the Microblaze, and the rest of the user modules are connected to the OPB channel. Figure 7.11 can also be viewed as the PBD paradigm of Figure 7.5 applied to the JPEG application for hardware/software mapping on one channel. In Figure 7.5a the current configuration (P) now contains RAM blocks from the functional level, a channel (OPB bus), a Microblaze processor, and standard devices (timers and programmable interrupt controllers). In Figure 7.5b the resulting mapping is shown. This configuration built in Simtek is called "software."

Figure 7.12 shows a refinement of the previous configuration which illustrates one possible solution to the second and third tasks, where two processors

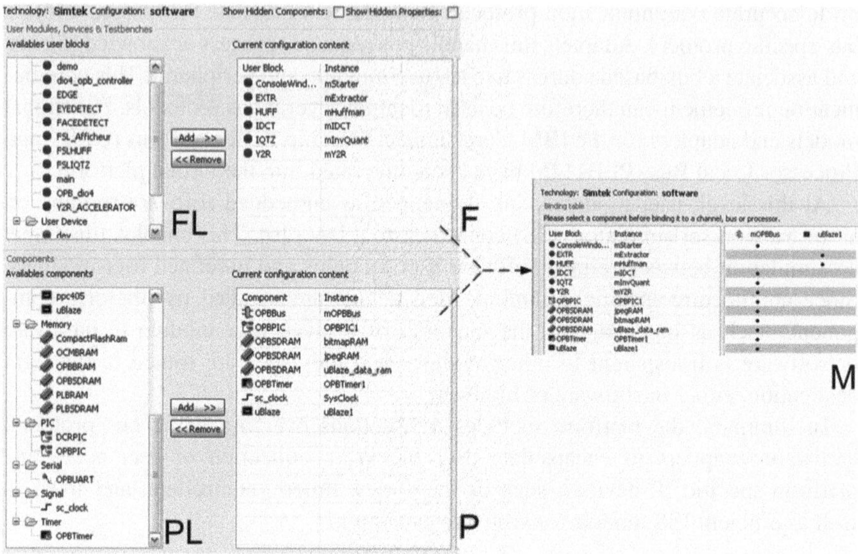

a) b)

Fig. 7.11 Hardware / software mapping with one channel

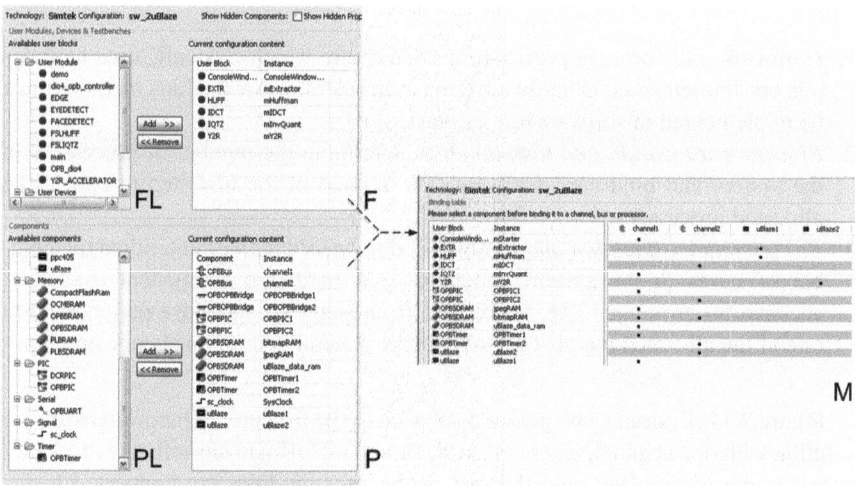

a) b)

Fig. 7.12 Hardware / software mapping with two channels

(Microblaze ISS) and two channels (OPB) are used. The first Microblaze is connected to channel1 with the mInvQuant as a hardware accelerator, while mExtractor and mHuffman are executed on the first MicroBlaze. The second Microblaze is connected to channel2 with mIDCT as a hardware accelerator and mY2R executed on the second MicroBlaze. Figure 7.12 can also be viewed as the PBD paradigm of Figure 7.5 applied to the JPEG application for hardware/software mapping with two channels. In Figure 7.12a, the current configuration (P) now contains RAM blocks from the functional level, two channels (OPB buses), two Microblaze processors, standard devices, and bridges (OPB-OPB). In Figure 7.12b the resulting mapping is shown. This new configuration is called "sw_2uBlaze."

Figures 7.11 and 7.12 showed how to modify the architecture (platform) to better suit the application constraints. But designers can also modify the application (functionality). For instance, consider the mY2R. One possibility is to split the user block into subblocks: mY2R is executed on the Microblaze and a hardware accelerator (mY2Raccelerator) is created specifically to compute the color conversion (R, G, and B values). The accelerator is connected directly to the Microblaze through two FIFO channels named FSL1 and FSL2. Fast Serial Link (FSL) is the FIFO communication mechanism used on an eventual Xilinx implementation. This point-to-point connection reduces communication latency between modules and consumed power. Performance also becomes deterministic. Indeed, data is transferred directly to the recipient without going through a shared channel and this eliminates both complexity and latency induced by bus arbitration and control mechanisms. This refinement of configuration "sw_2uBlaze" is illustrated in Figure 7.13, where the PBD paradigm of Figure 7.5 is applied to the JPEG application for hardware/software mapping with two channels and point-to-point communications. In Figure 7.13a the current configuration (P) contains the same components as in Figure 7.12a, but the current configuration (F) contains a new user module (mY2Raccelerator). In Figure 7.13b the resulting mapping is shown with a master/slave point-to-point communication between mY2R and mY2Raccelerator and vice versa. This new configuration is called "Split_Y2R."

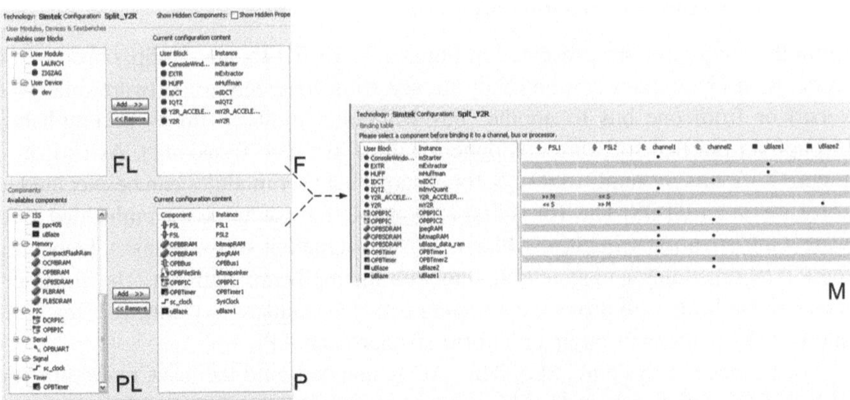

a) b)

Fig. 7.13 Splitting user blocks

When more than one communication mechanism is available (e.g., bus or point-to-point communication through FIFOs), a unified application programming interface (API) may be used. In Figure 7.13, putFSL and getFSL may be replaced by ModuleWrite and ModuleRead, respectively, to obtain a single interface from the programmer's point of view. When the mapping is done, point-to-point communication may be specified between any pair of user modules (hardware/hardware, hardware/software, or software/software) through the direct link manager. Figure 7.14 summarizes the different types of communication supported by the unified API with ModuleWrite and ModuleRead.

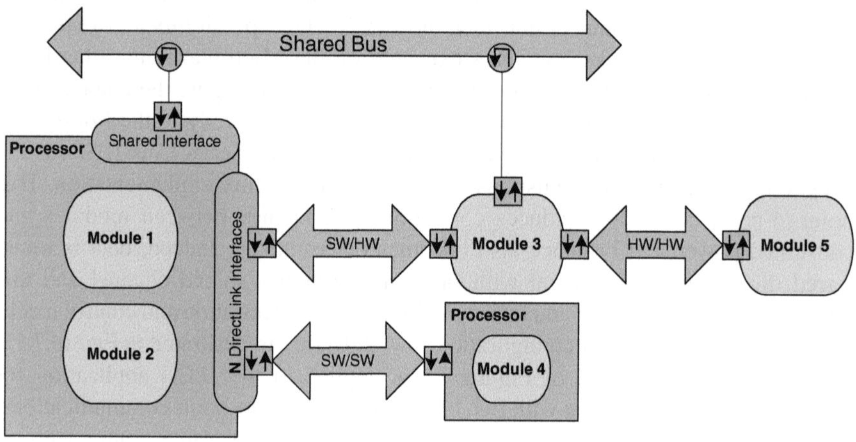

Fig. 7.14 Types of communication

7.3.2.2 Moving Modules Between Hardware and Software Partitions on a Multi-bus Architecture

From the configurations presented in Figures 7.11b, 7.12b, and 7.13b, observe that SystemC user modules can be easily moved from hardware to software (and vice versa) or from one bus to another bus without recoding. This is accomplished through an internal software component named Tor (for Translator). As shown in Figure 7.15, this SystemC to RTOS Tor ensures that user modules can be executed in software on an RTOS. The Tor is first responsible for executing the embedded software's initialization sequence, which consists of creating software tasks, launching the RTOS scheduler, and synchronizing with the hardware platform. The Tor interface's second role is to provide a mapping between standard SystemC 2.2 functions and RTOS functions through a platform-standardized API.

For instance, a SystemC SC_THREAD is mapped onto the taskCreate function of this API, which can further be mapped to, for instance, the OSTaskCreate of μC/OS-II [27], an RTOS that is supported. Finally, the Tor provides a communication manager, which establishes connections between software user modules and platform hardware. Using specific read/write methods, the Tor responds to task requests for communication with other software or hardware user modules. RTOSs are compiled with the appropriate HAL.

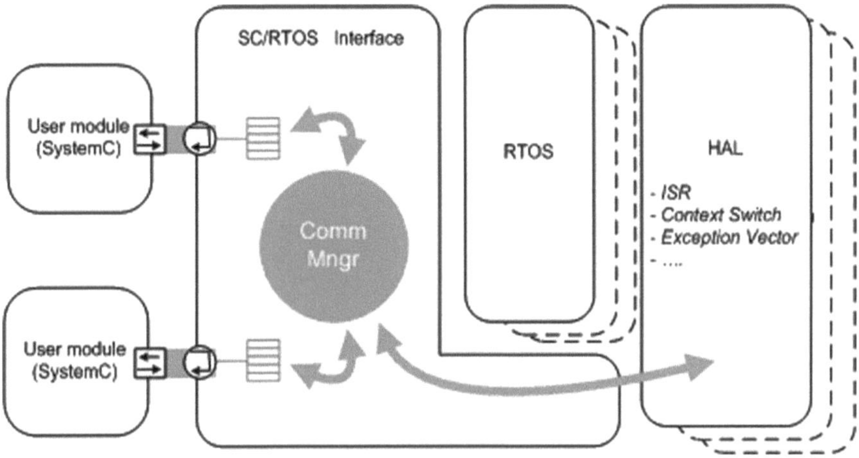

Fig. 7.15 Software environment to connect software SystemC user modules to the RTOS [26]

As mentioned above, the coding style of the user modules is close to the SystemC behavioral style and timing annotations can be added. However, these annotations are ignored if the module is mapped into software, because timing information is already provided by the cycle-approximate (or cycle execution) execution of software on an ISS.

7.3.2.3 ISS Abstraction

The Microblaze [28] core model was developed using two different abstraction levels [29]: a high-level non-pipelined version and a lower level pipelined model.

Non-pipelined Microblaze Model

The non-pipelined Microblaze implementation is based on an instruction-accurate C model used by GDB and wrapped in a SystemC wrapper. The processor model is divided into four main components: the ISS core, a GDB-compatible debug component, and two-bus adapters for bus and memory accesses.

Figure 7.16 shows the algorithm implemented in the only thread of the model. The model is mostly untimed except for multi-cycle operations as well as bus and memory accesses (data and instructions). This keeps a certain amount of information on execution time in order to guarantee thread and method ordering and to obtain a rough idea of the performance. To approximate the parallelism between the instruction fetch pipeline stage and the other pipeline stages without implementing a pipeline model, this sequential algorithm executes instruction decoding, single-cycle instructions and register write-back in zero simulation time.

For bus and memory accesses, a SystemC bus interface controller was implemented in order to abstract such accesses from the main processor model and to

```
thread() {
    while(1) {
        Fetch()
        Decode()
        Execute()
        If ( branch with delay ) {
            Fetch()
            Decode()
            Execute()
        }
        ProcessInterruptsExceptions()
    }
}
```

Fig. 7.16 Non-pipelined ISS main SystemC thread

allow insertion of a cache model. This component is instantiated twice, one for the instruction-side bus and one for the data-side bus.

The way the model is designed makes it extremely easy to add GDB [30] support to it. The debug component interprets the commands coming from the debugger and calls the appropriate methods on the ISS. If the command is a register access, the debug component directly reads the registers of the ISS. On the other hand, if the command is a run, stop, or step command, it is forwarded to the ISS and processed in the main loop before the instruction fetch takes place.

Because the model uses only one SystemC thread and its operations are executed sequentially, it cannot access both the instruction bus and the data bus concurrently. This imposes a limit on the performance that can be achieved with this model.

Five-Stage Pipeline Microblaze Model

Although the Microblaze model described in the previous section is instruction accurate with respect to the Xilinx Microblaze, it is far from achieving cycle accuracy. A more accurate model, better suited to detailed performance analysis is thus required. This model is closer to the architecture described by Xilinx [28] in that it implements a full five-stage pipeline with hazard detection and data forwarding.

The goals were to implement a near-cycle-accurate model without dramatically extending the simulation time and to use this model for software performance analysis as well as for functional software verification. A classic five-stage pipelined architecture [31] using high-level SystemC primitives was chosen to reproduce the Xilinx Microblaze behavior as closely as possible without falling into a complex low-level RTL description.

Figure 7.17 depicts the simplified architecture of the pipeline: five sc_methods/sc_threads each implementing a stage of the pipeline and separated by channels to send the results to the next stage and signals to send control information (branching and forwarding).

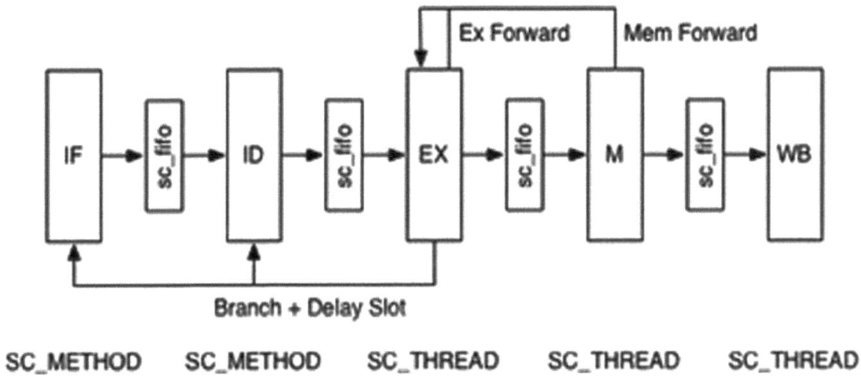

Fig. 7.17 Pipelined ISS model

The first two stages (instruction fetch and instruction decode) are implemented using SC_METHODs. SC_METHODs are useful when combinatorial behavior is needed because they can be sensitive to changes in signals or events and are in general faster than SC_THREADs.

SC_METHODs were chosen for these two stages to enable them to react immediately when a taken branch is detected. If the instruction in the ID stage is a delay slot instruction, it is passed on to the execute stage on the next rising clock and if not, the instruction is flushed. The same goes for the instruction fetch stage: if a branch is taken or an interrupt is detected, this stage either immediately starts transferring the new instruction or waits if there a pending delay slot instruction. The latter case happens when running from an external memory instead of the faster Microblaze LMB (Local Memory Bus).

Since the instruction fetch is a method instead of a thread, the memory access has to be performed in a separate thread. This enables the model to use TLM function calls while maintaining a good performance level.

The three remaining stages are implemented as SC_THREADs sensitive to the rising edge of the clock. This simulates a clock cycle without the burden of RTL level logic. For the execute stage, each instruction is executed in zero time and a SystemC wait() statement is added for multi-cycle instructions to simulate the execution delay. The memory stage accesses the LMB, OPB (Open Peripheral Bus), and XCL (Xilinx Cache Link) directly using TLM calls and finally, the write-back stage writes back to the destination register (if necessary) and waits for the next clock cycle.

In a pipelined architecture, inter-stage registers are used to propagate the data from one stage to the next. In the SystemC model, these registers are replaced by FIFOs with a length of 1. The use of FIFOs between each stage reproduces the inter-stage registers with a more flexible approach. If the next stage is still working on its instruction, the current stage will block until the FIFO is empty (in the case of SC_THREAD-based stages). Additionally, if the previous stage is not ready, the stage can simply block, waiting for the FIFO to be filled. Therefore, synchronization between stages is handled automatically.

The model also supports forwarding: for each instruction which uses a destination register, the execute and memory stages write to their respective forwarding structures. The forwarding is then handled in the execute stage to avoid the one-cycle latency which would be incurred by the FIFOs if it was handled in the decode stage.

The memory model for the pipelined Microblaze is the same as the non-pipelined model: two bus adapters are instantiated for the instruction-side and the data-side buses. Since the model is now pipelined, both buses can be accessed concurrently.

Finally, FPU instructions can be integrated in these different implementations of the Microblaze if required.

7.3.3 Analysis

The analysis part of the framework is based on a co-monitoring architecture that seamlessly and non-intrusively monitors both hardware and software user module computations and communications [32]. Indeed, the co-monitoring yields the same kind of performance data for every user module independently of the hardware/software partitioning of the system. Moreover, it does so without disrupting the simulation timing and without requiring any changes in user module source code when these modules are moved from hardware to software or vice versa. (Section 7.3.2.2). It also yields additional software performance data on processor load, instruction counts, and RTOS execution times. It is based on the general framework illustrated in Figure 7.18.

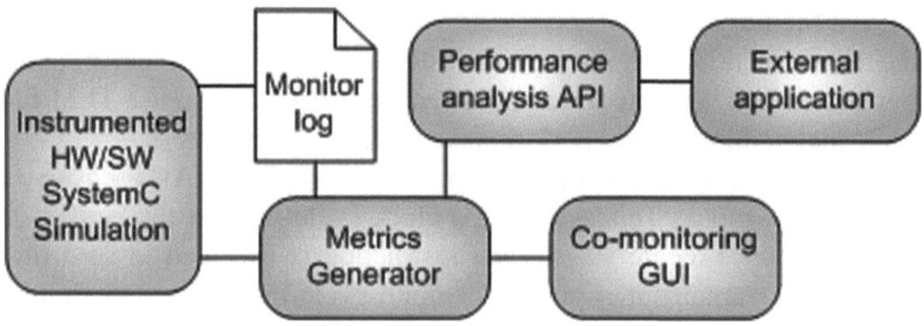

Fig. 7.18 General co-monitoring framework

First, the virtual platform is instrumented in order to send event notifications to a central SystemC monitoring component. This component processes the events and assembles co-monitoring records; it then forwards these records to a log file, or to a metrics generator outside of the SystemC simulation. This metrics generator computes performance metrics and statistics over these records and sends the results to a graphical user interface (GUI), or to another program using the performance analysis API.

Different approaches must be taken for the instrumentation of hardware and software user modules. The difference is due to the fact that hardware modules can do arbitrarily complex operations without affecting the SystemC simulation clock time (though they do take non-zero wall clock time) whereas software instructions are executed cycle-by-cycle on an ISS. To preserve timing accuracy within the simulation, we must ensure that the simulation clock time is not distorted by the monitoring of software user modules. Figure 7.19 gives a general picture of the platform instrumentation. Dashed lines represent monitoring components' lines of sight.

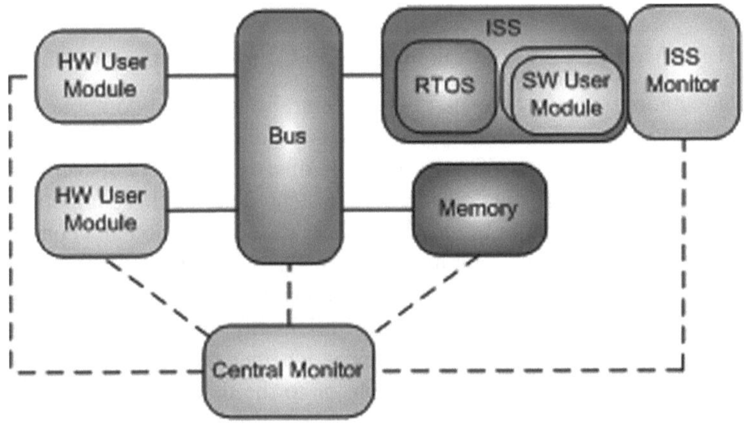

Fig. 7.19 Platform instrumentation

In order to minimize the amount of code inserted into platform models, generic instrumentation macros were implemented. This also allows co-monitoring to be easily enabled or disabled at compile time. These macros are responsible for collecting data on significant events and passing that information to the central monitoring component, which generates and timestamps the monitoring records presented in the following sections.

The instrumentation of hardware user modules is straightforward, since the relevant macros can be directly inserted into their respective platform models. Thus, bus models and memory models are instrumented by their designer whereas this is done automatically for hardware user modules through the instrumentation of the virtual platform's API functions. This instrumentation also takes advantage of timing annotations that are present in user modules, whether they were inserted manually, through a behavioral synthesis tool or via estimation techniques.

As explained above, software user modules cannot directly call monitoring macros, since doing so would mean executing additional software instructions and disrupting the simulation's timing. Each ISS is instead given an ISS monitoring component which calls the macros on behalf of the software user modules. Because the ISS monitor is itself a SystemC component, it can fulfill its role without distorting the simulation clock time.

The ISS monitor tracks the ISS program counter to detect when the execution flow reaches an entry or exit point of a context switch function or interrupt service routine. The ISS monitor then takes action depending on the event that occurs. For instance, if a context switch is requested, the ISS monitor retrieves the IDs of the software user modules being switched in and out. It thus keeps track of the user module currently being executed by the ISS. If an interrupt service routine is called, the ISS monitor records that the current module stopped its execution. At the end of the ISR, it records that the stopped module resumes its execution or that another module starts executing.

7.3.3.1 User Module Computation and RTOS Computation

Each computation record, generated by the central monitoring component, represents a user module or RTOS computation phase, and gives the beginning and end time of the computation. For software computations, such a record also contains the number of instructions, broken down by type, executed during this computation. An example of this can be seen in Figure 7.21 where the computation and communication of module mY2R for configuration "sw_2uBlaze" (Figure 7.12) is shown. The information on instruction frequencies by type can be used to estimate the degree of parallelism as well as the data- or control-orientation of user modules [33]. For RTOS computations, the record also indicates whether the computation is due to an interrupt service routine, context switch, or to the default idle task. Note that these are the only cases where record contents differ according to the hardware or software location of user modules.

The monitoring of task and RTOS execution also gives data on the total load of each processor and on the contribution of each module and of the RTOS itself

```
Data for module mY2R:

0.325518920 seconds of computation (89.265493%): 4824450 ALU instructions (43.912381%),
1217296 control instructions (11.079888%), 4944791 memory instructions (45.007731%)

780 bytes read from module mIDCT in 0.037098550 seconds
12 bytes written to module mIDCT in 0.000007740 seconds
98312 bytes read from module bitmapRAM in 0.089883170 seconds
65536 bytes written to module bitmapRAM in 0.058429600 seconds

Communication data for bus 1 initiated by module mY2R: 164256 bytes transmitted in
0.001807870 seconds (0.495764%)

Access data for memo ry bipmapRAM accessed by module mY2R: 163848 bytes from address 0
to 0x18020
```

Fig. 7.20 Instruction breakdown and communication per user module

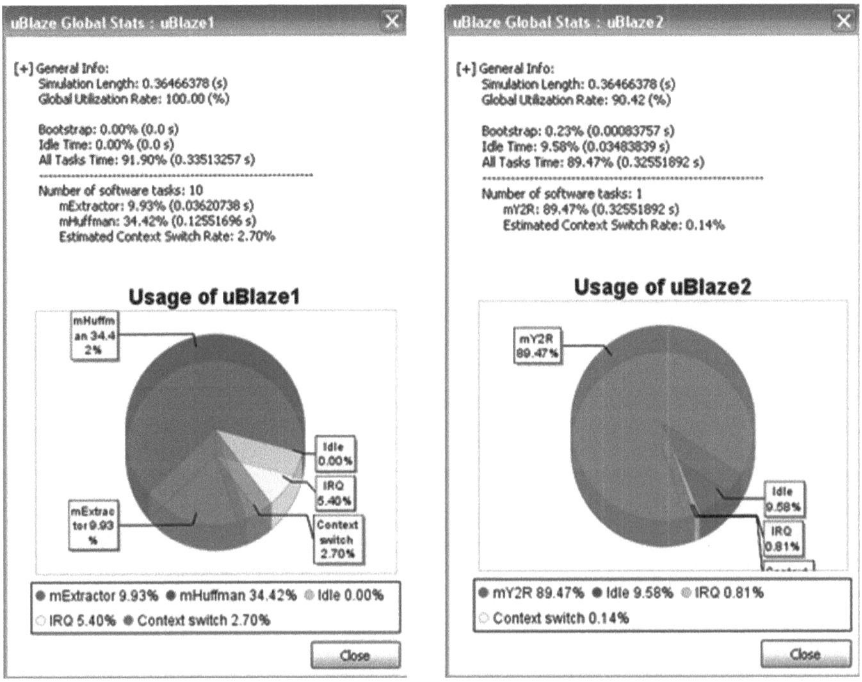

Fig. 7.21 Processor loads

to the load of each processor. Figure 7.21 shows an example of process loads for configuration "sw_2uBlaze" (Figure 7.12).

7.3.3.2 User Module Communication

The user communication record type provides information about completed end-to-end communications between a hardware or software user module and another component, which may be a second (hardware or software) user module. This information includes the identification of source and destination components, the start and stop time of the read operation at one end, and the corrresponding write operation at the other end, and the data length of the transfer.

Figure 7.22 summarizes the computation and communication performed by mInvQuant for configuration "sw_2uBlaze" (Figure 7.12).

7.3.3.3 Bus Usage

Bus models in the virtual platform are instrumented and call monitoring macros are used to gather information on bus usage. This bus instrumentation produces a series of bus transfer records. Each record contains information about the transfer of a data packet on a bus, such as the identity of the bus, the time at which the packet arrives

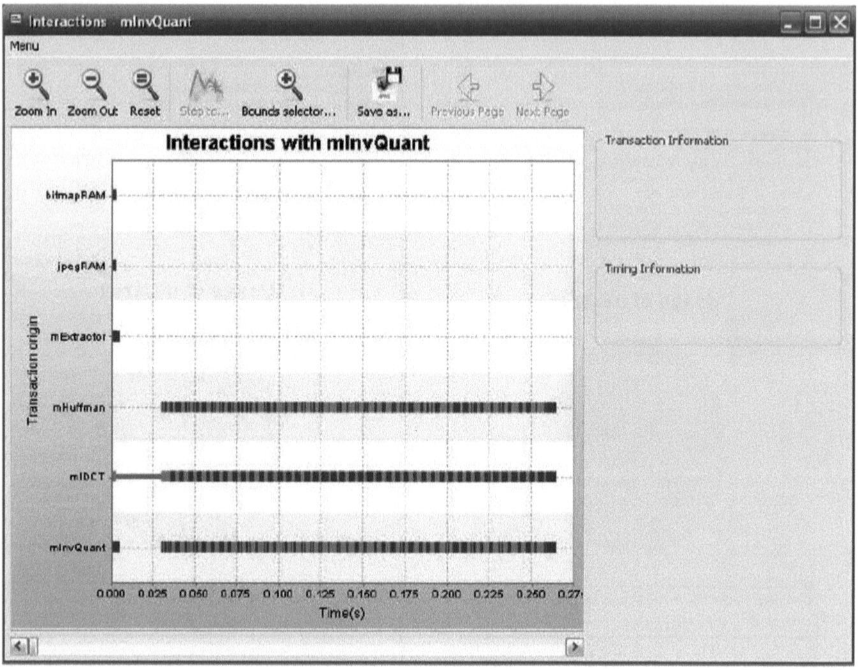

Fig. 7.22 Interactions with mInvQuant

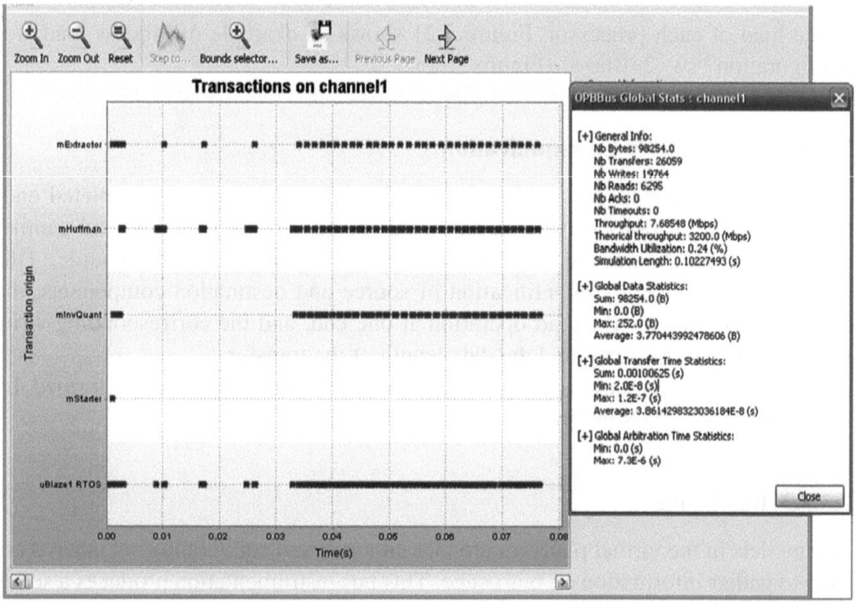

Fig. 7.23 Example load on a channel

on the bus, the times at which the bus starts and ends the transfer, the source and destination components as well as their data length. Note that a single end-to-end transfer may generate more than one bus transfer, because the packet may need to travel on more than one bus to reach its destination (for instance, in multi-processor systems) and because the communication protocol may require an acknowledgment to be sent. Figure 7.23 shows an example of the load on channel 1 for configuration "sw_2uBlaze" (Figure 7.12).

7.3.4 Integration

Figure 7.24 summarizes the refinement flow through points A, B, C, D, and F of Figure 7.4. Figures 7.24a,b have been already described in Sections 7.3.1 and 7.3.2, respectively. The last step, from Figure 7.24c–e, is to build a methodology to achieve a path between points D and F of Figure 7.4.

At point F (Figure 7.4), hardware (muxes, ALUs, decoders, memories, etc.) is described in terms of register transfers executed at each clock cycle (i.e., the RTL), and software is a sequence of instructions including user target code linked with the RTOS. The transformation from a transactional level (Elix and Simtek) to RTL level is referred to as hardware/software co-synthesis. It is considered one of the most challenging tasks in embedded design involving system-on-chips.

For the case of an implementation based on the Xilinx Virtex II-Pro FPGA, an implementation layer is already provided by the EDK (Embedded Development Kit) from Xilinx. EDK is an application for designing embedded programmable systems based on the Virtex II-Pro. This pre-configured kit includes all the tools and IP required for designing the implementation level of a hardware/software system. Therefore, one can perform the path between points D and F in three steps: interface synthesis, behavioral synthesis, and platform synthesis.

7.3.4.1 Interface Synthesis

This step makes application modules ready for behavioral synthesis. The generic communication protocol (Section 7.3.2) is refined down to a cycle- and pin-accurate RTL protocol, as shown in Figure 7.24c. A TLM–RTL transactor [22] is generated for each hardware user module. Transactors that support different abstract data types, different bus widths, and non-blocking semantics can be generated. An RTL–BCA transactor allows communications with the BCA bus adapter and its bus. This transactor also sends status information from the bus adapter to the TLM–RTL transactor.

7.3.4.2 Behavioral Synthesis

This step refines the computations of application modules and their TLM-RTL transactors to an RTL implementation, as shown in Figure 7.24d. Forte Design Systems' Cynthesizer has been integrated into our methodology to perform this behavioral synthesis [34]. Cynthesizer can synthesize C/C++ and SystemC arithmetic and logic

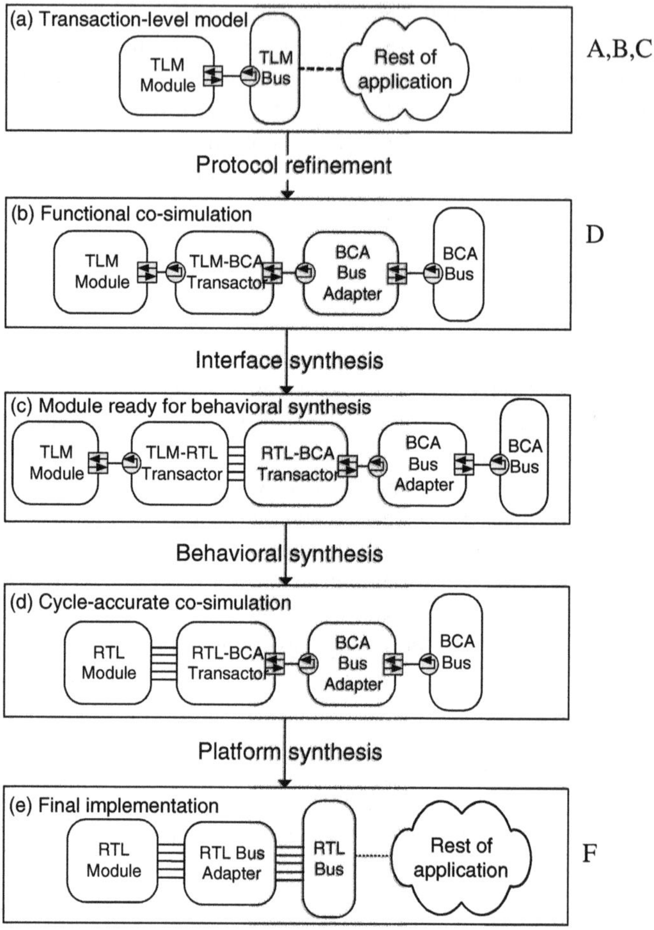

Fig. 7.24 Refinement flow through points A, B, C, D, and F of Figure 7.4

operations, control flow, loops, function calls, templates, and arrays as well as user-defined structures. Pointers are supported to the extent that they can be statically dereferenced at synthesis-time and that they do not manipulate internal data representations. Dynamic memory allocation is not supported. Both TLM application modules and generated transactors must therefore follow these rules.

Several options of Cynthesizer are utilized in order to optimize the synthesis of user modules and generation of RTL code: function in-lining, constant propagation, loop unrolling, common sub-expression elimination, unused bit trimming, and array flattening. Behavioral synthesis can be configured to minimize either area or latency.

With this automated communication refinement and hardware synthesis, we produced an RTL version of each of the five modules described in Table 7.2.

Table 7.3 Hardware resources used by JPEG RTL modules

	Flip Flops	LUTs	Slice Max (Flip Flops, LUTs) div2	BRAMs	Multipliers
mY2R	650	1,063	532	0	6
mInvQuant	2,672	2,447	1,336	1	3
mIDCT	9,674	18,684	9,342	0	44
mHuffman	3,511	2,790	1,395	2	3
mExtractor	1,183	2,274	592	0	3

Logic synthesis of each RTL module was performed with the Xilinx Embedded Development Kit (EDK) 9.1, targeting the Virtex-II XC2V2000-FF896 FPGA. Table 7.3 shows the hardware resources used by each user module when implemented as hardware accelerators, measured in flip-flops, lookup tables (LUT), and multipliers and block RAMS (BRAM).

Generated modules may use substantially more hardware resources than hand-coded versions. This area difference is explained by the fact that hand-coded versions are optimized to the specific characteristics of the module whereas the automated generation relies on generic refinement operations. In particular, the separation of communication and computation in the codesign methodology and automated flow might hinder the generation of an optimal implementation for modules (such as HUFF or IDCT) which transfer and process structures larger than the bus width.

However, for certain hardware/software architectures, a substantial part of the chip area is consumed by IP such as processors, buses and peripherals, not by hardware modules of the JPEG application. In such circumstances, the design productivity gained by automating communication refinement and hardware synthesis can more than offset these overheads. Since modifications are easier to make in TLM than in RTL code, such automated refinement is also well-suited to iterative and rapid system prototyping. However, some manual refinement might be necessary for the final production version of hardware-dominated applications with tight area constraints.

7.3.4.3 Platform Synthesis

After the behavioral synthesis step, all hardware modules have been synthesized as RTL hardware. This step is supported with a wizard called GenX available in the Space Codesign platform. It replaces all platform functional models of buses, adapters, processors, memories, and peripherals by existing RTL IP blocks, as shown in Figure 7.24e. These IP blocks are supplied by the target platform, except for the bus adapters which were implemented for the IBM CoreConnect OPB. Platform synthesis produces a full RTL implementation ready for logic synthesis on the target platform. Note also that software generation is completely automatic. The user software modules (tasks) and the RTOS are compiled for the embedded processor. The μC/OS-II RTOS [27] has been selected. μC/OS-II offers all the advantages of a micro-kernel: task preemption, a priority-based task scheduler, and an interrupt system.

7.4 Experiments

The experiments with these tools and example are discussed in two sections. First, two simulation models of the Microblaze processor (pipelined vs. non pipelined) are presented and tested using the mIDCT and mY2R modules. Next, we show different configurations of the JPEG decoder and we illustrate how the methodology allows the determination of the best ones.

7.4.1 Pipelined vs. Non-pipelined Models

Three memory configurations (BRAM, SDRAM, and SDRAM with cache) were tested using mIDCT and mY2R [29]. Each test, composed of a benchmark and a given memory configuration, was executed first on a real Microblaze soft-core on FPGA (Table 7.4) and afterward in simulation (Tables 7.5 and 7.6). The target hardware was the Xilinx University Program evaluation board and was composed of a Virtex-II Pro VP30 FPGA, 512 MB of DDR SDRAM (PC2100), and a wide variety of inputs/outputs. The tests were performed for a Microblaze 6.00 configured for maximum performance (five pipeline stages, barrel shifter, hardware multiply and divide, FPU and up to 32 KB of instruction and 32 KB of data cache) and running at 100 MHz. All standard output was redirected to the UART and viewed on HyperTerminal in Windows.

Table 7.4 Xilinx Microblaze – FPGA (100 MHz)

Memory configuration	Benchmark	Time (cycles)	Wall clock (s)
BRAM	mIDCT	9,406,362	0.09
	mY2R	2,136,361	0.02
SDRAM	mIDCT	141,633,320	1.42
	mY2R	48,074,779	0.48
SDRAM+Cache	mIDCT	8,120,972	0.08
	mY2R	2,357,424	0.02

Tables 7.4, 7.5, and 7.6 present the results obtained for the Microblaze on the FPGA and for the two models. The cycle time and wall clock time are presented for each test. For the simulated models, two additional columns are shown: the simulation speed in cycles per second and the error. The error is computed by comparing the cycle time of the simulated processor to the cycle time of the Xilinx FPGA Microblaze, using the same memory configuration for both.

Table 7.5 Non-pipelined microblaze model

Memory configuration	Benchmark	Time (cycles)	Wall clock (s)	Speed (kHz)	Error (%)
BRAM	mIDCT	13,204,510	34.0	388.4	40.4
	mY2R	2,704,072	8.0	338.0	26.6
SDRAM	mIDCT	159,616,705	361.0	442.2	12.7
	mY2R	52,949,991	124.0	427.0	10.1
SDRAM+Cache	mIDCT	12,399,339	41.0	302.4	52.7
	mY2R	3,125,142	9.0	347.2	32.6

Table 7.6 Pipelined microblaze model

Memory configuration	Benchmark	Time (cycles)	Wall clock (s)	Speed (kHz)	Error (%)
BRAM	mIDCT	10,270,991	72.0	142.7	9.2
	mY2R	2,071,325	15.0	138.1	3.0
SDRAM	mIDCT	153,846,188	591.0	260.3	8.6
	mY2R	49,853,028	211.0	236.3	3.7
SDRAM+Cache	mIDCT	8,775,232	61.0	143.9	8.1
	mY2R	2,436,881	17.0	143.3	3.4

First, results show that a lot of variability between the memory configurations is obtained for the non-pipelined ISS, while the pipelined version is more stable and closer in performance to the Xilinx Microblaze. Indeed, the error of the non-pipelined model (Tables 7.5 and 7.6) varies considerably, from 10% (SDRAM as memory configuration) to 52.7% (SDRAM + cache as memory configuration). This large variation means that this model gives only a rough estimate of the expected performance of the software on the target FPGA. On the other hand, the pipelined model is much more reliable as its error is almost constant for different memory configurations of the same benchmark. When compared across two benchmarks, its error varies between 3.0% in the case of the Y2R running on the BRAM to 9.2% for the IDCT also running on the BRAM (this indicates that there is an extra unforeseen latency when running computationally intensive code). This confirms that the pipelined model is well suited for performance evaluation, since several performance-enhancing techniques that are present in the Xilinx Microblaze and implemented in the pipelined model are not implemented in the non-pipelined version.

Also, high error values obtained for the non-pipelined model (52.7%, 40.4%, etc.) are mainly due to the sequential execution of each instruction preventing the inherent parallelism between instruction fetch, execution, and data access (Section 7.3.2.3). Indeed, the non-pipelined model does not support the overlap of instruction and data LMB requests which, in the pipelined model, can reduce the effective memory latency to one cycle instead of two. Moreover, the non-pipelined model does not implement the instruction fetch buffer. In the pipelined model, this buffer is useful when multi-cycle instructions and memory accesses are executed or when data hazards are detected. Thus, when the pipeline is stalled due to these conditions, the fetch stage can still retrieve the next instructions and store them in its fetch buffer. When the condition is resolved, the instructions that follow can be issued at the rate of one instruction per cycle. This can be an advantage when making data accesses to peripherals or external memories.

Another observation emerging from the collected data tells us that running a program from a slow memory lessens the advantages of pipelining since the processor has to wait to fetch the next instruction. This is supported by the error seen when running the benchmarks from the SDRAM only (when the cache is disabled). The estimated cycle counts of the non-pipelined and the pipelined version come very close to the performance of the Microblaze on the FPGA. Still, the accuracy of the pipelined model is consistently better than the non-pipelined one. In the case of the mY2R, the pipelined model is off by only 3.7% while the non-pipelined shows an error of 10.1%. But on average, the pipelined model is only 1.85 times more accurate than the non-pipelined model.

Finally, because of the minimal logic needed for the non-pipelined model, it is expected that the pipelined model will be significantly slower to simulate than the non-pipelined version. If we compare the speedup between each benchmark/memory configuration and average these, the non-pipelined model is 2.2 times faster than the pipelined model. This may seem like a large performance gap, but the speed of the pipelined Microblaze is still three orders of magnitude faster than RTL-level simulation, which only runs at around 100–200 cycles per second [35].

The conclusion that can be drawn from this data is that even though the two models are not cycle accurate, they each have their own usage. The non-pipelined model is quick to simulate and is used for fast debugging and verification while the pipelined model is used to more accurately model the performance that can be expected when the software will run on the FPGA.

7.4.2 Architectural Exploration of the JPEG Decoder

In this experiment, the goal is to find and verify a hardware–software configuration (partition) that satisfies the specification constraints, while maximizing the use of software. Table 7.7 presents nine configurations for the processing of a 128 × 128 pixel image, for which we provide execution time (Table 7.8), processor load, and hardware resources (Table 7.9). Building a new configuration takes only a few minutes. Non-pipelined MicroBlaze processors have been used in this experiment and assume no more than one processor per channel (bus).

Also, results for two levels of abstraction are displayed in Table 7.8: (1) point D and (2) point F of Figure 7.4. In (1) each hardware user module is at an approximately timed level, where the approximation is based on the synthesis results of Table 7.3. In (2) each hardware user module is at the SystemC RTL level, being the

Table 7.7 Description of configurations 1 – 9

Configuration	Description
1	Channel 1: SW={} HW={mExtractor, mHuffman, mIDCT, mInvQuant and mY2R}
2	Configuration 1 + point-to-point communications
3	Channel 1: SW={mExtractor, mHuffman, mInvQuant and mY2R} HW={ mIDCT }
4 (see configuration "software" (Figure 7.10b))	Channel 1: SW={mExtractor, mHuffman} HW={ mIDCT, mInvQuant and mY2R}
5	Channel 1: SW={IDCT} HW={ mExtractor, mHuffman, mInvQuant and mY2R}
6	Channel 1: SW={mExtractor, mHuffman,mInvQuant} HW={ mIDCT and mY2R}
7	Channel 1: SW={mExtractor, mHuffman} HW={} Channel 2: SW={mInvQuant} HW={} Channel 3: SW={mY2R} HW={IDCT}
8 (see configuration sw_2uBlaze (Figure 7.11b))	Channel 1: SW={mExtractor, mHuffman} HW={mInvQuant} Channel 2: SW={mY2R} HW={mIDCT}
9	Channel 1: SW={mInvQuant} HW={mExtractor, mHuffman} Channel 2: SW={mY2R} HW={mIDCT}

synthesis results of Table 7.3. The error column shows the accuracy of the approximately timed level (SystemC time at point D compared to SystemC time at point F), while the speedup column shows the speedup of the approximately timed level compared to a partial RTL level (wall clock time at point F divided by wall clock time at point D). Note that (2) represents a partial RTL version since software user modules are simulated by a non-pipelined MicroBlaze processor and communications are transaction-based bus cycle accurate. Therefore, the speedup would be larger if

Table 7.8 Execution time results

Configuration	SystemC time (s) point D/point F	Wall clock time (s) point D/point F	Error (%)	Speedup
1	0.031358840/ 0.04645864	3/141	32.5%	47
2	0.026011600/_	2/_		
3	0.896685260/_	282/_		
4	0.221537760/ 0.24016962	93/1,626	7.8%	17
5	0.52408192/_	161/_	7.6%	21
6	0.562002440/_	175/_	0.7%	19
7	0.362496740/_	351/_		
8	0.364663780/ 0.32088554	244/2,096	8.5%	8.5
9	0.374150650 /_	199/_	1.6%	20

Table 7.9 Processor load and resource results

Configuration	uP usage (%)	IRQ/context switch (%)	HW estimate (slices)	Memory usage (KB)	Bus usage (# transfers)
1	N/A	N/A	17,664	292	82,071
2	N/A	N/A	17,664	292	68,459
3	100	3.11/11.83	13,670	111	95,616
4	89	8.71/3.90	16,896	112	96,661
5	100	4.33/2.82	14,746	111	94,874
6	100	25.11/1.82	10,138	111	160,429
7	P1: 100 P2: 100 P3: 89.84	P1: 8.17/3.44 P2: 16.99/5.73 P3: 0.02/0.01	18,278	278	105,078
8	P1: 100 P2: 88.35	P1: 5.40/2.70 P2: 0.98/0.17	17,971	196	101,940
9	P1: 100 P2: 89.47	P1: 41.06/5.07 P2: 0.81/0.14	19,354	197	121,965

we compared it to a version completely in RTL (that is, replacing the non-pipelined version of the MicroBlaze and the communication by an RTL implementation).

We started with a simulation and verification of the JPEG application using Elix (based on the configuration of Figure 7.10b), where the crossbar is untimed. A SystemC simulation shows that 0.024749560s are required for the processing of a 128 $\times$ 128 pixel image. Thus 40 images per second is an initial upper bound on the system throughput.

Next, we simulated a hardware solution (Configurations 1 and 2 of Tables 7.7 and 7.8) that also confirms the potential performance of 40 images per second. The main difference between the configuration of Figure 7.10b and configurations 1 and 2 above is the introduction of the OPB channel (rather than a crossbar) that enables simulation of the behavior of an OPB bus (e.g., arbitration, timeout, and concurrency details) more accurately. Configuration 1 uses an

OPB channel for the communication, while configuration 2 uses an OPB channel and three bidirectional point-to-point communications: (1) mExtractor/mHuffman, (2) mHuffman/mInvQuant, and (3) mInvQuant/mIDCT. As no FIFO is required for the communication, the hardware resources used by configurations 1 and 2 are similar, while both bus usage and execution time decrease for configuration 2.

Once we introduce the MicroBlaze (Configurations 3–9) performance decreases by a factor of 10. Indeed, in the best scenario (Configuration 4) 0.221537760s is necessary for processing a 128 × 128 pixels image – i.e., a throughput of about four images per second. This confirms that the MicroBlaze is not well suited for this kind of data-intensive processing. As mentioned above, our goal is not to optimize the JPEG algorithm and processor resources.

Configurations 3 – 6 present different mappings based on one OPB channel. Configuration 3 shows that a complete software solution (with the exception of mIDCT used as hardware accelerator) is not efficient in terms of performance. Also, configuration 5 in which an IDCT is selected as a software task (while the rest of the JPEG application is in hardware) requires 0.52408192s to process an 128 × 128 pixel image, with a processor load of 100%. Considering that the IDCT is also the module that requires the largest number of slices when implemented as hardware accelerator, configuration 5 is the one that requires the minimum resources among the nine configurations but with non-optimal performance. Finally, configurations 4 and 6 show that mHuffman and mExtractor represent a good compromise in terms of processor load.

Next, configuration 7 suggests a solution distributed on three processors (excluding the IDCT) and with three bidirectional point-to-point communications as described above. If three images per second is an acceptable performance, configuration 7 represents an interesting solution that maximizes the use of software. Unfortunately, the use of three buses and three processor cores is costly in terms of resources, but not that much compared to a pure hardware solution (configurations 1 and 2).

Configurations 8 and 9 offer a hardware–software tradeoff based on two buses and two processor cores. For instance, configuration 8 is similar to configuration 7 with one less processor. Configuration 8 could also offer better performance in terms of memory usage and power consumption.

7.5 Conclusion

This chapter has presented a virtual platform which was defined to be cycle approximate. It is possible and efficient to explore different design partitions and mappings by starting with TLM simulations in SystemC, rather than starting implementation directly at the RTL level. The technology offers an environment to explore hardware/software partitions, by using drag and drop partitioning in the development environment. We applied an automated communication refinement and hardware synthesis to a JPEG application as an experimental demonstration. This illustrated the relative ease of exploring multiple configurations.

Two simulation models of the Microblaze processor at different speed-accuracy levels were presented and tested on two modules of the JPEG application. It was shown that the two models presented each have a distinct purpose and utility. A high-level model is perfect for early development and testing. It enables the programmer to develop code on an ISS that behaves like the FPGA-based Microblaze while simulating at a fast rate (442,200 cycles per second). The lower level model, while slower to simulate than the high-level model, is useful for performance evaluation and code profiling. Both models can then be combined in a multi-processor environment to speed up the simulation. This effectively enables the developer to use a progressive refinement methodology for the software as development proceeds.

We have also shown how a rich set of performance data from an embedded system simulated in a hardware/software codesign environment may be generated. In particular, the tool shows how the activities of hardware and software user modules can be seamlessly and non-intrusively monitored inside and across partition boundaries and how performance metrics and statistics may be efficiently extracted from monitoring records obtained from the instrumented model.

We have also described the refinement of transaction-level communications to pin-accurate protocols and the generation of synthesizable hardware from system-level specifications. These steps are integrated into the ESL methodology.

From nine configurations of the JPEG specification, it was possible to effectively identify three interesting configurations: (1) a hardware solution that maximizes the performance, (2) a software solution based on three processors that can be rapidly implemented, since only an IDCT hardware accelerator is required, and (3) a hardware–software solution based on two processors, less costly than solution 2 in terms of memory. Note that solutions 2 and 3 are much less interesting in terms of performance when compared to solution 1.

For the future, the developers are working on an automatic-mapping system. More precisely, they are developing a parameterized characterization and trace-based approach that helps evaluate alternative mappings during the design space exploration stage. From the execution trace one can generate a scheduling graph (e.g., who will communicate with whom, when, and how much) from which one can estimate the system performance. This approach will be faster than the conventional simulation approach by automatically finding promising configurations.

An important feature that is missing in this methodology, and currently under development, is a metric for energy consumption. The goal is to characterize implementation platforms (e.g., Xilinx platforms) in terms of the dynamic power consumption that accounts for the largest portion of the total consumption in digital circuits. The platform will be decomposed in different blocks: BRAM, processors, peripherals (timers, PIC, etc.), clock domains, IPs, etc. Then, it will characterize the consumption of each block for a range of potential clock cycles and the result will be annotated in the virtual platform (at the TLM level). Finally, there will be an analysis method for the scheduling graph to estimate the energy consumption of the whole system.

References

1. Brian Bailey, Grant Martin, Andrew Piziali. ESL Design and Verification, Elsevier, 2007.
2. http://www.spacecodesign.com
3. Brian Bailey, Grant Martin, Thomas Anderson. Taxonomies for the Development and Verification of Digital Systems, Springer 2005
4. Bill Murray. "Virtual Platforms – A Reality Check", www.scdsource.com, December 2007.
5. Bill Murray. "Virtual Platforms –A Reality Check, Part 2", www.scdsource.com, January 2008.
6. Frank Schirrmeister, Shay Benchorin, and Filip Thoen. "Using Virtual Platforms for Pre-Silicon Software Development", White paper, Synopsys, May 2008.
7. David Maliniak. "Virtual Platforms Technology 101", http://electronicdesign.com/Articles/ArticleID/21129/21129.html, April 2009.
8. Lukai Cai, Daniel Gajski. "Transaction Level Modeling: An Overview," in First IEEE/ACM/IFIP International Conference on Hardware/ Software Codesign and Systems Synthesis, Newport Beach, CA, 2003, pp. 19–24.
9. Felice Balarin, Yosinori Watanabe, Harry Hsieh, Luciano Lavagno, Claudio Passerone, Alberto Sangiovanni-Vincentelli. "Metropolis: An Integrated Electronic System Design Environment", IEEE Computer, 36(4): 45–52, 2003.
10. Douglas Densmore, Roberto Passerone, Alberto Sangiovanni-Vincentelli. "A Platform-Based Taxonomy for ESL Design", IEEE Design and Test of Computers, 23(5): 359–374, September–October 2006.
11. Frank Vahid, Tony Givargis. "Embedded System Design, A Unified Hardware/Software Introduction", John Wiley & Sons, 2001.
12. Adam Donlin. "Transaction Level Modeling: Flows and Use Models", Proceedings of CODES+ISSS'04, Stockholm, Sweden, pp. 75–80, 2004.
13. Gilles Kahn. "The Semantics of a Simple Language for Parallel Programming," in IFIP-74, North-Holland, pp. 471–475, 1974.
14. Prakash Panangaden, Eugene W. Stark. "Computations, Residuals, and the Power of Indeterminacy," in Proceedings of the 15th International Colloquium on Automata, Languages and Programming. Lecture Notes in Computer Science. Vol. 317, T. Lepistö and A. Salomaa (Editors). Springer-Verlag, pp. 439–454, 1988.
15. Erwin A. de Kock, Wim J. M. Smits, Pieter van der Wolf, Jean-Yves Brunel, Wido M. Kruijtzer, Paul Lieverse, Kees A. Vissers, Gerben Essink. "YAPI: Application Modeling for Signal Processing Systems," in 37th conference on Design automation, Los Angeles, California, pp. 402-05, 2000.
16. Pieter van der Wolf, Erwin de Kock, Tomas Henriksson, Wido Kruijtzer, Gerben Essink. "Design and programming of embedded multiprocessors: an interface-centric approach," in International Conference on Hardware/Software Codesign and Systems Synthesis, Stockholm, Sweden, 2004, pp. 206–217.
17. Kai Huang, David Grunert, Lothar Thiele. "Windowed FIFOs for FPGA-based Multiprocessor Systems," in IEEE 18th International Conference on Application-specific Systems, Architectures and Processors, Montreal, Canada, pp. 36–42, 2007.
18. Robert Kim Yates. "Networks of Real-Time Processes," in Proceedings of the 4th International Conference on Concurrency Theory, Hildesheim, Germany, pp. 384-97, 1993.
19. Edward A. Lee. "Modeling Concurrent Real-Time Processes Using Discrete Events," Annals of Software Engineering, 7(1–4): 25-45, 1999.
20. http://www.jpeg.org
21. Xilinx Virtex. Virtex-II Pro Platform FPGA Handbook. San Jose, CA: Xilinx, Inc., 2002.
22. Thorsten Grötker, Stan Liao, Grant Martin, Stuart Swan. System Design with SystemC, Kluwer, 2002.
23. Eclipse – an open development platform. http://www.eclipse.org/
24. Sudeep Pasricha, Nikil Dutt. "On-Chip Communication Architectures", Elsevier, 2008.

25. IBM. The CoreConnect Bus Architecture," IBM, Armonk, NY, 1999.
26. Jerome Chevalier, Maxime de Nanclas, Luc Filion, Olivier Benny, Mathieu Rondonneau, Guy Bois, El Mostapha Aboulhamid. "A SystemC refinement methodology for embedded software," IEEE Design & Test of Computers, 23: 148–158, 2006.
27. Micrium. "μC/OS-II," Micrium Inc., www.micrium.com
28. Xilinx. "Microblaze Processor Reference Guide - Embedded Development Kit EDK 9.1i", Xilinx, 2007.
29. Sébastien Fontaine, Luc Filion, Guy Bois. "Exploring ISS Abstractions for Embedded Software Design," dsd, pp. 651–655, 2008 11th EUROMICRO Conference on Digital System Design Architectures, Methods and Tools, 2008
30. Free Software Foundation, Inc. "GDB: The GNU Project Debugger", [http://sourceware.org/gdb/], 2008.
31. John L. Hennessy, David A. Patterson. Computer Architecture: A Quantitative Approach, 3rd edition, Maurgan-Kauffman, pp. A1–A79, 2002.
32. Laurent Moss, Maxime De Nanclas, Luc Filion, Sébastien Fontaine, Guy Bois, Mostapha Aboulhamid. "Seamless Hardware/Software Performance Co-Monitoring in a Codesign Simulation Environment with RTOS Support," in Design Automation and Test in Europe Conference, Nice, France, pp. 876-81, 2007.
33. Martin Holzer, Markus Rupp. "Static Code Analysis of Functional Descriptions in SystemC," Third IEEE International Workshop on Electronic Design, Test and Applications, pp. 243-248, 2006.
34. Forte. "Cynthesizer User's Guide For Cynthesizer 3.3," Forte Design Systems, San Jose, CA, 2007.
35. Tero Rissa, Adam Donlin, Wayne Luk. "Evaluation of SystemC Modelling of Reconfigurable Embedded Systems," Design, Automation and Test in Europe, 2005. Proceedings, pp. 253–258 Vol. 3, 7–11, March 2005.

Chapter 8
Transaction-Level Platform Creation

8.1 Introduction

Chapter 5 looked at the creation of a system-level virtual prototype that was used primarily for software and system-level functional verification. It addressed many of the issues associated with adding major architectural elements, such as processors, buses, and memories. Many of the issues associated with HW/SW partitioning were also discussed in Chapter 6 and 7. These chapters took a top-down approach to the problem. They started from the highest level of functional description and refined the models by adding information associated with the major architectural decisions. In this chapter we will introduce a flow in which the hardware is modeled incrementally and integrated into a system that may then be used for hardware verification, more detailed architectural exploration, and software integration. These platforms are created at the behavioral transaction level and often referred to as transaction-level platforms (TLPs).

In this chapter we will look at the creation of such a platform, how it can link with software, and discuss the adaptation of the hardware models to include concepts such as timing and power. These will be used to make further trade-offs in the architecture and micro-architecture of the blocks. The TLP model that is created can represent the actual hardware, rather than just the functionality of it. Chapter 9 will carry some of these concepts further by looking at the hardware implementation path using high-level synthesis, but in this chapter we will focus on the manual transformations and the integration of blocks into a transaction-level platform.

The goal with any model is to contain just enough information such that the relevant decisions can be made. If the model does not contain enough information, then the accuracy or fidelity of the results obtained is unreliable. If it contains too much information, then performance suffers to the point that it may not be possible to extract enough relevant data to correctly make decisions. Thinking back to the definition for ESL, it is the use of appropriate abstractions to reach comprehension and to improve the success rate of an implementation.

B. Bailey, G. Martin, *ESL Models and their Application*, Embedded Systems,
DOI 10.1007/978-1-4419-0965-7_8,

If we need to make architectural decisions about the hardware, then we need something that can adequately model the flow of information through and among the major architectural blocks in the system. This is already somewhat different from the prototypes that were introduced in the previous chapters, in that this prototype starts to create the structure of a solution, rather than containing just a functional decomposition of it. This is an important distinction and one that is essential to most hardware implementation flows.

The reason for this is that the structure creates a model hierarchy which is an essential element in all of the hardware description languages. It is this hierarchy that enables multi-abstraction modeling, since the structure can be preserved while substituting blocks at different levels of abstraction with the aid of interface models, usually called transactors. When changes in this structural hierarchy are made, we lose some ability to comprehend the transformations that are made or the ability to execute tools that cross this boundary without a lot of tedious and error-prone name mapping. For example, if a design becomes flattened during a synthesis operation, then equivalence checking would have to work out, or be told, all of the name mappings between the pre-and post-synthesized models. It also becomes impossible to replace part of the post-synthesized design with an abstract model because the interfaces have been removed. The interfaces thus become useful contracts between the blocks.

In this chapter we will also examine the concept of positive and negative verification. While many people may be familiar with block-level verification, those blocks have to be integrated into the system. In many cases these blocks may be IP blocks that come from a third party. While a block may have no known bugs, this is not a sufficient requirement for it to work correctly within the context of a system. The block must match the expectation of the system specification.

This creates several different verification needs. The use of such transaction-level prototypes can ensure that the interface specifications are correct before embarking on the implementation of the blocks. In this way it eliminates the possibility of integration errors, and this is highly beneficial in terms of time and quality. When IP blocks are being reused, or brought in from outside of the organization, it provides the platform within which they can be evaluated and integrated early in the development process, rather than waiting until the end to find such problems. It also provides a suitably fast platform such that the low-level software interfaces can be exercised and verified.

Another important aspect of such a platform is that it supports incremental refinement and verification. In this chapter we will show how timing and power information can be annotated onto the platform without the necessity to modify the functional model. This allows many choices to be evaluated and verified within the system-level context. When architectural problems are found late in an implementation, it can create severe project delays while the impacts of those problems are assessed. Changes that are necessary in one block often influence many of the surrounding blocks. The notion of immediate verification of decisions provided by a hardware platform means there are fewer potential problems to be found down the road to implementation.

8.2 Transaction-Level Modeling Comes of Age

Transaction-level models have existed for quite a long time. Most of these were written in C or C++ and had proprietary interfaces or APIs which could be used to connect them to other models or with tools. Most of these early prototype models were of processors with a few attached peripherals and possibly a memory subsystem. Connecting one of these models into a platform normally required the creation of a custom wrapper that would translate one API into a different API. This made it a costly proposition to create such a platform and slowed down the execution of the model.

In 1999 the first version of SystemC [1] appeared and after an industry battle replaced the other C-based languages as a standard language for modeling hardware at a higher level of abstraction than RTL. But a language is not enough. It provides a common execution environment, but that does not mean that models can communicate with each other. It was still necessary to create wrappers, although these wrappers were now written in SystemC rather than in C. For a more complete discussion about SystemC, refer to Chapter 3.

The next development was the creation of a transaction-level modeling (TLM) standard. This conceptualized the separation of communications and computation and started to build some standards for the communications aspect. The early version did not go far enough to achieve model portability. Then in 2008, a second version of the TLM standard (TLM 2.0) was introduced and this created model portability primarily for address-mapped communications (while other types of communications are possible, the standard does not provide utility functions to make it easy). While this standard is far from perfect, it made it possible to bring models in from multiple sources without the need for translation wrappers and this in turn enabled the creation of the type of platform model being described in this chapter.

Some of the standards behind this transformation were discussed in Chapter 2 for IP-XACT and Chapter 3 for SystemC and TLM. Other aspects of these standards will be discussed in more detail in various sections of this chapter.

Mentor's background in system-level capture and analysis tools came with their acquisition of Summit Design. That company has had a long and very eventful history that has included many acquisitions, including those with Viewlogic which itself was acquired by Synopsys, reformulated and divested, to create Innoveda, which was then split when acquired by Mentor, such that Summit once again became an independent company, until finally, Mentor acquired Summit itself in 2006.

All of this means that Mentor acquired a long history of knowledge and expertise in the creation of this kind of system and when coupled with the emerging standards enabled them to create an interesting platform for hardware exploration and refinement. With the many other ESL tools that Mentor has in their product portfolio, they have the capabilities to create a complete integrated ESL flow.

As one measure of interest in the new TLM 2.0 standard, John Aynsley, the CTO of Doulos said in a February 2009 webinar [2] that their company, which conducts SystemC training, had seen an increase in the popularity of their SystemC courses since the TLM 2.0 standard had been delivered. Maybe more importantly, all three of the major EDA companies are behind it, as well as some of the more specialized ESL tool providers, such as CoWare and JEDA. To have this kind of universal acceptance in the EDA industry is a rare event.

8.3 Model Abstractions

In this section we will formally introduce the models and abstractions that will be used in this chapter.

8.3.1 Terminology

When discussing transaction-level platforms, one item of confusion for many people comes from the terminology defined by the Open SystemC Initiative (OSCI). In their earlier works, such as the development of the TLM 1.0 standard, they defined a number of abstractions within the concept of transaction-level modeling. The highest level of abstraction was the programmers view or PV level of abstraction. This was intended to portray what a software engineer would require from a hardware model, in that timing was essentially unimportant but bit-level accuracy was required in registers or other data elements which were to be accessed by the software.

A second level of abstraction was defined that was called programmers view with timing or PVT. At times this is also referred to as PV+T. This acknowledged that some software, such as drivers, or other low-level functions that deal directly with the hardware, would require a more detailed knowledge of time, even though complete accuracy was not required. A good example of this would be to consider an instruction set simulator (ISS) that was defined to be instruction accurate. This means that while the timing associated with an instruction was known and accurate, the exact placement of read and write cycles within the instruction was not.

A third level was called bus cycle accurate or BCA. This had complete timing down to the notion of a bus transfer, such that all operations in software could be defined down to the exact order that they appear in the real system. Continuing with the previous example, this would mean an ISS that is completely accurate in terms of when its bus operations would occur. Most ISS models fall somewhere between these later two levels of timing accuracy, in that most instructions would be bus cycle accurate except under certain conditions caused by exceptions within their architecture that may produce small errors in the timing of their operations. To correct these would require complete modeling of the internal architecture of the processor, thus defeating the purpose and benefit of having an abstract model. For all practical purposes, such an ISS model can be thought of as "cycle accurate."

Several other abstraction views were also proposed such as the architects view and verification view, but none of these were widely adopted by the industry. The programmer's views, on the other hand, was adopted in the industry.

With the introduction of the OSCI TLM 2.0 standard [3], these terms were largely abandoned in preference of two new terms. The first of these is loosely timed (LT) and the other is approximately timed (AT). At the same time, the TLM 2.0 standard is primarily directed toward memory-mapped busses, which means that it is of direct support to the hardware/software interfaces rather than a more generic peer-to-peer connection. (See Chapter 3 for more details on TLM 2.0.)

In neither case are the terms LT and AT strictly defined, and thus models described as either of these may differ from those obtained from other people. This is part of the problem associated with putting together a complete and consistent virtual prototype. Most platform models that include ISS models for their processors would have been defined as PV models in the TLM 1.0 specification and as LT models in the TLM 2.0 specification. Similarly, the PVT models of TLM 1.0 are somewhat similar to AT models as defined in TLM 2.0. While this is not a strict one-to-one correspondence, it is close enough for our discussions.

Within the Mentor Vista environment, the notions of timing are aligned with the AT and LT notions of timing. However, Vista provides a much more flexible timing philosophy over what has been defined in TLM 2.0 as it supports both AT and LT as a continuum of timing possibilities. Vista also supports an untimed mode, in line with the original intention of the PV model. The combination of the PV untimed with the AT and LT timing is designated as PV+T.

8.3.2 Model Taxonomy

The programmer's view (PV) model is a pure functional model, which is sometimes referred to as a perfect model, in that it assumes infinite resources or bandwidth and that computation and communications happen instantly. Models within the Mentor Vista product, which is being featured in this chapter, may also have some notions of sequencing that emanate from models such as processors. This level of timing is appropriate for the loosely timed (LT) level of abstraction as defined by TLM 2.0. The taxonomy diagram for this is shown in Figure 8.1. See Chapter 1 for the taxonomy key. While this kind of model is fine for the verification of some aspects of a system, timing is often one of the most difficult and time-consuming aspects of a design to get right. Timing is an artifact of assigning functionality to physical

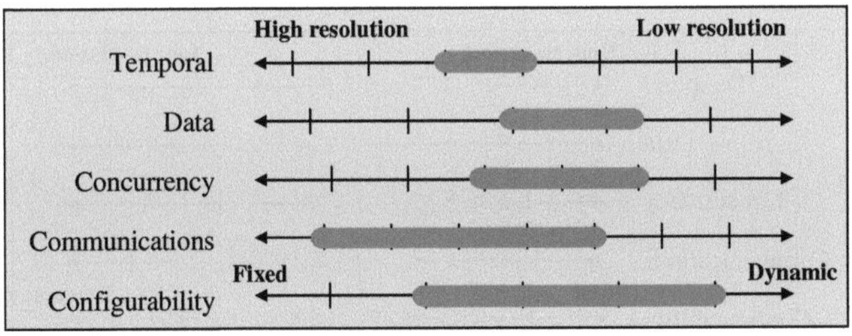

Fig. 8.1 PV models for Vista

entities, such as processors and buses, but also at a finer level of detail as well, such that the impact of adders and multiplexors is also considered.

The addition of these architectural elements can also affect behavior at a lower level of abstraction. For example, at the PV level, we may have a transaction that flows between two blocks. That transaction clearly has a protocol and a payload associated with it. In reality, both aspects of this transaction consume time and resources. The protocol will probably be implemented as a state machine in both the producer and the consumer and this may take multiple clock cycles. This adds latency to the transaction. The payload transfer also takes time and it is unlikely that the totality of it can be moved from one place to another in one go. It will need to be split into several pieces, with each piece being transferred as a mini-transaction (a burst). The lower level protocol may also have to get involved in the sequencing of each of the pieces of the payload, which also adds latency between each of the payload piece transfers.

While at the end of the transaction, the functionality that has been executed should be identical to the original PV intent, a lot of other functionality has been added in order to make the transaction a reality. At the same time, the addition of this timing information may have affected the order in which things happen at the system level and this in turn may affect the operation at the higher level as well. In other words, the functionality has remained unchanged, but the behavior is different.

A specification defines one function but may permit many possible behaviors, all of which are correct. As design choices are made, the range of possible correct behaviors that match the required functionality is reduced until only one remains – the implementation. Some people also see this as the necessary ambiguity that is contained in a specification. That ambiguity is what allows multiple design decisions to be made, so long as the requirements and constraints are also fulfilled.

Timing also indirectly affects other aspects of design. A wider bus will mean that more of the payload can be transferred in a single operation, but the wider bus will also mean increased cost of the final product or a larger chip area. Power may also increase. It thus becomes important to investigate the impact of various computation and communication architectures without expending the time to actually implement them. At the end of the exploration process with the Vista tool, the taxonomy diagram will look more like that of Figure 8.2.

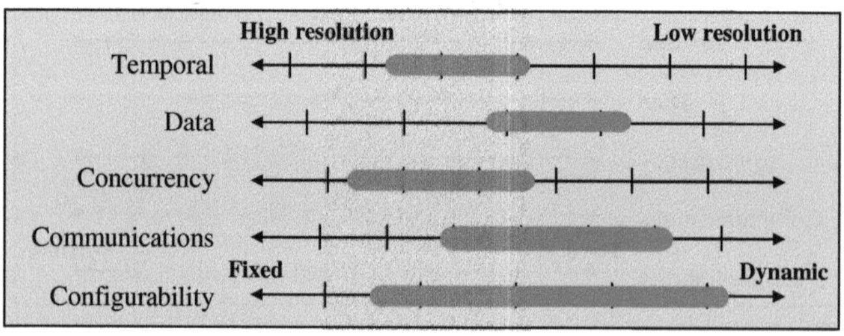

Fig. 8.2 Refined model from Vista

The biggest changes between the before Figure 8.1 and the after Figure 8.2 are that the timing has become more refined and this in turn has increased the potential levels of concurrency to include pipelining. While timing does not quite reach the cycle-accurate marker, it does extend toward this marker. Additional forms of communications have been defined such that the speed characteristics of buses have now been taken into account. We also see something that may be counterintuitive in that the communications types supported have been pushed toward the lower resolution end of the scale. This is because things such as dedicated point-to-point communications that represented abstract notions have been replaced with more accurate physical forms of communications. Overall, the model has become more refined.

8.4 Roles of the TLM Platform

A TLM platform provides an essential framework within which many of the essential design and verification tasks can be performed. The main functions that are performed using this framework are modeling, assembly, verification, validation, and analysis.

- **Modeling:** Models contain many parts and in the case of models for hardware implementation we can list them as
  - **Function:** This is the pure function of a model as defined in the industry standard taxonomy [4]. As such it contains no timing or other implementation artifacts.
  - **Interface:** The separation of interface and function is very important for reuse, not only in the case of a block that needs to be connected using several different interfaces but also when connecting models of different levels of abstraction. The interface provides both the separation and the means of connection for models in a virtual prototype.
  - **Timing and Power overlay:** Timing and power are artifacts of implementation, and while they do not form the core essence of a virtual prototype, they allow a peek down into the implementation domain. At the same time, they could be viewed as setting power and timing budgets that are expected to be met by an implementation. However, it is not ideal for the addition of this architectural information to require modifications to the functional model itself and thus needs to be kept separate, but connected to the model.

  Each of the pieces of these models must function together and in an ideal world would be completely independent of each other, such that the modification of any one piece, for example timing, would not require any changes in the others. This is sometimes referred to as aspect-oriented programming. The reason this is so important is for reuse, both in terms of reuse within the same design and within different designs. Within a design, it may be that multiple instances of a particular function are required, but with different interfaces or different performance characteristics. If the function of the block can be defined independently, then if any bugs are found in that functional description, it only has to be changed once and not in each of its variants. Reuse over time makes this even more important, especially since different implementation possibilities may result in completely different architectures being chosen. If the core functionality can be reused without modification, then it is more likely to provide a bigger time and effort savings for the entire design.

- **Assembly:** What is a system to one person is a subsystem to another. The task of integration and verification has until now been a process performed toward the end of the design and verification flow. (See discussion in Chapter 4 related to the breakdown of the V diagram.) Errors found at this time are very costly not only because of the delay they inject into the project but in some cases the widespread need for modifications that they can cause, invalidating much of the verification that was done on those blocks. Many companies have instituted programs to try to reduce the number of these kinds of errors that escape detection during the earlier phases of verification. One of the most effective ways to do this is to perform the interface verification task early in the design flow using a system-level virtual prototype, thus ensuring that the integrated functions work together to perform the desired system functionality.
- **Verification:** The verification of high-level models can have a huge impact on the total development time of a project [16]. Not only does it provide a golden model against which an implementation can be verified, or synthesized, but in the case of a manual implementation path, it can provide significant insights into the block such that the design team has to spend less time understanding the concepts while they are implementing the RTL. Having to "think" at the RTL level is slow and cumbersome, and rework is both time consuming and error prone. Being able to gain comprehension at the system level has been a huge productivity boost to some companies.
- **Validation:** The integration and verification tasks of the old V-diagram methodology, as discussed in Chapter 4, ended in the assembly of a complete system, and at that time, it became possible to perform validation, which is the determination of whether the specification is as intended. With older methodologies, complete system models were only available at RTL or lower levels of abstraction, and this meant it was only possible to perform very small simulation runs or run longer simulations for only small portions of the overall functionality. It was thus almost an expectation that emulation, physical prototypes, or the first chips produced were the ones on which validation would be performed and where performance, timing, or in some cases functionality errors were detected. Some of these errors were not errors in the implementation, but in the specification itself. Now with the introduction of system-level virtual prototypes it becomes possible to perform both verification and validation in the early stages of the design, thus ensuring that these kinds of errors are minimized or even eliminated. For companies that have historically assigned one chip spin to the validation phase, they could see a considerable cost and time savings by the adoption of a complete ESL flow using system-level virtual prototypes. This is on top of all of the other benefits that may be obtained.

- **Analysis:** The design process is one of looking at alternative implementation possibilities and selecting the one that best matches both the functional and non-functional needs, such as timing, area, and power. One way to do this is to perform an implementation and test it to see how well it matches the goals. A system-level virtual prototype provides the ability to do a number of quick what-if kinds of analysis that can allow a huge range of potential solutions to be explored without actually performing the implementation. At this stage in the methodology it thus becomes almost impossible to split the concepts of design and verification as they have become a very tightly knit pair of operations. It is equivalent to the scientific process which is based on the proving of a hypothesis. We first propose a hypothesis, which in the case of electronics design may be related to an architectural option. This may establish the timing of some functionality, which can then be analyzed within the context of the system along with the similar effects of decisions about other blocks. This in turn leads to the selection of appropriate implementation choices, unless it is found that in combination they do not yield the required performance or some other such criteria.

8.5 Contextual Verification

In the introduction to this chapter, we mentioned that one usage of the TLM platform was for verifying the implemented blocks in the context of the system. In the previous book *ESL Design and Verification*, [5] we also introduced the notion of positive and negative verification. Positive verification is used to ensure that a required functionality is present, whereas negative verification concentrates on the elimination of bugs. This may seem like a strange differentiation to make as you may think that the lack of bugs means that the necessary functionality is present. This is not necessarily the case. Similarly, if all system functionalities are present, then you may reason that you have finished verification. This is also not true.

When a block is verified it is verified against the specification for that block. We can ensure that there are no erroneous behaviors existing in the block that do not match the specification. If we achieve this, then we can say that no bugs exist in the block's implementation of the specification. However, we do not know that the specification is correct or that the specification as defined is what is required in the system. This is a validation function. So when we integrate a verified block into the context of a larger system we are

1. verifying the functionality of the larger system and
2. validating the specification against which the lower level block was implemented.

Consider a simple example. A processor is to be used in a system and this is to be obtained from a third party. Given that most processors come from very stable companies, who have spent enormous amount of time and money on the verification task, it can be reasonably expected that the processor will function as specified, especially if it has already been used in several designs. Does this mean that the processor will work in every design? The answer to this is clearly no. There are expectations about the capabilities that the processor will provide, the buses and interfaces that it will provide and the performance that it has. We can reasonably hope that the system architects will have made sure that all of this is true as part of the selection process. This is not so easy with arbitrary hardware blocks, especially if this is the first time that they have been used within a company. Is the partitioning of a specification into the individual specifications of the blocks foolproof? Not at all! This is why we perform system-level verification.

In this chapter, we are creating a TLM platform where we can ensure that the integrated system functions as specified. For a moment, consider the hypothetical possibility that we have developed a set of tests that ensure that 100% of the necessary functionality exists and nothing more. This also assumes that the specification was correct and complete. Now when a block has been implemented, it can be integrated back into the system model and those same tests executed. If everything passes, we have proven that the block performs every function necessary without any bugs. At first, it may thought that we need to do no further verification of the block. If we were to collect code coverage data for that block, we may expect to find that large portions of the code have not been executed. This is because the code is redundant as to its meeting of the high-level specification. All of that code could be removed without affecting its operation in any way.

There are some exceptions to this general statement, such as redundant logic for error detection and correction which cannot be adequately modeled at the high level. However, while these statements are generally true, it is also a very dangerous position to take. Part of the system is probably software and it is very likely that the software will change over the life of the product. A change in that software could cause completely different aspects of the hardware to be exercised. If this functionality has not been tested, or indeed eliminated, then we could be stuck with hardware that only works with specific versions of the software. This has happened to some companies in the past and it is a sign that they did not perform enough negative verification. Most of these companies relied on physical prototypes for much of their verification, where it is difficult to inject stimulus that was not generated by a real system. It can and does create some differences of opinion between verification experts when considering the verification of processor-based system.

If the system does contain a processor, then one school of though says that we should run the drivers and other software as the correct way to exercise the rest of the system. In many industries, they say that if it runs the software, then it works. The other school of thought says that this limits the input to the system too much and much better control can be obtained by injecting bus cycles directly, even if they do not represent typical or possible patterns. This too is a question of balancing positive and negative verification. Neither answer is completely right as both are a necessary part of the overall system verification.

The use of a TLM platform enables both types of verification to be performed in a fairly effortless manner and allows both a top-down verification flow, and a bottom-up flow to be intermixed. This can save large amounts of integration time that were previously performed on RTL implementations as it ensures the specification for those blocks was correct before the implementation starts. Engineers can also test their interpretation of a specification, by placing partially implemented blocks back into the system. Alternatively, stimulus can be captured from the system-level execution and applied to the blocks in a stand-alone fashion. Care should be taken anytime differences are seen on the outputs as these alternative behaviors must be put back into the TLM platform to see if they upset any other parts of the system. These differences can come about for several reasons and may be timing related or due to particular choices that an engineer makes where ambiguity remains in the specification.

For example, if a specification says that under conditions of congestion, packets may be dropped so long as no packets above a particular priority are dropped, the decision about which packets to drop is arbitrary and a different decision than the one represented in the high-level model is not incorrect, just different.

We will expand on the way in which these issues can be addressed more fully in Sections 8.6.2 and 8.6.3.

8.6 Creating Models

Functional models are built primarily for speed since their primary target is software execution. The PV or LT levels of abstraction contain only just enough timing information to ensure the correct functionality can be presented. Most methodologies in use for this today use a blocking function call approach, which is fast (1,000–10,000$\times$) and relatively simple to write (10–100$\times$) compared to RTL. However, the exact accuracy of these models can vary between different applications. For example, when used for applications software development, almost no timing information is necessary and thus the model does not have to reflect the actual structure of the solution. However, if the model is being used to port an operating system, knowing if a bus transfer will use individual reads and writes or if a DMA transfer can be used is important. Thus this model would have to mimic some aspects of the implementation structure, even if it does not implement the details.

We also have to consider the nature of the software. While the above statements may be true for control-oriented applications, it may not be the case for data-plane software which will often have real-time deadlines associated with it, such as audio/video encoding or decoding or wireless baseband functions. In this case both high execution speeds and accurate timing are necessary.

The important characteristic of these models is that they are bit-true in terms of their behavior and have accurate register interfaces for all of the components. These are necessary such that binary code can be run on these platforms without modification or recompilation. However, there is no guarantee that they will execute exactly the same as an RTL model unless the models also exhibit sequence correct behavior. This last requirement is problematic because unless the software performs correct synchronization itself, then this requirement is almost impossible to obtain without full cycle-accurate models. A small change in timing could mean an interrupt or some other asynchronous event will be detected and registered at a different time which can result in grossly different behavior. These cycle-accurate models run at 1/100th the speed of a functional model and make it impossible to run any significant amounts of code. As an example Carbon Design Systems reported the following performance ranges for the models they provide for ARM processors [6] (Table 8.1).

Table 8.1 Relative model performance

Modeling style	Internal timing	Bus timing	Performance
Instruction set accurate (ISA)	Instruction cycle	None	10–100 MCPS
Cycle Approximate	Cycle count	Yes	~2 MCPS
Cycle accurate	Pipeline accurate	Yes	100s kCPS

Some systems do exist that can switch between operational modes (interpreted, compiled code, statistical etc.) for different parts of the software execution, or can gather performance statistics so that they come closer to mapping to actual clock cycles while maintaining a higher level of overall performance. An example of this was discussed in Chapter 6.

There has thus been a lot of debate in the industry about the appropriate amount of timing information to include in a model [7]. The methodology described in this chapter will demonstrate how multiple levels of timing accuracy can be put into a model without having to make significant modifications to the underlying functional model. The timing can either be defined in a declarative manner or it can be automatically extracted from an RTL model and annotated onto a functional model.

8.6.1 Model Refinement

Adding timing to a model has been attempted in multiple different ways. Gajski and colleagues [8] defined two independent aspects to the modeling domain, namely computation and communication. Each of the aspects is defined by three models at different levels of implementation detail. Each of these aspect models can be combined together to populate the entire chart as shown in Figure 8.3.

The most abstract level describes an untimed functional model A, also commonly understood as a specification model. Adding architectural information and the correlated functional timing behavior produces the refined model B, called a component assembly model. The bus-arbitration choice allows one to introduce a first notion of real communication timing and results in the bus-arbitration model C. From this model the designer could choose two ways to reach the final goal. Either way, via the behavioral bus-functional model D or the cycle-accurate computation model E, leads to the final RTL implementation model F.

One implementation that somewhat mimics this modeling structure is from the Open Core Protocol (OCP) International Partnership (OCP-IP) [9]. This organization creates standards for intellectual property (IP) with the goal of being able to

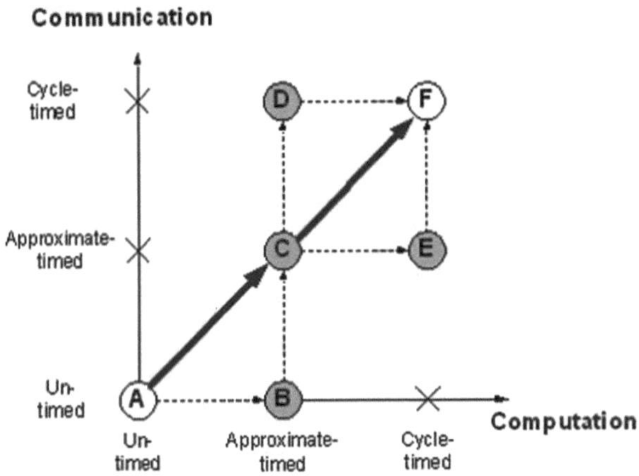

Fig. 8.3 Gajski system modeling graph

connect them together in a "plug and play" manner. It does this primarily through interfaces or sockets. As with TLM 2.0, it also concentrates on the processor bus interface, although other interfaces such as test and debug are also covered.

OCP [10] defines three levels of abstraction as shown in Figure 8.4.

Communication Accuracy		Data Accuracy	Timing Accuracy	Addressed Design Problems
domain specific MoC		tokens	causality, partial event ordering	algorithm design
TLM	TL3	packets	total event ordering, burst-level annotation	functional specification, generic architecture exploration
	TL2	burst of words	burst-level and word-level	exploration of OCP based architecture
	TL1	word	cycle accurate	100% cycle accurate performance profiling
RTL		bit vector	cycle accurate	synthesis

Fig. 8.4 OCP transaction abstraction levels

OCP does not define the timing model for the computation element at all, instead only defining that of the communications between models. In fact in a white paper [10] they state

> The first observation is that the abstraction level of the SoC platform model is mostly influenced by the abstraction level of the deployed bus model. More specifically, the accuracy and simulation speed inherent to the chosen model of the communication architecture determines the aptitude of a TLM simulation model for a specific use-case.

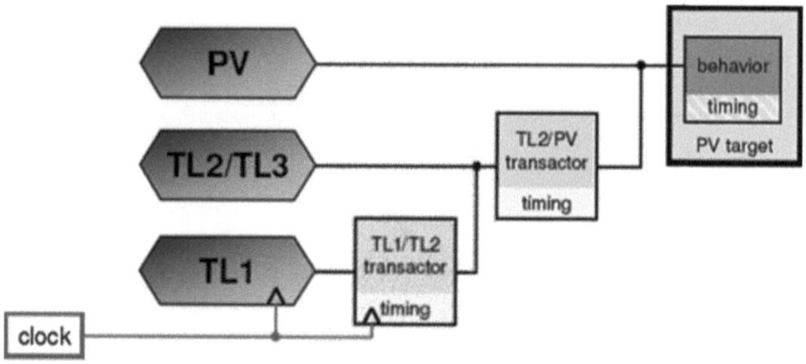

Fig. 8.5 OCP TLM methodology

The PV model as shown in Figure 8.5 communicates with the bus transactor models using a single interface method. Because of this they state that you must be *careful to construct the model such that timing can be introduced into the behavior as and when it is required.* Timing can either be implicit or explicit. Explicit means that the timing is modeled inside the functional model using events, synchronization, or other mechanisms provided by SystemC. Implicit means that the timing is annotated into the TLM calls. While implicit timing is their preferred method, it is not possible to fully define timing in this way. Implicit timing can be defined in a number of ways, including *a configurable constant value, a statistical delay value using configurable constraint randomization or a delay calculated from the internal state and the actual data.* The document goes on to show some of the limitations of this approach when dealing with concepts such as pipelining and arbitration.

Ghenassia and colleagues [11] promote an approach where each PVT model is a combination of an untimed PV model that may have delays annotated or a PV model in combination with a stand-alone timed model. With the annotation approach, (which they state is useful when the structure of the untimed model matches the structure of a micro-architectural model) annotations are simple wait statements related to the computation time of a specific functionality. These delays are put into the PV model and used in conjunction with an untimed TLM model for communications.

In the stand-alone timing model, the timing behavior is modeled such that delays are computed within the timing model. It thus becomes a high-level analytical model without functional information. When timing behavior depends on functional behavior, the timing model can be controlled externally through a trace stream generated by the functional model. This is depicted in Figure 8.6.

The untimed PV model is executed until it reaches a synchronization point. Then the timing model starts and computes the delays associated with the data transfers that the PV model generated. This includes both computation and communication delays. When completed, control is returned to the PV model again.

One advantage of this methodology is that the PV and timing models can be developed independently and one PV model may have several timing models that each represents a different micro-architecture. One problem with this is that it may be over-accurate for the PV mode and under-accurate for the PVT mode.

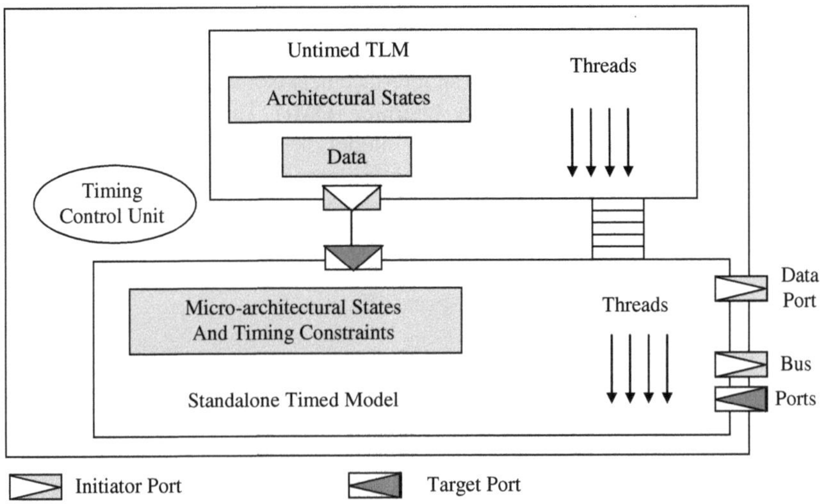

Fig. 8.6 Stand-alone timed model

This approach may make too many compromises so that it can operate in both modes. It would be slower than it could be for PV only and not have enough fidelity for accurate T.

All of these approaches suffer in the ways in which the functional model and the timing model interact and this in turn limits the accuracy of the timing models that can be created. In addition, the ways in which delays are distributed in the model can make it difficult to see the real timing policies that may exist within a model. For these reasons, Mentor decided to try a different way to define timing and the ways in which the models communicate. This approach, which is based on declarative methods of defining timing, will be more fully described in Section 8.8.

8.6.2 Multi-abstraction

There is one further requirement for a TLM platform that must be met before it can be universally accepted as part of an ESL flow. That is the ability to integrate models at different levels of abstraction. This could be because either the only model available for a particular component is at the RTL level or the TLM platform is going to be used as a test harness for blocks as they get integrated and it is desired to run the system-level tests on this multi-abstraction set of models. While this cannot depend on a single standard, each of the major HDL languages provides an interface layer that enables an RTL model to communicate with a C model. For VHDL (VHSIC hardware description language), this interface is called VHPI (VHDL Procedural Interface); for Verilog it is called PLI (Program Language Interface); and for SystemVerilog, a new interface has been defined called Direct Programming Interface (DPI). All simulators that support these languages should provide these interfaces. We will concentrate on the SystemVerilog DPI which is the one used by Vista and is also the only one that has high enough performance to be useful for a virtual prototype at this level of abstraction.

In addition, TLM 2.0 also creates an interface between a model and the contents of its memory called Direct Memory Interface (DMI). This means that while the memory may be instantiated in a memory model, it is possible for a processor or other hardware block to directly access the contents of that memory without having to perform all of the bus transactions necessary to access this in the way that real hardware would. This is very useful in the creation of a loosely timed virtual prototype, to load large quantities of information into memory or to provide faster execution of certain non-interesting aspects of a prototype.

8.6.2.1 SystemVerilog DPI

The SystemVerilog DPI interface was discussed in Chapter 4. It is used here in co-operation with the Open Verification Methodology (OVM), a class library used for the verification of designs at the RTL level. The suggested methodology for OVM (and indeed for VMM, another similar methodology as discussed in Chapter 4) is to use transaction-level communications between the testbench and the model being verified. A transactor model is used to bridge the abstraction gap between the two. When used with a virtual prototype, that transactor is no longer required as the inputs and outputs of the model are already at the transaction level. This means that the same testbenches can be used across both levels of abstraction.

However, nothing in real life is that simple. The first thing is that we are mixing not only multiple abstractions but also multiple languages since we now have the testbench written in SystemVerilog and the model in SystemC. Many of the problems and limitations of this were discussed in Chapter 4. These problems do not manifest themselves with the more abstract PV modeling style since the recommended way of writing models does not necessitate time advancement.

However, this does mean that care must be taken when dealing with testbenches coupled to timed SystemC models or when the system model is required to operate as the reference model for RTL verification. Some mixed language simulators attempt to make this as easy as possible. These issues will be discussed in the following sections.

8.6.2.2 Transactor Models

As previously stated, a transactor is a particular type of model that sits between two blocks that are modeled at different levels of abstraction. The role of a transactor is to translate high-level transactions on one side of the model into lower level transactions on the other side. It also performs the reverse function of monitoring low-level transactions and converting them back into the appropriate high-level transactions.

In Chapter 2 a detailed description was provided for the most common type of transactor, the bus interface model, which is the one that exists between the hardware and software domains. This type of model takes bus requests on one side and outputs a sequence of signal operations on the other side that represent the hardware implementation of the bus protocol. When performing a bus operation, such as a read, it will also monitor the data bus and at the right time capture the information it sees on the signals that represent the bus and return this as the data payload for the bus read operation (Figure 8.7).

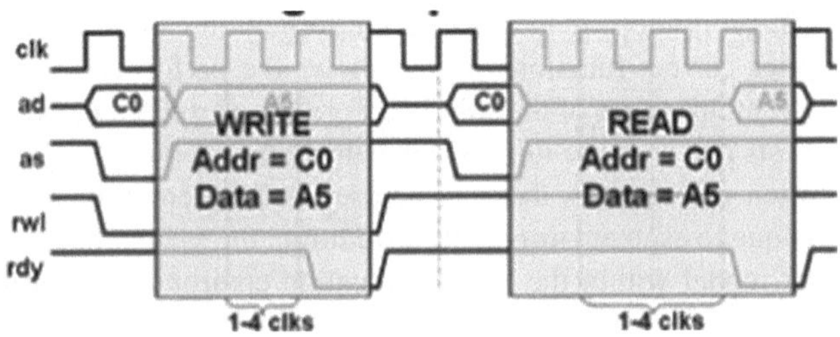

Fig. 8.7 Bus transactor operations

While the previous example converted between a transaction-level model and an RTL model, a transactor can exist between any two levels of abstraction.

With the use of transactors, which can be stacked, it thus becomes possible to put together platform models in which many different levels of abstraction exist, a concept that is essential to being able to verify a block implementation within the context of a system model.

A different type of transactor model that is becoming more common today bridges between the software model executing in a simulator and a physical hardware model. In this context, software model means everything that is modeled in software at any level of abstraction. We have seen these for some time when connecting a simulation model to an emulator. Here the transactor is split into three primary pieces, namely the software model, a communications infrastructure, and a synthesizable model that will reside in the emulator. The Accellera standard co-emulation modeling interface (SCE-MI) is the interface that most vendors use to provide a common infrastructure so that models need only be written for one such system and will operate on all of the others.

In Chapter 5 we also saw some examples of transactor models that bridge between the software model and actual physical devices that virtualize the functionality that is being reflected in the model, such as virtual USB or Ethernet ports. We can thus see that over time, an increasing number of model types can be integrated together to form a prototype, many of them being hybrids between software models and hardware models that mimic the necessary functionality in a variety of ways.

8.6.3 Verification

While there is a desire to have a single testbench that covers every aspect of verification and validation, the reality is that this is not possible today. The testbenches constructed for RTL, even though modeled at the transaction level, do not have some of the necessary capabilities to be suitable for the creation of system-level testbenches. This is primarily because they assume a timed DUT model. At the same time, as was discussed in Section 8.5, there are some differences between positive and negative verification, and these are often manifested in terms of different testbenches.

At the system level we need to put through long sequences of data that mimic actual operation. From these long sequences information can be gathered that not only demonstrates correct functionality but shows the performance of the system that relates to the architecture of the solution. For RTL verification, we need shorter more targeted tests that exercise the specific behaviors of a block.

There are often several tests that can be used in both environments, and some tests that are unique to each environment. In addition, the system model can be used as the reference model within the RTL verification environment, but may require some special interconnect between the models to ensure that each model executes the same functionality where multiple valid functionalities exist.

8.6.3.1 Verifying the System Model

There are four primary ways to verify the system, namely

1. Using traffic generators to create streams of transactions that mimic the typical or extreme loads that the system may encounter under various types of operation. The actual software would be run on any processors within the system and thus test both the hardware and the software.
2. Running the system in a real environment where the peripherals are connected to actual data streams through the use of speed gaskets or some other mechanism to compensate for the speed disparity between the model and the real world.
3. Running test software on one of more of the processors that causes specific activity to take place in the hardware. This clearly tests the hardware only, but can target specific areas of the hardware functionality in a more focused manner.
4. Removing the processors from the system, leaving only bus-functional models (BFM) into which specific bus transactions are driven. This allows highly focused execution of the hardware unconstrained by the limitations of the processor or software.

In many cases a combination of all these verification methods will be used and it should be noted that cases 1 and 2 roughly correspond to the notions of positive verification, and cases 3 and 4 correspond to negative verification as discussed earlier in Section 8.5. It is thus important that at least one method from each class be chosen for complete system verification.

Running test software was discussed in some detail in Chapter 4 and will also be the focus of the example later in this chapter. Running the system model in a real environment was also discussed in Chapter 5.

Traffic generation can take two primary forms. The first of these is the equivalent of a directed testing approach, where a specific scenario or sequence of data is injected into a port of the system. Figure 8.8 shows a simple example for a traffic generator on an ARM cortex A9 bus.

```
void Cortex_A9_pv::thread_simple() {
        unsigned int Size = 1;
        unsigned int wData [Size];
        unsigned int rData [Size];

      wData[0] = 47;

        cpu_master1_read (0x10,   rData, Size);

        wait(1, SC_US);
        cpu_master1_write(0x301,  wData, Size);
        cpu_master1_write(0x1010, wData, Size);
        cpu_master1_read (0x1010, rData, Size);
        cpu_master1_read (0x301,  rData, Size);
```

Fig. 8.8 A directed traffic generator

In the example above, we can see specific read and writes being performed on the bus. No checking of the data is conducted as it is assumed that this function will be performed by other components of the verification environment. In a complete system test, it may also be necessary to synchronize activities across multiple traffic generators to create the complete scenario.

More typical would be a traffic generator that randomly creates transactions, such as the one shown in Figure 8.9.

In this example, the address, data size, and data are all randomized. Five hundred transactions are created with a random time interval between them. It should also be noted that not everything is left purely random. For example, the address is kept in range of the actual address map and if more reads than writes are desired, this can be achieved by setting the appropriate value for `ReadPercent`. This generator also does the checking of the data as it keeps track of the data that was written to each memory location. However, it does not check that data was not inadvertently written to other locations, so it is possible that this will not find all errors. It is also possible that locations that were written to will never be read or that locations will be read before they have been written to.

It is also possible to create traffic generators in languages such as SystemVerilog or *e* that have sophisticated constrained random generation functions or to utilize the SystemC verification library.

8.6.3.2 Using the System Model for RTL Verification

Once a system model has been created, many companies wish to use this to verify their implementation. If the implementation is synthesized from the system model, then this corresponds to equivalence checking between the two models. It is thus imperative that full functional verification was performed on the system model as no additional verification is being performed on the RTL model. These two models are not independently created models and so a comparison made between them is not very useful.

```
void Cortex_A9_pv::thread_random() {

    saDistribution *randAddr = saCreateDistribution(AddrDist);
    saDistribution *randData = saCreateDistribution(DataDist);
    saDistribution *randSize = saCreateDistribution(SizeDist);
    saDistribution *randWait = saCreateDistribution(WaitDist);

    saDistribution *randM1M2 = saCreateDistribution("uniform 0 100 79");
    saDistribution *randRnW  = saCreateDistribution("uniform 0 100 47");

      //First: write some data to snoop, save it in "local memory",
      //then, check that they are identical
      unsigned u=0;
      unsigned count = 500;
      for (; u<count; u++) {
        saLogMessage("%s: stage: %u\n", (const char*)(name()), u);
        unsigned int cacheIndex = (randM1M2->getNextInt() < Master1Percent) ? 0 : 1;
        unsigned int length = randSize->getNextInt();
        unsigned char data[length];
        unsigned int address = randAddr->getNextInt() % (memory_size - length - 1);
        unsigned int readOrWrite = (randRnW->getNextInt() < ReadPercent) ? 1 : 0;
        switch(readOrWrite) {
          case(0) : //write
          {
        //create some data
            for (unsigned j=0; j<length; j++) {
              data[j] = randData->getNextInt();
            }

            memcpy(memory + address, data, length);
            switch(cacheIndex) {
              case(0) : { cpu_master1_write(address, data, length); break;}
              case(1) : { cpu_master2_write(address, data, length); break;}
            }
            break;
          }
          case(1) : //read data from cache
          {
            unsigned char shouldBeData[length];
            memcpy(shouldBeData, (unsigned char*)(memory + address), length);
            switch(cacheIndex) {
              case(0) : { cpu_master1_read(address, data, length); break;}
              case(1) : { cpu_master2_read(address, data, length); break;}
            }
            for (unsigned j=0, jCount = length; j<jCount; j++) {
              assert(shouldBeData[j] == data[j]);
            }
          }
        }
        wait(randWait->getNextInt()*10, SC_NS);
      }
}
```

Fig. 8.9 Random traffic generator

If the RTL implementation is constructed by hand, then the system model is an independent reference against which the RTL model can be verified. While many people may not consider this to be an ESL flow, there are several companies that have found that this has considerable value.

The reason for this is that it increases comprehension about a system which is an important aspect of the ESL definition. That increased comprehension leads to the right architecture being chosen for the implementation and fewer bugs being injected into the system due to incomplete and ill-defined specifications.

One example of this was relayed to the authors in a telephone conversation with Dave Zalatimo from TI. They developed the system model because they had some problems with performance on previous designs since this aspect of the implementation could not be verified until the RTL was ready. Instead they created a C++ model and performed extensive verification on this model using SystemVerilog testbenches. They used a similar approach to the one described in this chapter where they started with an untimed model and incrementally added timing as the control path in the design became more refined. The end result was that their RTL came in ahead of schedule and had far fewer problems than in previous designs.

While in cases such as these, the system model and implementation are not completely independent, the fact that extensive verification is first conducted at the system level means that many of the problems that would have resulted from problems in the specification have already been resolved. This aspect of verification has thus already been dealt with and is thus redundant to perform again.

Bringing together a C, C++, or SystemC model with a SystemVerilog or *e* testbench is not without its own difficulties. Verification methodologies, such as VMM and OVM, do provide some guidance and support for this integration. However, the interface needs to be constructed in certain ways to make this possible and these requirements may influence how the system model is constructed. The biggest issues relate to data representations across the boundaries as well as time advancement issues.

In the case of SystemVerilog and the DPI interface that was discussed in Chapter 4, the data types are restricted to those that are native to SystemVerilog. While this does cover many of the simple data types, others need to be handled with care.

While a number of methodologies exist that recommend adding a `tlm_fifo` between all interface points, this can be tedious because SystemVerilog does not support templatized functions or function overloading as SystemC/C++ does. Reference [12] outlined an interface methodology in a Mentor whitepaper, where a SystemC component is created for each input or output port that exports a function to be called from SystemVerilog. These functions either put or get a transaction from SystemC. A `svOpenArrayHandle` data type is used so that SystemC can extract the size and data value for each element.

Packet creation in SystemVerilog is shown in the code snippet below:

```
//SV Transaction
class Packet;
  rand int pkt_id;
  int len;
  rand int data[];
  constraint data_c
{ data.size()=='MAX_LENGTH; };

  function void post_randomize();
    len=data.size();
  endfunction
```

```
  function new(int id=0, int l=0, int d[]=null);
    pkt_id=id;
    len=l;
    if(d) data = new[len](d);
  endfunction

  function Packet clone();
    Packet tmp=new(pkt_id, len, data);
    return tmp;
  endfunction
```

The SystemC equivalent looks like the code snippet below:

```
//SystemC Transaction
class Packet {
public:
  int pkt_id;
  int length;
  int * data;

  Packet(int p_id = 0, int len = 0, int *d=0) :
      pkt_id(p_id), length(len)
  {
    data = 0;
    if(length>0) data = new int[length];
    for (int i = 0; d&&data && i < length; i++)
{
      data[i] = d[i];
  }
....
```

The next step is to create the SystemC components that export the put/get functions as shown below:

```
// For each input port create a SC_MODULE
// with the right put DPI function
SC_MODULE(stim) {
sc_port<tlm::tlm_blocking_put_if<Packet>>
   stim_p;

   SC_CTOR(stim):
    stim_p("stim_p") {
#ifdef MTI_SYSTEMC
      SC_DPI_REGISTER_CPP_MEMBER_FUNCTION("dpi_put", &stim::
  dpi_put);
```

```
#endif
    }
  int dpi_put(int i, int len, svOpenArrayHandle p) {
    int * dd = (int *)svGetArrayPtr(p);
    Packet * pkt=new Packet(i, len, dd);
    stim_p->put(*pkt);
    pkt->print("PUT");
    return 0;
  }
};

// For each output port create a SC_MODULE
// with the right get DPI function
SC_MODULE(resp) {
sc_port<tlm::tlm_blocking_get_if<Packet>>
   resp_p;

   SC_CTOR(resp):
    resp_p("resp_p") {
#ifdef MTI_SYSTEMC
      SC_DPI_REGISTER_CPP_MEMBER_FUNCTION("dpi_get", &resp::
  dpi_get);
#endif
    }
  int dpi_get(int* i, int * len, svOpenArrayHandle p) {
    Packet *pkt=new Packet;
    resp_p->get(*pkt);
    pkt->print("GET");
    *i=pkt->pkt_id;
    *len=pkt->length;
    int * dd = (int *)svGetArrayPtr(p);
    memcpy(dd,pkt->data,pkt->length*sizeof(int));
    return 0;
  }
};
```

Now connect the modules through a tlm_fifo as shown below:

```
// For each input/output port connect the
// above modules through tlm_fifo
SC_MODULE(sc_top) {
  simple s _mod;
  stim s;
  resp r;
```

```
  tlm::tlm_fifo<Packet> stim2simple;
  tlm::tlm_fifo<Packet> simple2rcvr;

  SC_CTOR(sc_top) :
    s_mod("s_mod"),
    s("s"),
    r("r"),
    simple2rcvr("simple2rcvr"),
    stim2simple("stim2simple")
  { s.stim_p (stim2simple);
    r.resp_p(simple2rcvr);
    s_mod.in_p (stim2simple);
    s_mod.out_p(simple2rcvr);
  }
};
```

Finally, it is possible to import and use the DPI function in the SystemVerilog testbench to put and get data to and from SystemC as shown below:

```
//SV Put Port
import "DPI-" context
    task dpi_put(input int pkt_id, input int len, input int data[]);
class stimgen;
  task run();
    Packet pkt=new;
    for(int i=0; i<10; i++) begin
      Packet tmp;
      assert(pkt.randomize());
      tmp=pkt.clone();
      display("SV putting %p",
          tmp.data['MAX_LENGTH/2]);
      dpi_put(tmp.pkt_id, tmp.len, tmp.data);
    end
  endtask
endclass

//SV Get Port
import "DPI-SC" context task
    dpi_get(output int pkt_id, output int len, output int data[]);
class responder;
  task run();
    for(int j=0; j<10; j++) begin
      Packet pkt;
      int i, l, d['MAX_LENGTH];
      //fixed size array required for dpi_get
```

```
        dpi_get(i, l, d);
        pkt=new(i,l,d);
        display("SV getting %p", pkt.data['MAX_LENGTH/2]);
      end
    endtask
endclass
```

One further problem with languages based on C++ is the name mangling nature of compilers. To overcome this it is necessary to declare any function that will pass across the interface as an `extern"C"`. This of course applies to any SystemC model.

8.7 Timing

As discussed in Section 8.4, Mentor's Vista product separates function and timing by allowing timing to be added to a model as an independent layer that can be changed without any changes to the original model. This is useful in many different ways. For example, if a particular block is instantiated multiple times in a design, the same PV model can be used with different timing overlays, each representing a different implementation option. This is even more crucial when considering reuse over several designs. Being able to preserve the function without also being forced to adopt the same implementation decisions used in the last design makes it much more reusable.

In the creation of this methodology, Mentor had the following goals:

1. Enable writing the PV models in a fast and concise functional style.
2. Introduce the T part in a way that will leave the PV code mostly unchanged.
3. Have a clear, well-defined interface between the PV and T parts of the design.
4. Enable the system to work in two modes: PV and PV+T with the option to change between modes at run time. The T part existence should have a negligible performance overhead when the simulation runs in a PV mode.
5. Allow a T coding style that provides transaction accurate timing together with achieving good simulation speed.
6. Avoid, as much as possible, the duplication of code of the PV in the T code.
7. Encourage the evaluation of different micro-architectural implementations of the same PV system.
8. Enable one-to-many mapping of PV to T models.

Most approaches to this problem in the industry use a concept similar to that shown in Figure 8.10a. In this approach, timed transactions are fed into the T model. Here multiple transactions may be combined into a single transaction, which is then fed into the PV model. After a transaction has been created by the PV model, it is fed into the timing model which adds appropriate timing and may possibly split the

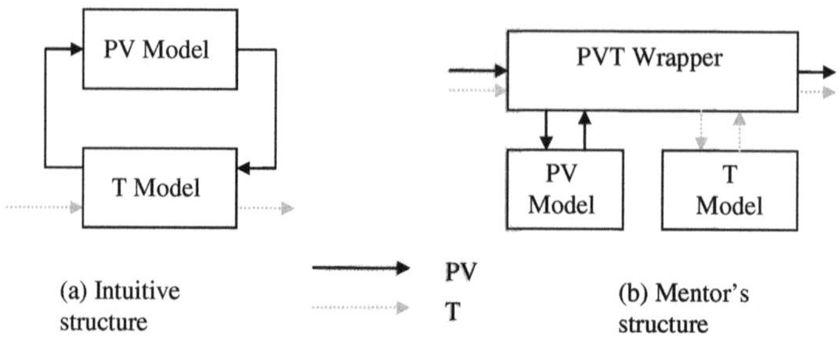

Fig. 8.10 Adding timing to a PV model

transaction into multiple transactions on the outgoing timed interface. As previously mentioned this has problems in the accuracy of timing related to pipelining and other hardware effects that can be modeled.

The Mentor approach is composed of three elements as shown in Figure 8.10b. The PVT wrapper is automatically generated from within the Model Builder tool, so the user only has to create two models.

Mentor recommends that the PV model is written using a function and block style. By function they mean that the model is written in the methods on the ports in a sequential style – without breaking it to threads. Block style means that the model does not break transactions into a sequence of smaller transactions. Using this style of modeling makes it fast to write the models and achieves maximum LT simulation performance.

The order of the outputs from the sequential PV code can be different from the order that is scheduled from the timing model associated with a concurrent implementation, and in general this is not desirable. To reconcile this difference in ordering, the PV code runs without consuming time but the wrapper queues the PV calls internally and schedules them as (T) transactions under the control of the timing model.

The problem with this scheme is that the PV model needs all of the data immediately in order to continue its sequential execution. To solve this problem the wrapper gets the values immediately for the PV model through the TLM2.0 DMI. This makes it possible to create timed models that are accurate and yet still retain the ability to run the PV models in LT mode.

The PV model executes until it either reaches a synchronization point or computes the results in zero time. The T model performs the computations related to how the output should be timed, including how the data should be chunked. The wrapper then schedules the transactions. Many of these may be associated with a single PV call if, for example, a bus transfer would have been split into multiple pieces. It can start producing this output timing data even if the incoming transaction has not yet been completed. This enables pipelining to be emulated.

An example of this will be shown in Section 8.7.5. This model of operation is quite new but has some resemblance to one of the features of the late lamented VCC created by Cadence's Alta Group [13].

There are also several possibilities to consider with data types. The fastest are obviously the native C++ data types, but the problem with these is that they are limited to the concepts that exist in the software world, such as int and byte. Additional data types are necessary to represent specific hardware concepts. Another division of Mentor defined a package of fixed-point data types, originally with a high-level synthesis motivation, which are fully defined in Chapter 9. This includes variables with arbitrary bit-widths and according to Mentor executes at high speed and has been proven to be accurate and consistent across all of the data types. Finally there are the data types defined as part of SystemC itself. While these may be slower than the other two types of data, it is necessary for Vista to support them so that arbitrary SystemC models can be run. Vista supports all of these datatype methodologies.

8.7.1 Timing Policies

Traditionally, the languages used to design and verify electronic systems have been imperative languages. This means that they have a set of instructions that are executed one after the other. An example of this is the C language. While the hardware description languages (HDLs), such as Verilog and VHDL, added concepts of concurrency, the imperative approach is still used within blocks. More recently, assertion languages have been introduced that enable functionality to be verified using either formal or dynamic methods. These assertion languages, such as property specification language (PSL) [14] and SystemVerilog assertions (SVA) [15], are based on a different programming concept called declarative programming. Declarative programming defines paths of functionality from inputs to the expected outputs. It is often possible to define complex behaviors in very a compact style using these languages. Mentor has adopted a similar approach to the definition of timing. Rather than attempting to define each element of timing as it applies to a functional model, they allow the user to define timing policies that are related to transactions on ports or between ports. Thus the user does not have to worry about how these policies would be written in a language such as SystemC as the Mentor tool suite does this automatically via code generation.

Four timing policies are supported, namely delay, sequential, split, and pipelining. Each of these will be explained below. These timing policies are attached to ports and thus some are appropriate for master ports, i.e., those that initiate a transaction, slave ports, i.e., those that respond to transaction requests, or between ports. This last category of policies allows dependence and sequence to be defined such that it is possible to say that after two consecutive writes, an additional delay will be added or that if a read follows an interrupt, then it will have a modified delay.

8.7.2 Delay

This is the simplest of the timing policies and represents computation time. It can be used for many different aspects of a model, such as the time that it takes to access a

Fig. 8.11 Delay timing

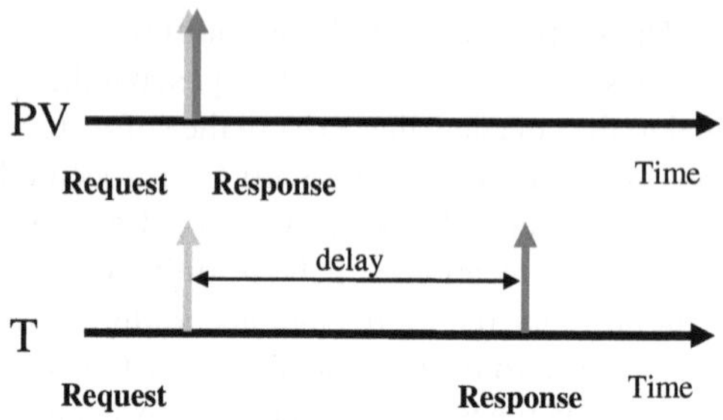

piece of data after a read request has been made, or the processing time associated with a piece of data. This timing policy can be applied to any slave port and is specified in a number of clocks rather than as elapsed time. It is also possible for the delay to be defined as a variable, or an expression containing variables, such that the delay can be dynamically calculated rather than a static number (Figure 8.11).

An example of the policy defined in declarative textual form would be

```
data_delay_policy ON slave WAIT STATES 2//
The latency of read and write transactions on port slave is 2 clocks
```

8.7.3 Split

A split policy applies to master or outgoing ports. It defines how data is to be split into a sequence of T calls. Note that the actual data in the PV call remains untouched. This is used to model buses that have a defined width or other policies that may require that the data be transferred in a particular way. It can also represent buffers at a model's output that restrict the size of the output bursts. Thus if a 128 byte data packet has been produced and needs to be passed across a 4 byte bus, then at least 32 consecutive writes would be required.

Three parameters are associated with this policy, namely the size, the latency, and the bus protocol. In our example, the size would be 4. The latency defines the time in clock cycles between the writes. Thus if the latency were defined as 2, then a delay of 2 clocks would be inserted after the completion of the previous write. It should be noted that the completion time of the first write may be delayed by the timing policies of the receiver. Further, if multiple transactions are utilizing the same port and have been interleaved, as will be described in Section 8.7.5, then additional time may pass between the write transactions associated with this PV call.

The latency thus represents the minimum time that can elapse between T calls for this PV transaction. In addition to this chunking, it is possible that the protocol used on a port, which is the third parameter to this policy, may impose further constraints, such that chunk sizes must be a power of two or to require address alignment.

While the policy is called "split," it should be noted that what is really being defined is the burst size and the model will then perform any necessary splitting in the transaction to accommodate the size of the transaction being transferred and bus policies (Figure 8.12).

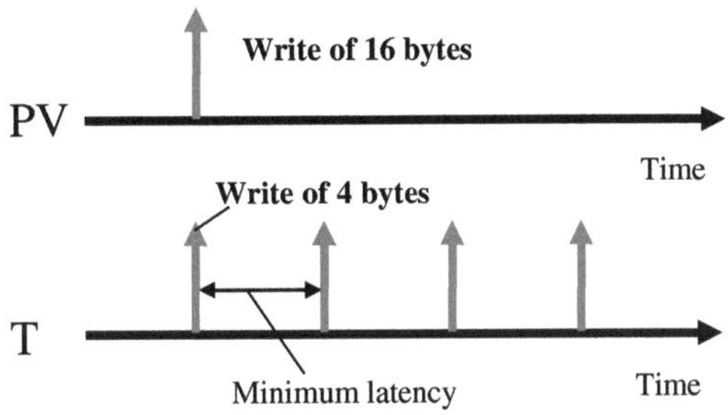

Fig. 8.12 Split policy

While these parameters allow many split policies to be defined, there may be times when this is not sufficient. For these cases Mentor allows custom split policies to be written in C++. However, this is outside the scope of this chapter.

An example of the policy defined in declarative textual form would be

```
split_policy ON master BURST SIZE bufferSize BUS ahb LATENCY 3
// The traffic on the master bus (read and write) is done through AHB bus policy
//    with maximal bursts of size bufferSize (a parameter of the model)
//    and 3 clocks delay between bursts
```

Note that `bufferSize` is a variable rather than a constant value and was defined as an `mb_variable` in the PV model making it accessible to the T model. The naming convention "mb_" is that used by Mentor for all variables within Vista.

8.7.4 Sequential

The sequential policy defines timing between events. There is thus a *cause* and a possible *effect*. The cause does not make the effect happen, that is controlled by the PV model, but *if it happens*, then this timing policy will apply.

Consider a policy that says if we have two back-to-back writes, then the second write will be delayed. This policy will not be applied if a read follows a write. While this example talks about transactions on the same port, it can equally apply to transactions on different ports. One common example may be an interrupt being received on one port which could be followed by an interrupt acknowledge cycle on a bus. If the acknowledge happens, then this defines the minimum delay associated with that event (Figure 8.13).

The latency is calculated from the end of the cause T transaction. When defining the policy, events on specific ports within a PV model are referenced.

Examples of the policy defined in declarative textual form would be

```
sequential_policy CAUSE slave.write EFFECT master.read DELAY 20
// The master read starts after the end of a register write in 20 clocks
sequential_policy CAUSE slave.write EFFECT end_write_int.write LATENCY 0
// The end-write interrupt occurs after a register write
```

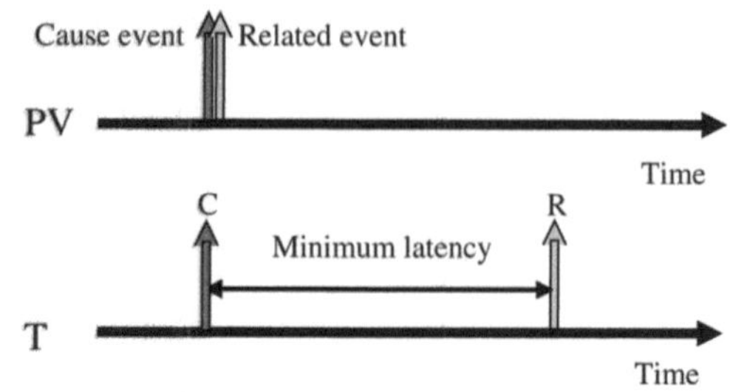

Fig. 8.13 Sequential policy

In the second example, we can see that a latency of 0 has been defined which means that it happens within the same clock cycle.

8.7.5 Pipelining

Pipelining can be used in two major ways. The first is pipelining through a model and the second is pipelining on a port, which is sometimes also called interleaving. We will start by considering pipelining through a model.

Pipelines are a very common element of many functional blocks, but one that is very difficult to put into a model without modeling its micro-architecture. Changing the micro-architecture of a model can require considerable rewriting and is why few people have performed micro-architectural exploration in the past. Vista is able to mimic the effects of pipelines within the timing shell and this capability makes Mentor's Vista product fairly unique among the PVT platforms. Its ability to do this relies on the way in which PV and T transactions are fed through the system separately (Figure 8.14).

The Pipeline policy has two parameters, namely, the amount of data that can be consumed by the pipeline and the delay between the consumption of the data and production of results. Consider that a block of 128 bytes is sent to a model. We define the pipeline as being able to handle 8 bytes at a time. Thus when 8 bytes would have been available on the inputs, based on the timing of the associated T transactions, the first set of data is free to move into the pipeline. Then the pipeline is expected to produce results after the given latency.

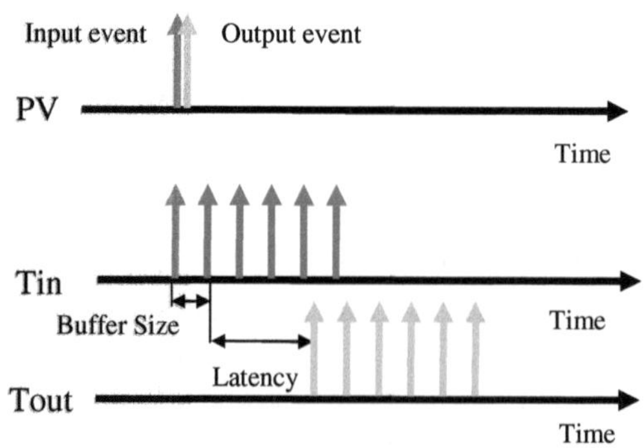

Fig. 8.14 Pipeline through a block

An example of the policy defined in declarative textual form would be

```
pipeline_policy CAUSE master.read EFFECT master.write DATA DELAY pipelineSize
        LATENCY 0
// The master writes are pipelined to the master reads after a pipelineSize data
// delay and zero clocks delay (pipelineSize is a parameter of the model)
```

Note that this example also shows the use of a PV parameter – `pipelineSize`.

Often a pipeline cannot consume data as soon as it is available. It requires a recovery period. This could be because the pipeline clock rate is lower than the surrounding logic or it takes multiple clocks per stage in the pipeline. You can get this effect when the delays between the data element on the output are greater than that of the input. The policy will apply backpressure when its buffer, or pipeline, is full.

The second way that pipelining can be used is to create interleaving on a port. This would be very common for a DMA block. In natural language it would say; start writing data on port Y after X bytes have been read from the same port.

In Figure 8.15 we can see that two PV events happen on the same port – a read followed by a write. After enough read calls have happened to fill its defined buffer size, it is free to interleave write calls associated with the contents of the buffer.

An example of the policy defined in declarative textual form would be

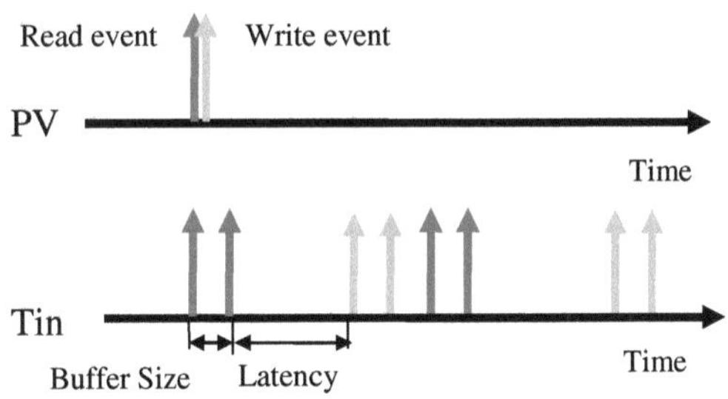

Fig. 8.15 Pipeline – interleaving

```
pipeline_policy CAUSE master.read EFFECT master.write DATA DELAY pipelineSize
        LATENCY 0
// The master writes are pipelined to the master reads after a pipelineSize data
// delay and zero clocks delay (pipelineSize is a parameter of the model)
```

8.7.6 Putting It All Together

At this point, we have presented the way in which timing policies can be expressed in a declarative manner, rather than writing the code yourself. You may be asking questions like – what is the big deal or why use policies. Let us consider the example for which we have been showing the timing policies – the FIR filter. The PV code for this is shown in Figure 8.16 (note that the header has not been included but can be downloaded from the book's web site – http://ElectronicSystemLevel.com).

As you can see, it quite neatly fits into a single page and is easily readable. The code even contains some diagnostic information to aid in understanding what the model is doing, so it could be made even smaller. Now if we apply the timing policies and generate the full timing model for this, it is quite a different story. Admittedly, the code is generated by a tool so it is not meant to be human readable, but the C++ code now contains over 2,500 lines and the header file is almost 1,000 lines. Not only would this amount of code by time consuming to write, but how much of it will need to be re-written if you want to apply a different timing policy?

Even this is not a full evaluation of the amount of code it takes to write a fully timed SystemC model since all of the timing policies are using the built-in routines (such as the ones that evaluate and perform the split policy) and are not included in the line count.

```
#include "FIR_pv.h"
#include <iostream>

// Constructor
FIR_pv::FIR_pv(sc_module_name module_name) : FIR_pv_base(module_name)
{
  cout << this->name () << " Constructor" << endl;
  irq.initialize (true);
  for (int i = 0; i < NUM_TAPS; i++)
    regs [i] = 0;
}

// Write callback for Trigger register.
// The newValue has been already assigned to the Trigger register.
void FIR_pv::cb_write_Trigger(unsigned int newValue) {
  ac_int<8> Data [NUM_TAPS];
  ac_int<8> Fir_Data [NUM_TAPS];
  ac_int<8> coeffs [NUM_TAPS];

  cout << name () << " @ " << sc_time_stamp () << " Trigger Callback" << endl;

  cout << name () << " Source=0x" << hex << Source
       << " Destination=0x" << Destination
       << " Size=" << dec << NUM_TAPS << endl;

  master_read (Source, Data, NUM_TAPS);
  for (int i = 0; i < NUM_TAPS; i++) {
    fir_filter (Data[i], coeffs, Fir_Data[i]); //  print (Data, Size);
  }
  master_write (Destination, Fir_Data, NUM_TAPS);

  cout << name () << " @ " << sc_time_stamp () << " (" << dec << sc_delta_count ()
       << ") Assert Interrupt" << endl;
  irq.write (false);
}

// Write callback for Acknowledge register.
// The newValue has been already assigned to the Acknowledge register.
void FIR_pv::cb_write_Acknowledge(unsigned int newValue) {
  cout << name () << " @ " << sc_time_stamp () << " (" << dec << sc_delta_count () << ")
Deassert Interrupt" << endl;
  irq.write (true);
}

void FIR_pv::fir_filter (ac_int<8> *input, ac_int<8> coeffs[NUM_TAPS], ac_int<8> *output ) {
  int temp = 0;
  int i;
  SHIFT:for ( i = NUM_TAPS-1; i>=0; i--) {
    if ( i == 0 )
      regs[i] = *input;
    else
      regs[i] = regs[i-1];
  }
  MAC:for ( i = NUM_TAPS-1; i>=0; i--) {
    temp += coeffs[i]*regs[i];
  }
  *output = temp>>7;
}
```

Fig. 8.16 FIR filter PV code

8.7.7 Timing Callbacks

While it may appear that there are only a few timing policies that can be set, it should be noted that these present a very flexible set of capabilities. In addition, for those who are not faint of heart, users may directly manipulate the event callbacks in a similar manner to the code generated by the policies.

Timing control is achieved through callbacks that are called at different stages of each transaction. For each port type three kinds of callbacks are called:

- Slave (Target) Port:
  - Transaction-Start – is called at the start of the transaction.
  - Delay policy – is called at the start of the data phase (if there is no distinct data phase it is called just after the Transaction-Start).
  - Transaction-End – is called at the end of the transaction.
- Master (Initiator) Port:
  - Transaction-Start – is called at the start of the transaction.
  - Split policy – is called at the end of a burst (used to set the delay of a subsequent burst).
  - Transaction-End – is called at the end of the transaction.

In the slave port's delay policy callback, the user can specify the time of the start of the actual data transfer and the delay incurred by the slave for each word (wait states). The callback gets a handle to the transaction. The Model Builder's API enables rescheduling the start of pending slave transactions forward and backward in time. Thus arbitration of concurrent transactions can be implemented. Here are some representative examples:

- The starting time of a transaction's pending data stage can be moved forward if the data stage of a transaction with higher priority arrives.
- The starting time of a transaction's data can be scheduled to infinity when its execution time is not known beforehand. It can be moved to a known time in a different transaction's ending callback.
- Moving forward the starting time of a transaction whose data has already started being moved has the effect of delaying the remaining data's time.

The same API allows arbitration of transactions that arrive on multiple slave ports (e.g., multi-port memory). The starting time of a transaction's data can be scheduled to infinity in the delay policy callback and its handle can be put in a data structure that belong to the model and rescheduled when the time comes.

The T model defines additional callbacks: sequential and pipeline policy. These callbacks are called for both slave and master ports:

1. Sequential policy – is called at the end of a transaction (named cause). It is used to set the delay to the start of a master transaction (named effect).

2. Pipeline policy – is called at the start of a transaction (named cause). It is used to set the delay to the start of a master transaction (named effect).

The callbacks get handles to the *effect* transactions. The Model Builder's API enables scheduling the start time of pending master transactions forward and backward in time.

8.8 Power

Power can be annotated onto a model in a similar way to timing. There are two principal kinds of power, static and dynamic. Static power models the leakage current that remains fairly constant through the time that power is applied to a model. This is thus defined by a single value per block. While this block receives power, its leakage power will be added to the total power consumption for the system.

Dynamic power relates to the switching currents in a system and is thus associated with transactions moving through a block. Power is specified for each port and transaction type and also is established per data word transferred and the time period over which it will be drawn. A FIR filter, for example, reads a word into the pipeline, and a computation relating to this data executes. The word is moved to the second pipeline stage. This process repeats as long as the word is in the pipeline. All the power consumed by this activity is associated with the original word, and it is consumed in a time interval of the number of clocks needed to propagate the word through the pipeline.

While the power policies are defined on the PV ports, it should be noted that the timing policies can affect the power profile. For example, if the pipelining policy is changed, this will affect the amount of time over which a word of data consumed power.

8.9 Creating a Model

The Vista Model Builder provides several ways in which models can be brought into a system. These include modeling it from scratch within Model Builder, importing an existing functional model that was written in SystemC while adding timing and power annotations, or extracting timing and power information from simulations of an RTL model. We will look at the Model Builder creation process first.

8.9.1 Using Model Builder

Model Builder provides a development environment that can take a lot of the drudgery out of creating a SystemC TLM model. It not only provides a pre-built infrastructure that ensures that models will work together in a consistent fashion but also deals with many of the complex language features. It can thus save considerable time in the creation of the models, provides documentation about the interfaces to those models, and when problems are found makes it quicker to correct them.

The first things that must be defined with any model are the ways in which it communicates with the outside world. In a SystemC world, this means the ports. Each port must have a name, a direction, the protocol it will use, and its width. Figure 8.17 shows a simple example where three ports are being defined in a DMA block. In this case it has a 32-bit wide master port on which it can write data, a 32-bit slave port from which it will read data, and a 1-bit interrupt line which it will raise when a DMA operation has completed.

Fig. 8.17 Defining the ports

The next things to define are the registers, memories, and any parameters to the model. Parameters are set statically and allow a model to be customized for each simulation run. This allows multiple scenarios to be created simply by loading different parameter files at the beginning of each simulation. These could represent different architectural choices that are being saved.

Registers and memory can be configured by setting their base address, size, and if they can be both written to, and read from. In the DMA example there are no memories, but several registers are defined as shown in Figure 8.18.

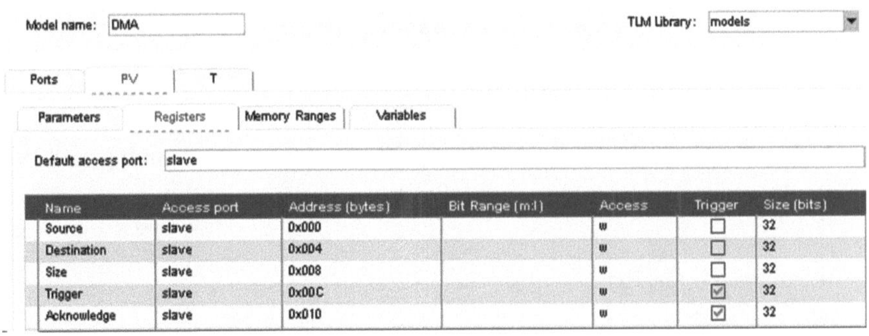

Fig. 8.18 Defining the registers

All registers have been defined as write-only 32-bit registers, so it is not possible for them to be read from the master port. It should be noted that two of the registers namely `Trigger` and `Acknowledge` have the trigger box checked. This means that callback function prototypes will be built for these. These functions will then be called whenever these registers are accessed.

At this point, Model Builder has all of the information necessary to build a template for the model including the convenience layer that wraps around the PV and T models. Into this template it also adds the read and write callback functions on each port. The code generated is fully compliant with the TLM 2.0 standard including both the forward and backward paths.

The functionality for the DMA is contained in the two register callback functions. This can easily be added into the templates that were generated.

```
void DMA_pv::cb_write_Trigger(unsigned int newValue) {
       unsigned int Data [0x100];
       cout << name () << " @ " << sc_time_stamp ()
            << " Trigger Callback" <<
       endl;
       master_read (Source, Data, Size);
          //DMA read - Master transaction embedded in slave
       print (Data, Size); // Print debug information
       master_write (Destination, Data, Size);
         // DMA write - Master transaction embedded in slave
       cout << name () << " @ " << sc_time_stamp ()
            << " Assert Interrupt" <<
       endl;
       irq.write (false); // Interrupt CPU upon DMA completion
}
void DMA_pv::cb_write_Acknowledge (unsigned int oldValue,
         unsigned int newValue) {
      cout << name () << " @ " << sc_time_stamp ()
      << " Deassert Interrupt" <<
      endl;
      irq.write (true); // De-assert Interrupt upon CPU
        interrupt acknowledge
}
```

Adding the power policies is done in a similar manner as shown in Figure 8.19.

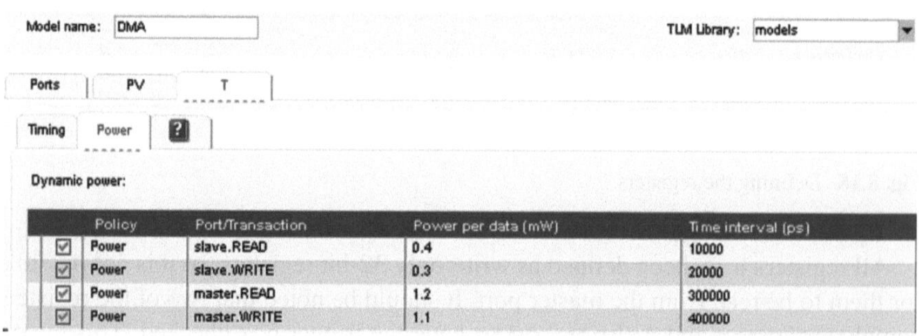

Fig. 8.19 Power policies

8.9.2 Synchronization

The timed model should run so that transactions appear on the interfaces at the correct time. Since the PV calls are blocking, the PVT wrapper can use wait statements in order to block the functional model. This would be one way to synchronize each transaction; i.e., block the PV code until the transaction finishes. Generally this does not work as demonstrated by the following scenario:

Assume that a model issues the following sequence of transactions: read1 followed by read2. In an implementation read1 and read2 may be issued concurrently and it is possible that read2 may finish before read1. If the code waits for read1, then read2 will not finish at the correct time. This situation is quite typical for modern processors. The processor does not wait for the sequential order of the reads and writes that it issues. It may issue a read, continue, issue another read, get the result of the second read, and wait to get the result of the first read only when it cannot continue without it.

To enable this type of scenario, the functional code runs in zero time, while the PVT wrapper obtains the data through the DMI interface. In TLM2.0, a model is able to get direct access to memory through DMI only if transactions to this are free of side effects, e.g., reads from memory are generally without side effects while reads from a FIFO probably are not. In the later case, the PVT wrapper has no option but to block the functional model until read transactions with side effects have completed.

In the case of the accesses that have side effects, it is possible that the timing accuracy may be impacted. However, an important observation is that if those accesses do cause side effects they are probably non-concurrent by nature, e.g., it would be pointless to write to a register that triggers a DMA before the corresponding size register is written. Processors that utilize a bus that enables out-of-order accesses have ways to force the sequencing of the accesses. They often do this for driver code that accesses memory-mapped registers. Thus blocking read operations with side effects will not adversely affect timing modeling in practice.

The model builder enables the user to specify, through policies, that a transaction's start is synchronized. When this happens the PVT wrapper synchronizes the current thread, i.e., it waits until the time that the transaction should start. This feature enables the support for time-dependent behavior modeling.

Here are two examples that support time-dependent behavior:

1. A thread is sensitive to a FIFO whose size is fixed. It synchronizes before getting each element from the FIFO. A slave port method pushes requests into the FIFO. Depending on the number of elements in the FIFO, the method can raise one of the following interrupts: BELOW-LEVEL, ABOVE-LEVEL, and FULL (an error indication).
2. A thread is sensitive to a number of FIFOs. When it is triggered, it scans all of the FIFOs and determines which of the waiting elements to process according to some arbitration algorithm. It synchronizes before getting an element. A number of slave port methods push requests into the FIFOs; data is lost when a FIFO is full.

8.9.3 Integrating 3rd party Models

There will be times when a model is available from a third party or being used for the first time within Vista. In these cases the functional model already exists, but not in a form ready to enable the addition of the timing and power layers. The easiest way to accomplish this is to create a model wrapper in the same way as described in Section 8.9.1. This will ensure that all of the correct functions appear on the ports of the model. The actual functionality can then be added into all of the applicable callback functions that were created.

A few other changes may be necessary within the model to ensure it works correctly within Vista. For example, any sc_events that are expected to be used for synchronization must be converted into mb_events. These calls have the same functionality but cause additional things to happen within Vista.

8.9.4 Model Abstraction

In some cases, legacy models exist and need to be brought into a system environment so that the impact of changes or additions to the system can be assessed. The timing and power profiles for an RTL model are more accurate than could reasonably be created in a system model.

Characterization of accurate power for preexisting processor and peripheral models and mapping it into a TLM representation would allow for improved definition of power domains and better allocation of power budgets across the system. Such power characterization would allow software developers to look at the power profiles resulting from the applications they apply to the system.

In addition, tools such as the ones supplied by Carbon Design Systems can take an existing RTL model and compile it into a high-speed software model that conforms to the OSCI interface specification. This enables them to be easily integrated into Vista as described in the previous section.

8.9.5 Building a System

We have now seen how models can be created and how they can be refined, but systems usually contain many models that are interconnected. In this case, all connections are TLM 2.0 connections and platform assembly is done in the Vista block diagram editor.

Since all of the connectors are compatible with each other, platform creation is a simple process of instantiating the blocks and connecting them in a one-to-one manner. An example of the connected platform can be seen in Figure 8.26. Once this has been done, the simulation executable which includes the OSCI SystemC scheduler can be built and run.

8.9.6 Navigating a System

Before we talk about navigating around a system, we should perhaps talk a little about the hierarchies that have been created in such a model. The obvious hierarchy is the one built using the Vista block diagram editor that ties many models together into a complex system. This is the classic structural hierarchy that is familiar to everyone who has ever handled a design at the RT level. We start at the top-level block and in it we can see the children. By pushing into a child, we see the structural components that it is built from and so on.

But this is not the only hierarchy that exists in a system model. We also have the functional hierarchy that is made of the threads and processes. This is the hierarchy that every software programmer will be familiar with. This hierarchy exists in two ways within the system platform. It exists in the software itself and it exists within the SystemC models that describe the hardware.

In a system-level environment, we need to be able to see all of these hierarchies and in the case of those that represent the hardware, they need to be interlinked such that anything in one view can be mapped into the corresponding entity in an alternate view. It should be noted that this is not always a one-to-one relationship.

Once a system has been constructed, any design browser, such as the one in Vista allows the design to be navigated in two basic ways. The first of these shows the actual code structure as depicted in Figure 8.20.

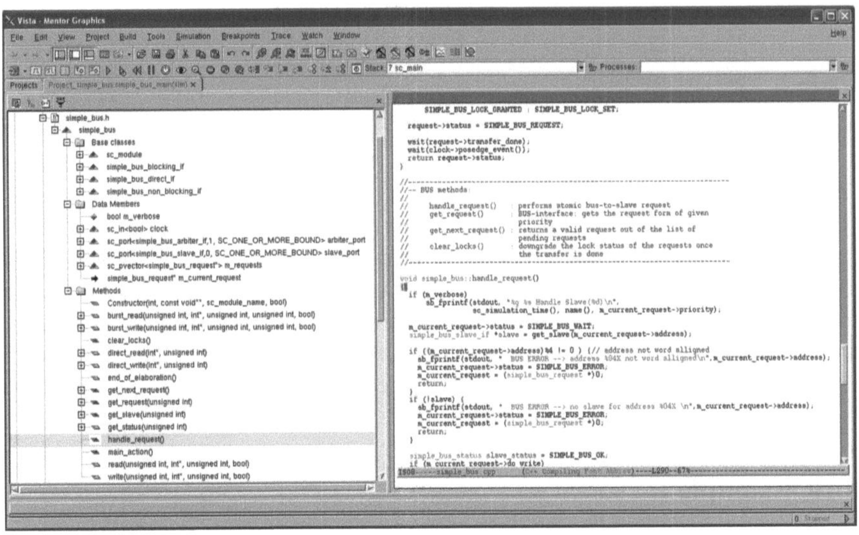

Fig. 8.20 Functional navigation

Here we can see a navigation panel on the left-hand side of the window showing the decomposition for the `simple_bus`. Expanding this we could extract more details about its base classes, data members, methods, etc. On the right-hand side we see the actual code that is associated with the items selected from the navigation panel. This allows quick and easy traversal of the code that has been constructed.

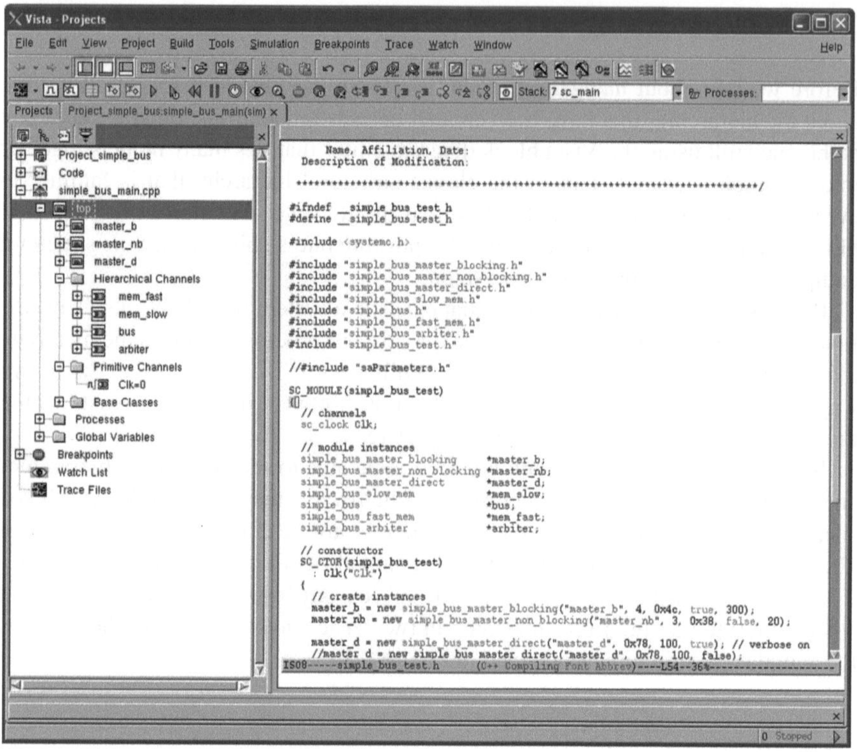

Fig. 8.21 Structural navigation

A second view is equally important that shows the design structure. This is depicted in Figure 8.21.

Using this view the design hierarchy can be traversed. Again the navigation panel can be seen to the left of the window and provides the traditional hardware-centric view of the system. The right-hand panel, as before, shows whatever is selected in the navigation panel. Vista provides easy ways to switch between these two views, and selections can be made in either of the panels and reflected in the other. So for example, if a particular section of code is selected, it can be identified in both the structural view and the functional view.

8.10 Example

In this section, we will use a fairly simple example to demonstrate some aspects of the flow and to show the kinds of analysis which can be performed. It should be understood that this example is not meant to be reflective of the typical designs on which a product such as Vista would be most highly valued since many of the concepts demonstrated are quite obvious to anyone familiar with processor-based design.

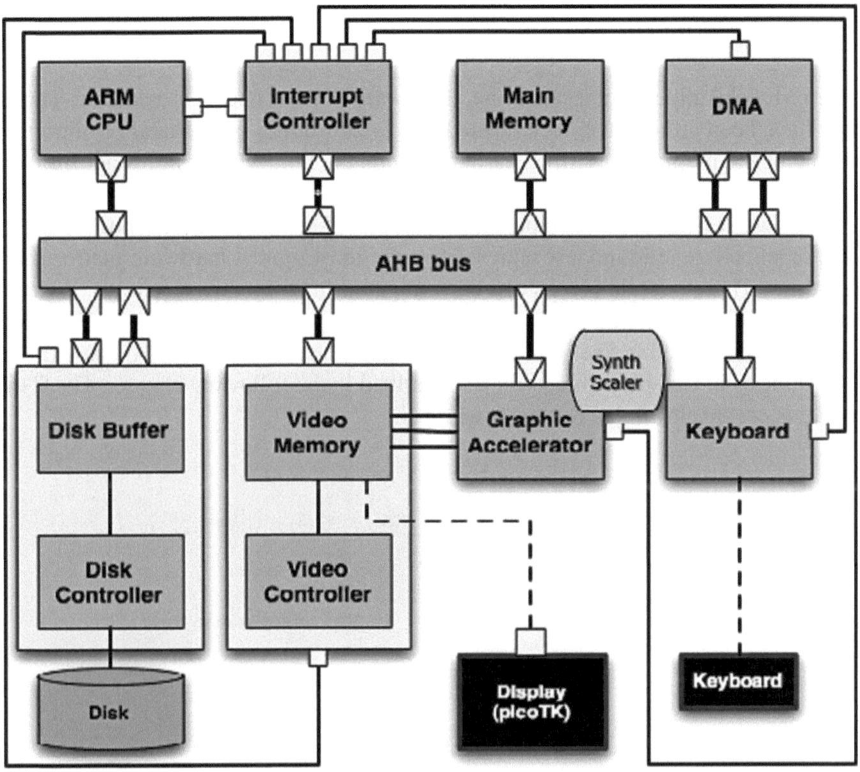

Fig. 8.22 Example block diagram

The design (Figure 8.22) is a single processor system, with some memory and a few peripherals attached to a bus. Also present in the system is an accelerator block for a graphics scaler. We will not be concerned with the actual design of the accelerator but will use it to show some of the design trade-offs that could be made.

The design has been set up to perform some very simple commands based on input from a virtual keyboard. First a frame of a picture can be loaded from disk into the main memory. Second, the picture is scaled in one of two ways. It can either be done in software or utilize the accelerator. The before and after pictures are displayed on a virtual terminal window so that the result can be assessed for correctness and quality. Finally, the result can be stored back onto disk at the end of the process.

In addition to the use of the accelerator, the design can also be invoked with either a generic processor or a specific processor. Three tools are used in the flow. First Vista Model Builder will be used to enter the necessary model description and to set the parameters for model creation. Then the Vista block diagram editor will be invoked to build the complete system from the models. Finally Vista will be used to perform analysis on the system.

8.10.1 Building the System

When Model Builder is first started up, the library of components is available. These may have been either have been created in Model Builder beforehand or imported by one of the methods described earlier. Here is the library view in Model Builder (Figure 8.23).

Here we see the list of components available. You will note that most of them are predefined generic models representing the most typical hardware platform IP functions. These generic blocks can be configured in many different ways. We can navigate through each of the blocks to get to the parameters, timing, and power policies as described in Section 8.9.1. The scalar block itself is a C++ block written for synthesis in Catapult which will be described in the following chapter. The C++ model is encapsulated in a SystemC wrapper.

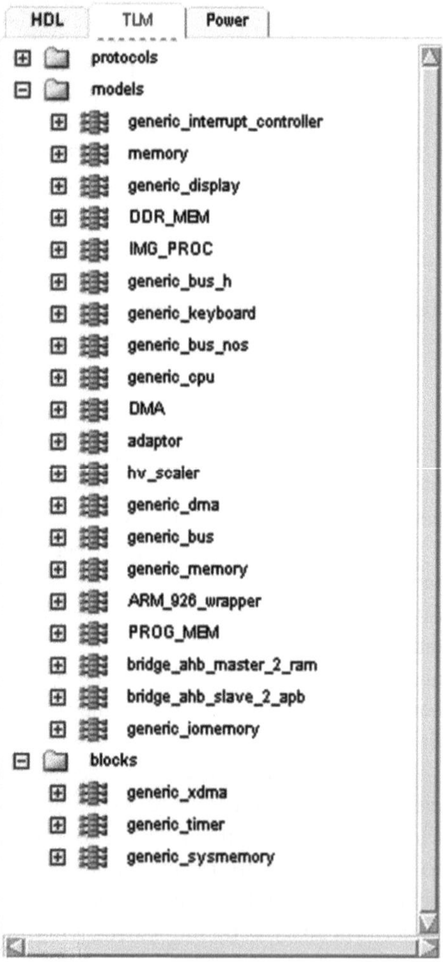

Fig. 8.23 Model Builder interface

In this example we will be concentrating on the DMA block which has the parameters shown in Figure 8.24 and the register set is shown in Figure 8.25.

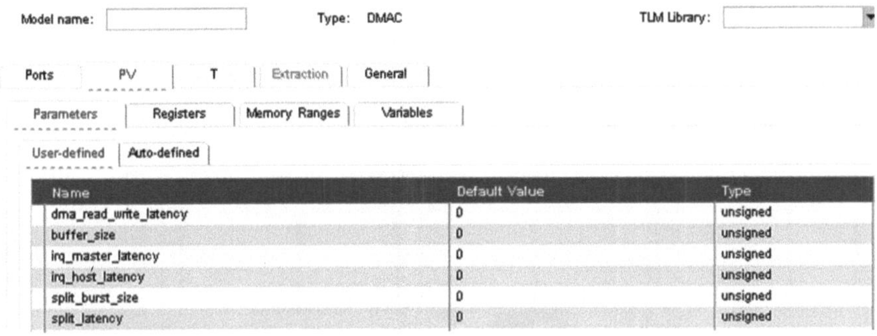

Fig. 8.24 Parameters for the generic DMA block

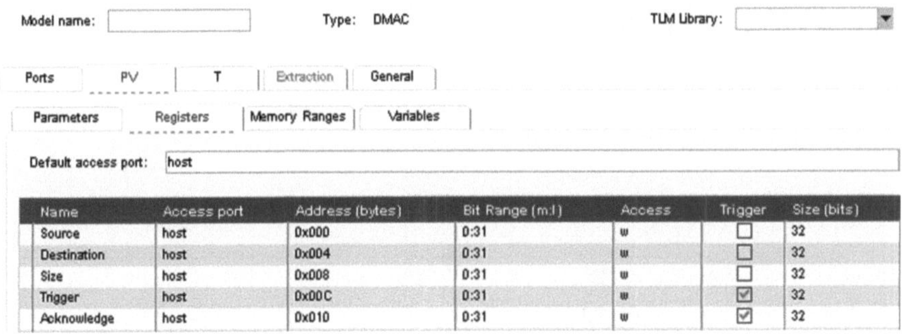

Fig. 8.25 Generic DMA register set

Once all of the parameters, timing, and power policies have been set for a block, the model can be generated. Either a PV model or a model with timing can be generated. In the next step, the platform is constructed. All connectors are TLM 2.0 interfaces, so connection of the interfaces should be an easy task. The resultant platform looks very much like the conceptual diagram introduced earlier (Figure 8.26).

8.10.2 Running the Simulation

When this step has been completed, the SystemC model can be built and this creates the simulation environment on which analysis can be performed.

The main function of this window is to control the simulation process. Signals, processes, and ports can be selected for incorporation into a waveform display. When the simulator is advanced, a traditional waveform is captured, except that

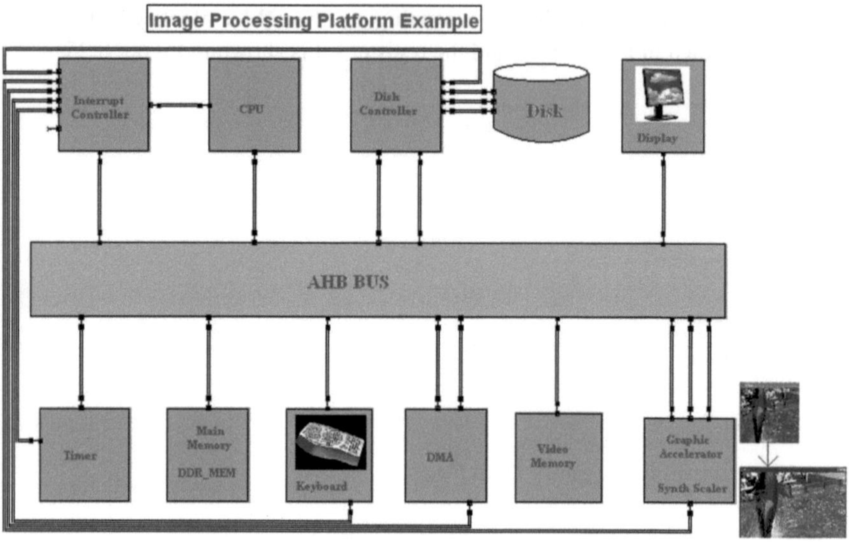

Fig. 8.26 Platform assembly in Visual Elite

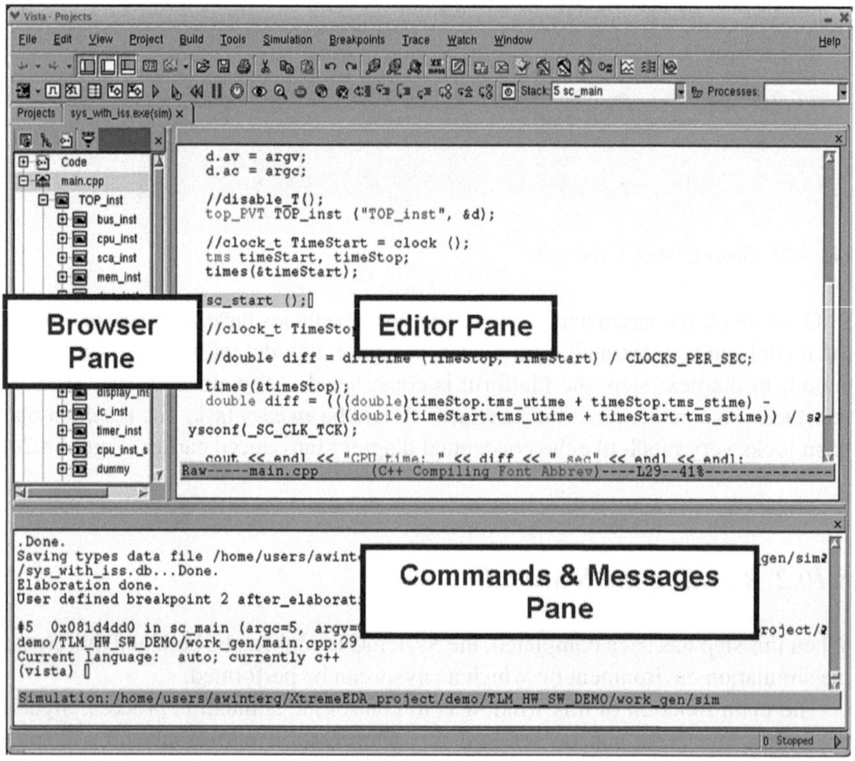

Fig. 8.27 Vista Browser

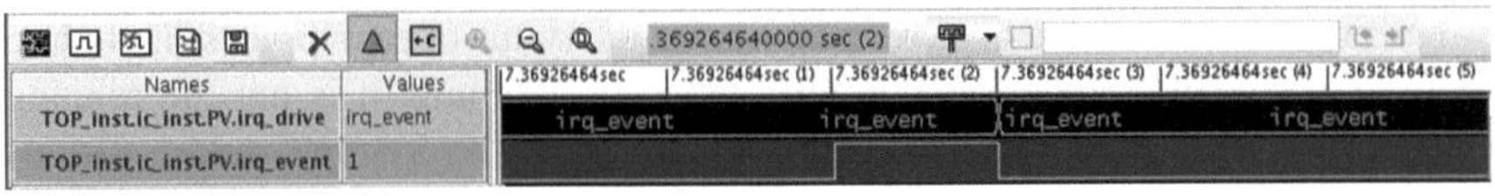

Fig. 8.28 Vista transaction trace

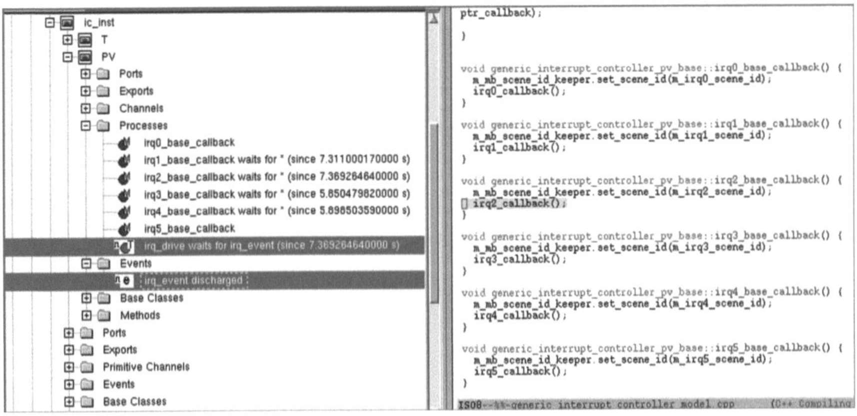

Fig. 8.29 Navigator showing cause of event

transactions are displayed rather than the more typical signals associated with an RTL simulation.

Events are coupled with the source that caused them, such that if an event or process state is selected and the cause button pressed, the main Vista window will be adjusted to show the code that caused the event to happen.

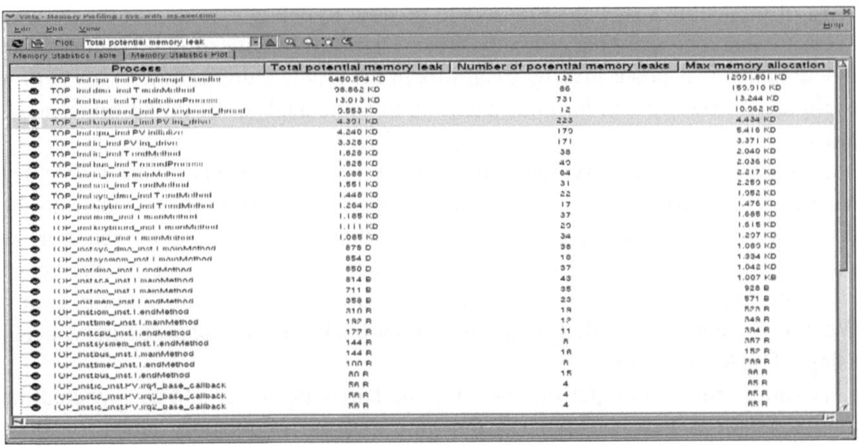

Fig. 8.30 Memory consumption

The processes list in the browser pane shows the status of each process including the reason why that process is currently blocked, assuming it is not the running process. This allows tracing through the execution of the SystemC code. Various other types of tracing are possible, such as the ability to track memory usage in all of the processes.

Transactions can also be viewed in a tabular list such as shown in Figure 8.31.

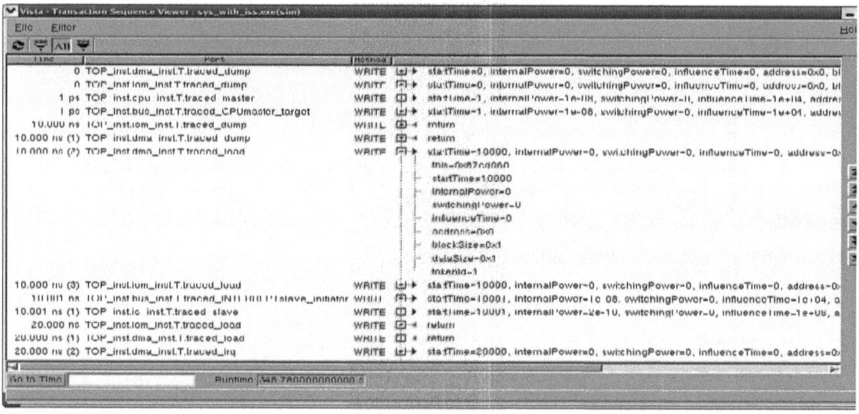

Fig. 8.31 Transaction listing

Each transaction can be expanded to obtain more information about the transaction such as start and end times, and associated data.

8.10.3 Analyzing the System

Most of the features shown so far would be used to debug a model, or the system, to ensure that it is operating correctly. But system-level verification is not just about functionality, it is about making trade-offs in the architecture or how the system is configured. Now we will turn our attention to some analysis of traffic flows in the model.

In Figure 8.32, we have selected the read and write events of the DMA that resulted from a run utilizing the software scaler. Vista makes use of color so the distinction between the reads and writes is not that easy to see, but the reads are in a slightly lighter gray. We thus see a set of reads followed by writes as the image is transferred into the video memory and then a long delay as the scaler operates. When completed, the scaled image is written back into the main memory, again using a DMA operation.

A similar run can be performed using the hardware scaler and as expected the time separations between the transfers into and out of memory are a lot close together, as shown in Figure 8.33.

Similar analysis can be performed to show the power profiles of each execution.

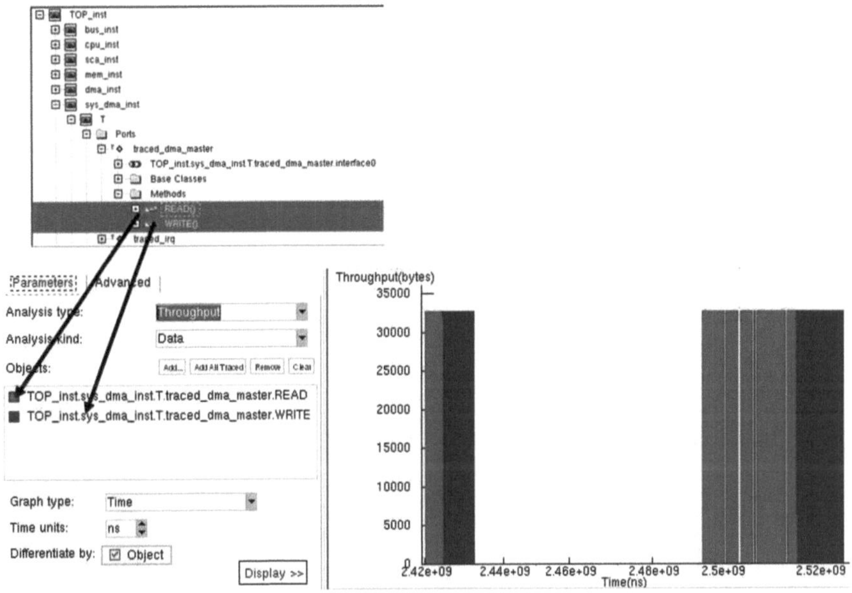

Fig. 8.32 Looking at the DMA in a SW run

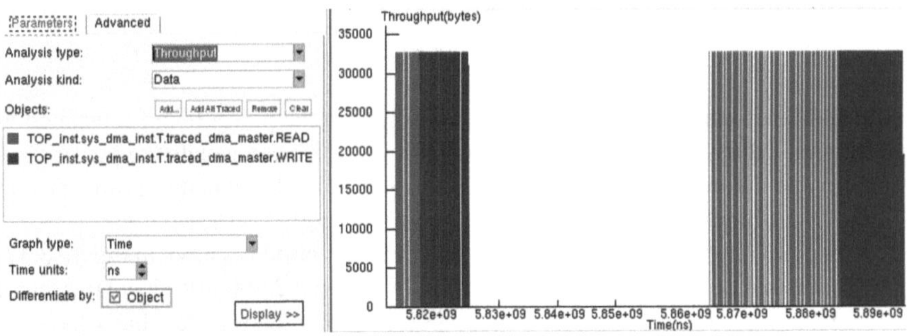

Fig. 8.33 DMA operations with HW scaler

It is clear from Figure 8.32 and 8.33 that the DMA block is buffering all of the data so that it first completes all of the reads and then performs the writes. This may require a much bigger DMA block due to its memory requirements. This could be adjusted by changing some of the timing policies on the DMA block. So, for example, we could add a pipelining policy as shown in Figure 8.34.

Vista supports incremental compilation, so it is quick to show the effects of a change, as shown in Figure 8.35.

While it is not clear from the figure, the reads and writes are now being interleaved which should result in a smaller implementation without having any impact on the time it takes to perform the transfers.

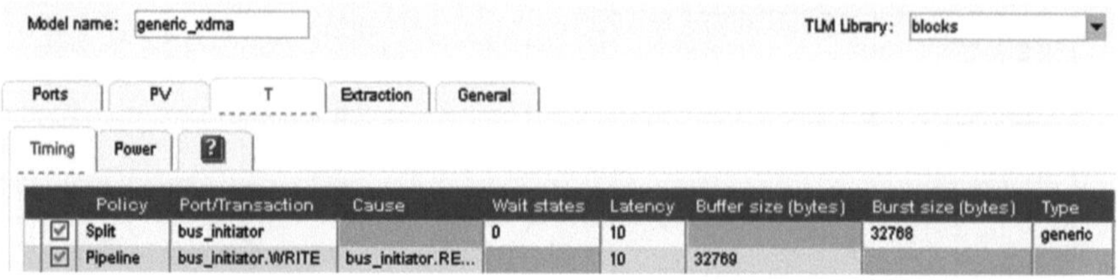

Fig. 8.34 Adding a pipelining policy

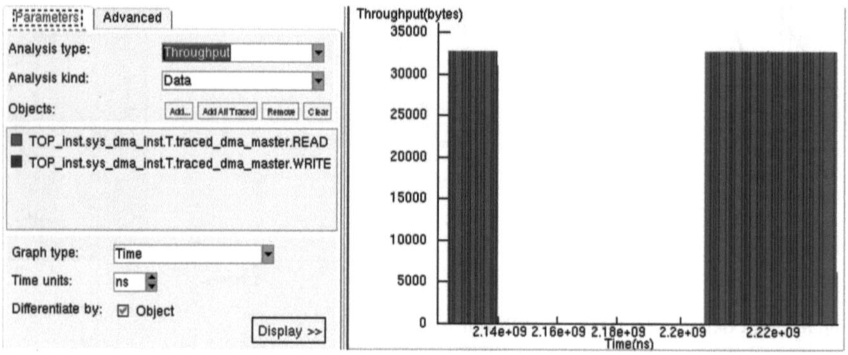

Fig. 8.35 Interleaved reads and writes in DMA

In addition to the time analysis, we can also look at the power being consumed in the system. Either this can be done at the top level or we can isolate the power consumption of the individual components. In Figure 8.36 the dynamic power profile for the DMA block before the pipelining policy was added is shown.

With pipelining added the impact on power consumption is shown in Figure 8.37. We can see that peak power has been reduced from about 2,000 mW down to around 800 mW, but the downside is that average power during the transfers has gone up a little from 500 to 800 mW.

8.10.4 Inserting an ISS Model

So far we have done all of the testing using a generic processor that has high performance and reasonable accuracy, but for fine-tuning the architecture we may need more detail, such as the detail that could be obtained from an instruction set simulator. For this capability, Mentor turns to its Seamless product line, which has long been the dominant hardware/software co-simulation environment in the industry.

Seamless has support for most processors in the industry and for many of them can use its XRay debug environment to control and analyze the processor operation. Multiple processors are supported in this environment. Seamless connects into the

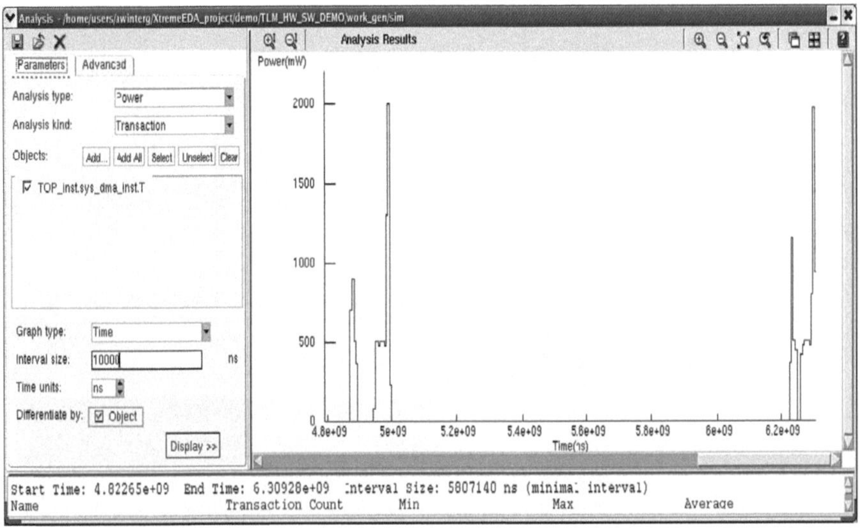

Fig. 8.36 Power consumption in the DMA

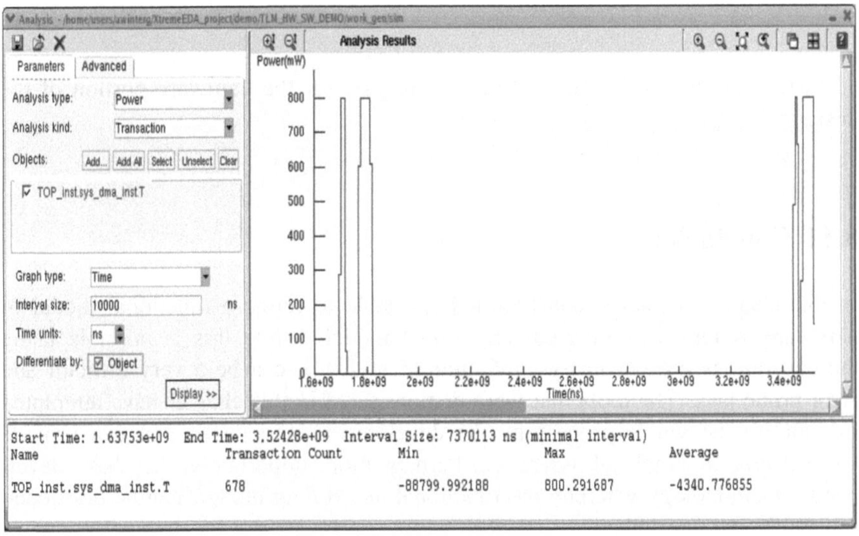

Fig. 8.37 Power profile with pipelining

Vista environment such that all of the rest of the system can continue to operate at the system level with no changes required.

With this environment, even greater analysis capabilities become available, such as being able to analyze cache configurations, to perform accurate profiling of the code, to perform heat profiling of memory and processor power consumption. Some of the analysis windows are shown in Figure 8.38.

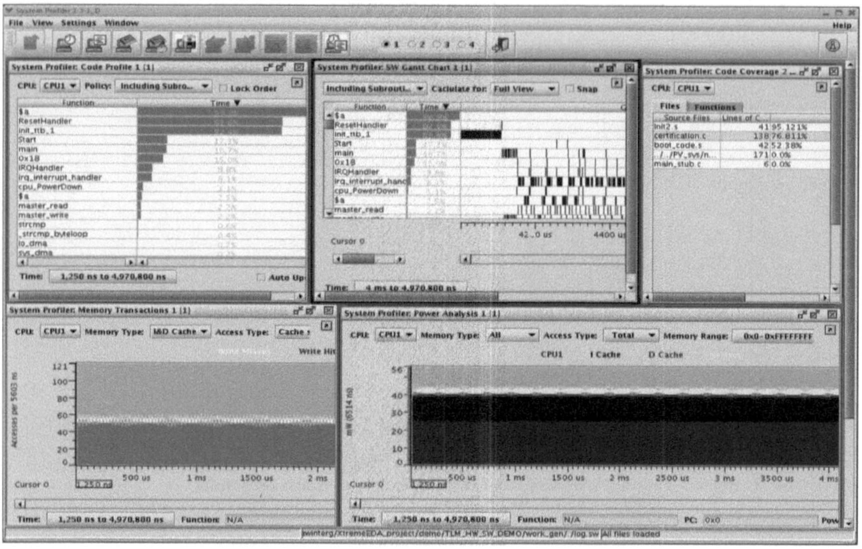

Fig. 8.38 Software analysis in Seamless

Vista continues to provide all of the analysis for the hardware portion of the design.

8.11 Conclusions

In this chapter we have seen how it is possible to refine a functional model of hardware system into an accurately timed model. While this is possible using the capabilities present in SystemC and TLM 2.0, it can be a very difficult and error-prone task. The tools that were demonstrated in this chapter have attempted to simplify the whole process by the development of an IDE that can provide some degree of graphical assistance. Perhaps more importantly, they have developed a methodology whereby the functional and timing models can be developed separately and interlinked.

They then went one step further in the development of a set of timing policies that can be attached to the ports of models and describe the timing in a declarative manner. These timing policies are easy and quick to define and the necessary SystemC code to make it happen is all generated automatically.

Once a system has been constructed it becomes possible to perform a number of analysis functions on the system that will be at a much greater level of detail and fidelity as compared to platforms that only have more limited timing capabilities.

References

1. IEEE 1666 LRM http://www.systemc.org/downloads/lrm/
2. Migrating to Transaction-Level Modeling in SystemC. Aldec/Doulos Webinar Jan 29th 2009
3. Bart Vanthournout. An Insider's View on the Making of the New TLM-2.0 Standard. http://www.dac.com/newsletter/shownewsletter.aspx?newsid=34
4. Brian Bailey, Grant Martin, Thomas Anderson (Eds.). Taxonomies for the Development and Verification of Digital systems, Springer, 2005
5. Brian Bailey, Grant Martin, Andrew Piziali. ESL Design and Verification: A Prescription for Electronic System Level Methodology, Elsevier Morgan Kaufmann, 2007.
6. Mixed-Abstraction Virtual System Prototypes Close SOC Design Gaps. SoCentral. April 2006 http://www.soccentral.com/results.asp?EntryID=18709
7. Bill Murray. OSCI TLM-2 Proposal – The Debate Begins. SCDSource 1/9/2008. http://www.scdsource.com/article.php?id=84
8. Lukai Cai, Daniel Gajski. "Transaction level modeling: an overview," in First IEEE/ACM/IFIP International Conference on Hardware/Software Codesign and Systems Synthesis, Newport Beach, CA, 2003, pp. 19–24.
9. OCP Organization Website http://www.ocpip.org/home
10. Tim Kogel, Anssi Haverinen, James Aldis. OCP TLM for Architectural Modeling. http://www.ocpip.org/socket/whitepapers/OCP_TLM_for_Architectural_Modeling.pdf
11. Frank Ghenassia (Ed.). Transaction-Level Modeling with SystemC, Springer, 2005
12. Shashi Bhutada. A Scalable Approach for TLM Across SystemC and SystemVerilog. Mentor Graphics white Paper.
13. William W. LaRue, Sherry Solden, Bishnupriya Bhattacharya. Functional and Performance Modeling of Concurrency in VCC, Concurrency and Hardware Design, Springer, LNCS volume 2549, 191–227, 2002
14. Property Specification Language Reference Manual http://www.eda.org/vfv/docs/***PSL-v1.1.pdf***
15. SystemVerilog home page: http://www.systemverilog.org/
16. Bill Chown. System-level validation increases design productivity and saves errors. Mentor Graphics white paper.

Chapter 9
C/C++ Hardware Design for the Real World

9.1 Introduction

In previous chapters we have seen how system-level virtual prototypes get created and how they are often partitioned into pieces intended to be implemented as either software or hardware. This partitioning was done to manage costs and performance and to optimize many other attributes of the system. For the parts of the system intended to be hardware, several choices still remain. It is possible that those functionalities can be implemented using off-the-shelf components, such as processors or DSPs, or as components that require small extensions in order to meet the necessary requirements, such as configurable and extensible processors. It is likely that there will be a few components for which dedicated custom hardware is required and we now need to refine those models in ways that make them more amenable to the transformation process that is often called behavioral synthesis or high-level synthesis [1].

This chapter will both talk about the manual transformations of the models that are necessary and also demonstrate the power of the commercial tools that are available. To get the most from those tools, the user must understand why this transformation is necessary and guide the tool toward the right implementation. High-level synthesis involves the manipulation of many interrelated aspects of the hardware generation process and the technology is not yet at a level where it can automatically find an optimized solution without expert guidance. This chapter aims to provide you with the knowledge that you need to be able to generate a good solution.

In order to demonstrate the process, we will be showing the Catapult® C Synthesis tool from Mentor Graphics [2]. As with the other demonstration chapters of this book, Catapult is not the only high-level synthesis tool on the market, but it is necessary to concentrate on one tool in order to explain more than just general concepts. If after reading this chapter you believe that high-level synthesis can provide you with the productivity gains that you are looking for, then you should perform a complete analysis of all of the tools available to you.

Mentor Graphics has a long history in the development of high-level synthesis. Their attempts to break into the RTL synthesis market with Autologic in the

B. Bailey, G. Martin, *ESL Models and their Application*, Embedded Systems, DOI 10.1007/978-1-4419-0965-7_9, © Springer Science+Business Media, LLC 2010

early 1990s were never successful, even though they tried several times. Synopsys' Design Compiler had a tight grip on the market and even significant improvements in algorithms or performance were not enough to shake the Synopsys dominance in the marketplace. Mentor has a business philosophy that it wants to be number one or two in a market and if it cannot achieve that, then it considers the market to be too difficult to navigate. Because of this, Mentor often looks for markets that are underserved by the other companies, and this led them to examine the FPGA synthesis market. Their equity purchase of Exemplar in 1995, and complete purchase in 1999, gave them a solid foundation on which to build other synthesis technologies. Mentor made their first attempt in the high-level synthesis market in 1997 with Monet. This was based on VHDL as its input language, a language that was used by many Mentor customers, and thus seemed to be a logical choice. They did attract a number of users, but not enough to warrant making it a permanent member of the Mentor tool line-up. Even though this product was not successful, Mentor went back to the drawing board, and the technology resurfaced as Catapult C synthesis in 2004. This revised product had been the result of quietly working with customers through the intervening period, and when the product was announced they already had a significant number of endorsements lined up for the launch and could claim 10 product tape-outs that had successfully deployed it. Now if you do a search for Catapult on the Internet, you will find piles of customer endorsements, certifications for ASIC libraries, or integrations with other related tools. For a more complete history of the progress of high-level synthesis in general, and other solutions that are available in the marketplace, the reader should refer to ESL Design and Verification [3] – Chapters 3 and 11 and [2].

Mentor has quickly become the market leader in this segment. Based on numbers from GarySmithEDA [4], Mentor had a 44% market share in 2006, a substantial lead over its closest rival. Over the period 2004 through 2007, the market had seen a 78% compound annual growth rate, making it one of the fastest growing EDA segments at that time. A later report in October 2008 showed that while the growth rate had slowed to 29.7% for the latest year, it is still growing at an impressive rate and has now reached a respectable size of $28.4 M in 2008 as shown in Figure 9.1. Market share figures are basically stable.

9.1.1 Chapter Overview

In this chapter we will examine how high-level synthesis fits into a complete ESL flow and how it integrates with the existing RTL flow. These tools require a new (more abstract), language in order to be both expressive enough and avoid locking a designer into a particular implementation architecture. For this chapter we will be concentrating on synthesis from a C++ description.

Synthesis, just like any design process, is a set of trade-offs that balance various attributes of an implementation, such as performance, area, and power. In order to guide a tool toward a desired implementation, an engineer needs to understand several concepts including the algorithms they are trying to implement, the fundamental

The growth is good, with 29.7% CAGR, but could be better. This is one of the main sub-applications that has a good chance of exceeding our forecast. *(See Figure E-6)*

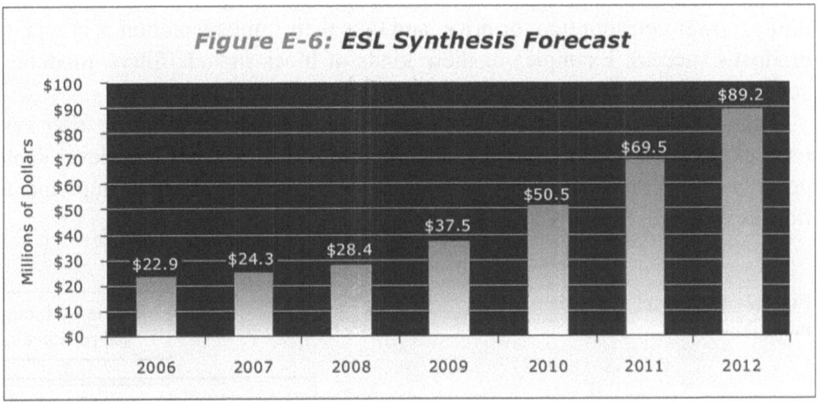

(Source: Gary Smith EDA, September 2008)

Fig. 9.1 HLS market growth – used with permission

capabilities of the tool, and a thorough understanding of hardware structures such as pipelining, memory architectures, and interfaces.

The chapter will introduce some of the basic feedback mechanisms that high-level synthesis tools provide to an engineer that will enable them to change both the architecture of the original input description and the micro-architecture of the solution that is generated. Together they allow the complete solution space to be explored and the best solution obtained. These concepts will be explored using several examples, including one that follows the development of a complex block that is part of most OFDM communications systems.

The design and synthesis flow has to fit into a corresponding verification flow. While this chapter does not fully develop that verification flow, it does show many of the hooks that it creates to enable efficient verification of the input description and to ensure the result of the synthesis process functionally matches the input description.

9.2 Where Does It Fit in an ESL Flow

In *ESL Design and Verification* [3] we talked about where high-level synthesis fits in a structural sense, namely after HW/SW partitioning and the appropriate analysis and verification steps. At this point, the partitioning choices have been made between hardware and software and we have a complete specification of the functionality to be implemented in hardware.

Within the hardware partition, several blocks will go through specialized implementation flows such as the processors, memories, and bus control systems. These blocks will not be discussed in this chapter. Almost all designs will have one or more pieces of hardware which are unique to that chip and these custom pieces often provide the competitive differentiation between it and the other products on the market. These custom portions can have a major impact on the product speed, utility, power consumption, or price, and thus their implementation is crucial to the product's success. Examples of these kinds of block include filters, modems, and codecs.

Thus the domain of high-level synthesis can be defined as being after system-level exploration and partitioning is complete and as a result high-level synthesis generates everything necessary to integrate with the existing RTL implementation flow as shown in Figure 9.2.

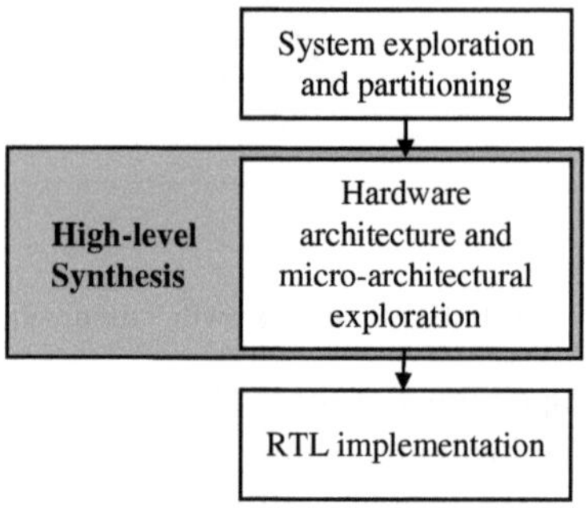

Fig. 9.2 High-level synthesis in the flow

In addition to its structural location in an ESL flow, there is a second aspect of where high-level synthesis fits in the flow which is related to the type of choices that have become fixed in the design. A specification should, in an ideal world, be completely devoid of any assumptions about its implementation. This means that we have the ultimate freedom to make any choices we want. It is the sum total of those choices that moves us from specification to implementation. Some choices are more important than others or have more far-reaching implications, such as which processor to use, or how to configure IP, or the memory architecture. Others, such as the way in which a clock tree is going to be balanced or routed are equally important for the success of the product, but do not have the same kind of follow-on consequences related to this decision. All of these decisions are basically architectural decisions, and it could be argued that at any level in the design process, a set of architectural decisions has to be made. It is thus important to know which architectural choices have already been made and fixed at the time that we enter the high-level synthesis stage. For a more concrete example of architecture embedded into a C description see Section 9.6 in this chapter.

Taking this to a slightly greater level of detail, we can see in Figure 9.3 that the input to the high-level synthesis process is usually a C/C++ description of the algorithmic blocks coded in a fixed-point style. Almost all of the verification to ensure that the input description matches the generated RTL is performed automatically meaning that virtually no verification is necessary at the RT level and the time can instead be spent ensuring that the chosen algorithm is correct.

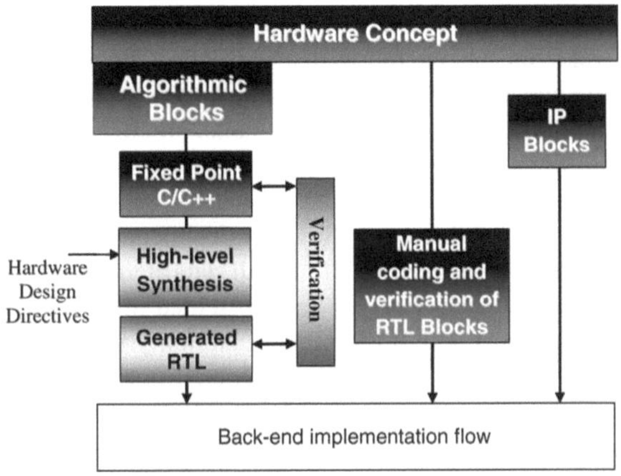

Fig. 9.3 High-level synthesis flow

If we were to compare this high-level synthesis flow to a traditional RTL flow, the major difference is where an engineer spends their time. In an RTL flow, a small amount of time is spent choosing an architecture and then the vast majority of the time is spent coding and verifying the RTL. Attempting to get the best optimization of the chosen architecture can result in several time-consuming iterations. Compare this to a high-level synthesis flow, as depicted in Figure 9.4. Now we spend most of the time performing architectural exploration and almost no time at the RTL level. Still, with the additional time spent in finding the right architecture, the total time spent in the entire process is considerably diminished.

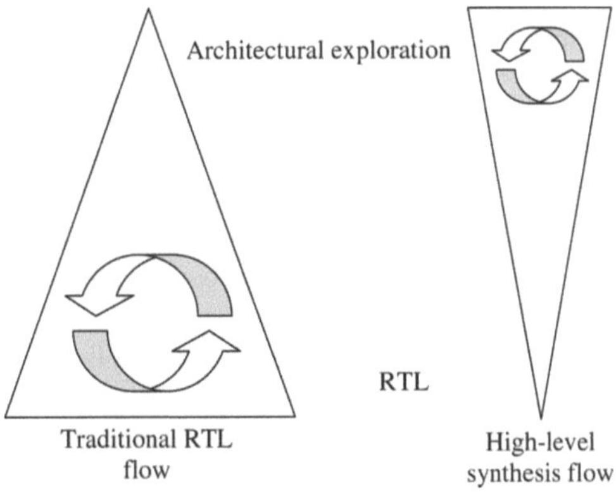

Fig. 9.4 Relative time utilization

9.2.1 Hardware Implementation Input

The input to the high-level synthesis process is ideally one in which operations have been ordered but not timed. By ordered we mean that we know when something will start and end only in relation to when other things start or finish. It is the high-level synthesis operation that will then refine this by adding an explicit notion of time and take us all the way down to a cycle-accurate model. In addition to the temporal changes that will be made, significant changes will also be made in the data domain. The inputs at the beginning of high-level synthesis are valued, but not typed. For example, the inputs will be integers or floating point values, where the values of them and the numerical accuracy are known. Floating point will be converted into fixed point and all data items will be transformed into an optimal format in which they can be represented in hardware. It is these two axes (temporal and data) within the model taxonomy that define abstraction and in this case we clearly have an abstract functional model, which is more fully defined in the industry taxonomy [5]. The key to the taxonomy can also be found in Chapter 1 of this book.

If we consider the other three axes of the ESL taxonomy (concurrency, communications, and configurability), the input model to the hardware synthesis process is usually a sequential description written in C or C++. In Chapter 1, we defined these languages as being incapable of explicitly expressing concurrency and communications. However, this does not mean that models written in these languages cannot support these capabilities. For example, the C language is designed to execute on a processor, which we assume operates sequentially. This defines the semantics of operation for the language. However, many modern day compilers will make use of processor features that enable more than one instruction to be executed in parallel. So long as the semantics are not modified, this parallel execution model is equally valid.

In Chapter 7, Section 7.2.5, we talked extensively about the qualities of a Kahn process network (KPN). In summary, KPNs have the desirable property of having a uniquely defined behavior which does not depend on the timing characteristics of its processes and communication channels. This means that we can express a system in a sequential language and it is possible to enable concurrency to be inferred. In addition, the insertion of concurrency happens in a completely deterministic manner so that even after synthesis the functionality is guaranteed to stay the same as the original sequential description. In addition, as we will discuss later in Section 9.4.3, the insertion of channel components that come from a library can be used to explicitly define concurrency.

So, while no concurrency exists explicitly in the input description it is possible, by conforming to a KPN, to extract concurrency in a safe manner. To the outside world, each block is a black box that will perform a function over time and return a result, usually to the calling parent. It will not be aware of other functions being performed outside of the block, unless they are explicitly made known to it through one of the block interfaces. Those interfaces are likely to be either point to point or buffered, perhaps with the inclusion of a FIFO to control data rates between blocks. The use of a FIFO would be the normal situation when using the KPN methodology. At the entry point to high-level synthesis, the configurability axis is completely undefined since this is a design requirement that will affect how the block is to be

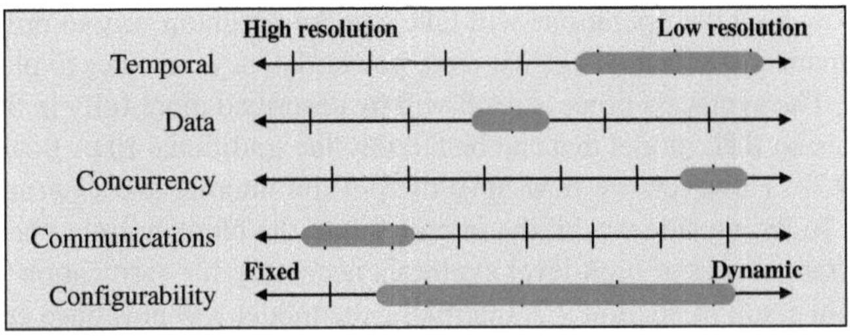

Fig. 9.5 ESL hardware input

implemented. So we know that it is at least design time configurable, and additional flexibility may be built into it during the implementation process.

We can thus map out the input to the hardware implementation process as shown in Figure 9.5.

During the manual transformation steps getting ready for the high-level synthesis flow, changes may be made in the description that affect the kinds of implementation that can be produced, but these changes do not in general change either the abstraction of the description nor do they change any of the ESL attributes of the model. The one exception to this is that a refinement is done in the data types, such that floating point numbers are changed into fixed-point types and in many cases integers will also be converted into fixed-point data types with a specified size. Thus the input to the high-level synthesis process is more accurately reflected by the taxonomy chart as shown in Figure 9.6.

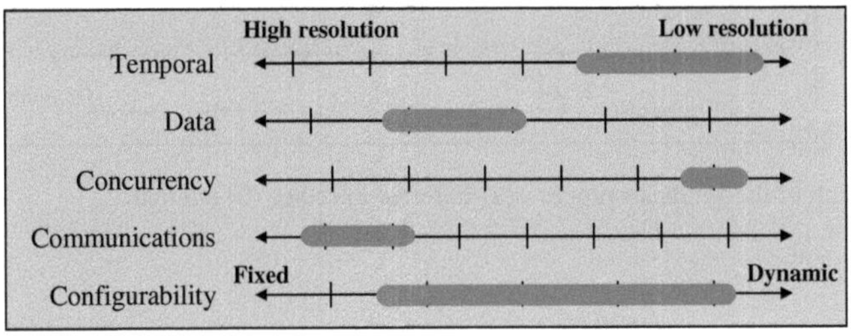

Fig. 9.6 High-level synthesis entry

9.2.2 High-Level Synthesis Output

The main purposes of high-level synthesis are to define a feasible time schedule that meets constraints and to optimize usage of computational resources. What had been ordered events in the input description can now be related to the exact point in time, in terms of clock cycles, at which they will occur. In order to do this, it has to perform functional allocation, timing estimation, and many other tasks necessary to work out how long the operations will take and the optimum way to order them so that performance is balanced against area, power, latency, or other implementation constraints. The synthesis process itself will be described more fully in Section 9.4. The output is an RTL model that can be fed into the traditional EDA flow, as shown in Figure 9.7b. However, we now have to split the internal and external views of the model. To the outside world, the interface into the block has not changed. This is an important aspect of high-level synthesis necessary for verification which will be covered in detail in Section 9.7. Internally, the model will now have concurrency and the communications used to couple those concurrent blocks together. These two models are shown in Figure 9.7a, b.

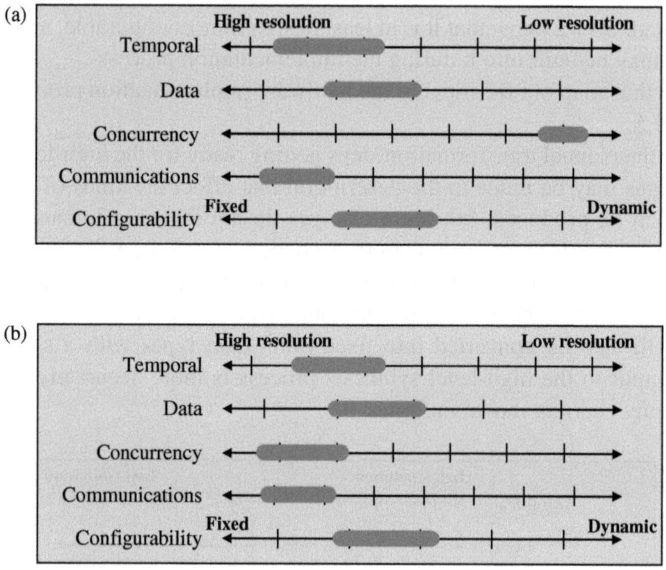

Fig. 9.7 Result of the synthesis process. (a) External interface (b) internal

9.2.3 Verification Models

In addition to producing RTL models, which is the primary goal of the synthesis process, many high-level synthesis tools also produce a high-performance cycle-accurate C, C++, or SystemC model of the hardware which can be used for verification purposes. This model reflects the scheduling that was performed in the synthesis process, but does not reflect the hardware structure that was created. Thus it does not worry about issues such as resource sharing or the actual form in which a function will be implemented. This model will often be 10× or faster than the comparable RTL model and can be plugged into ESL tools or used as a replacement for RTL in system-level simulations. This model has the same external taxonomic view as shown in Figure 9.7a, but internally the model can be anything and does not necessarily reflect the actual device in any way. It is thus not useful to draw the diagram for this.

9.2.4 Other Uses for the Input Model

There are other ways to use these abstract models of the hardware, many of which are an integral part of the ESL flow, such as the benefits of being able to integrate hardware and software together, or higher performance simulation which enables longer and more complex scenarios to be fed through the models. The synthesis process also enables some other flows to be created that may have additional benefits. While we will not be talking about the synthesis processes being targeted at FPGAs, it can become an integral part of the ASIC/ASSP flow, where the FPGAs are used as prototypes that can either enable faster system-level evaluation or have the ability to feed live data into parts of the system, if the FPGA clock rates are sufficiently close to the ASIC speeds, or those necessary for a particular application. This subject was discussed in more detail in Chapter 5 on system-level virtual prototyping.

9.3 Why C/C++/SystemC

With the prevalent RTL methodologies in use today, it is only possible to explore a very small area of the possible architectural space. This is because of the time and effort it takes to specify all of the details necessary to implement such a solution, including the encoding of any state machines, constructing the necessary datapaths and pipelines, and performing enough verification to know that the unit is working correctly. With that amount of investment, it is impractical to look for other potential solutions unless the chosen one does not meet the requirements or is shown to be problematic in its eventual implementation. Thus any optimization work that is performed tends to be associated with local optimizations and making sure that the best implementation is found for the architecture selected.

A design team may have even less flexibility when IP blocks are selected. These may have been implemented by a different team, at some time in the past. At that time, different requirements or known architectural possibilities may have led them to implement the block in what would now be considered to be an inferior architecture. Thus there is no generalization that is possible as to the IP performance as it is possible that an IP developer, having more experience with that particular protocol, or codec, may have selected a better architecture than a less educated implementer may have done. On the other hand, the developer may have in-depth domain expertise that means their implementation is superior to what the ordinary design team can accomplish. The point is that the architectural choices are more limited with that solution.

While RTL is a good implementation language, it is not suitable for design exploration, nor is it a particularly good language to enable the full benefits of reuse. To solve both of these issues and to achieve higher design productivity, there is a need

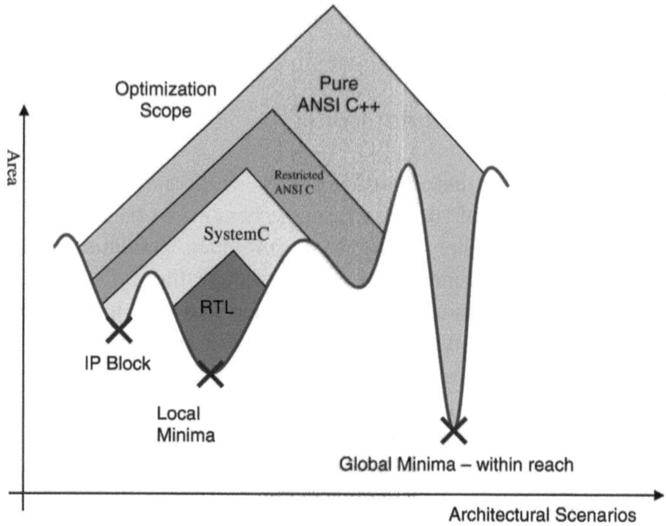

Fig. 9.8 Optimization scope

to migrate to higher levels of abstraction, but the choice of language can influence the benefits that will be obtained (see Figure 9.8).

When considering a migration to higher level languages, it is very important to consider the impact that a language may have in enabling the exploration of the full range of possible implementation solutions. In addition, there are trade-offs that should be considered in terms of the amount of architectural information that should be embedded in the source code, rather than existing in separate files, or in comments or pragmas. The more architectural information that is embedded in the file, the closer it is to an implementation, and if reuse is ever contemplated, it may not represent an ideal set of architectural decisions for that new application. This is to a lesser extent the same problem that existed for the fixed IP solution, although in this case there is still a large implementation space that it covers. The counterargument to this is that mistakes are more likely if there is a separation of information that is intimately related. The industry went through the same dilemma with assertions a few years ago. They could either be defined directly in the place they were to be used or they could be described in separate files and then attached into the design code during simulation.

One popular high-level language is SystemC. This language has been discussed in detail in Chapter 3 and its usage demonstrated in several other chapters. SystemC extends ANSI C++ by adding a class library that describes more traditional event-driven simulation semantics common to the existing HDL languages, such as Verilog and VHDL. These extensions allow the explicit modeling of time, concurrency, and bit types that are more likely to be found in hardware implementations. In the C language data types such as integer and byte are the only ones available. This explicit control over concurrency, as found in SystemC, could be seen as one of the trade-offs that were illustrated in the previous section. By describing the concurrency in the design file itself, we can see the direct impact that it has in the code, but this requires additional time and effort to code. That in turn may restrict the kinds of architectural design solutions that can be explored in the implementation process and in its eventual flexibility when reused. SystemC also adds class definitions for more abstract modules, channels, interfaces, and transactions.

While SystemC is a superset of the C++ language, the entire language is not generally available for high-level synthesis, and the embedding of some of the architectural aspects of a solution is seen by some to limit the scope of the optimization solutions available to the synthesis tools. It is thus possible that the use of SystemC will enable better solutions to be found, but some architectural avenues have been blocked off unless some amount of rewriting is done to explore the other parts of the solution space.

For years, companies have used ANSI C for constructing system-level models. Indeed, C code is often used to create reference implementations of new algorithms – e.g., for new communications or multimedia standards. The attraction is a language that is almost universal among the engineering community and has lots of free development systems and compilers. Once compiled, a model can be run on any suitable machine and will often execute at orders of magnitude faster than RTL (up to 10,000× faster) when compared to the eventual speed it is expected to operate at once implemented in hardware or run on a target processor. It also has the advantage that the architectural decisions are not embedded in the code. The problems though are quite considerable, including the lack of bit-accurate modeling types, the difficulties in enabling reuse, and a messy coding style. However, for reference implementations of algorithms that just need to be functionally correct, this is a good starting point.

What is needed is a solution that solves these problems and at the same time does not constrain the architectural exploration space. For Mentor and their Catapult synthesis system, that was a difficult choice. C++ is a modern language that both builds on the capabilities of C and does retain full compatibility with modules that may be written in SystemC. It has all of the same advantages in that it is a compact language and is object oriented. In order to solve the bit-accurate data types issue, Mentor created an extensible library package that makes arbitrary bit-widths possible, along with the ability to apply almost any form of rounding, saturation, or truncation policy. This will be described in more detail in Section 9.4.6.

A less important consideration is that of archival quality. C++ is a stable language for which many compilers, debuggers, and other tools exist. The chance that a model written in C++ will become obsolete is very low. Having said that, there is no industry agreed upon synthesizable subset for C or C++, so it may be possible that a model written today may not be compatible with future changes in the tool. This means that although the model inputs will still execute or simulate in the future, they may not synthesize if the tools change.

On the other hand, SystemC is still a young and incomplete language. It has already gone through a number of changes. It is possible that a model written today may not maintain the same functionality in the future. In addition, there is no industry standard synthesizable subset for SystemC although work has started to define this.

A similar situation existed with RTL synthesis before the synthesizable subsets were standardized. Each vendor would accept different constructs or subsets of the language. Similarly, the languages (Verilog and VHDL) have been updated over time and some features of the original languages have been deprecated, meaning that some models may have to be updated in order for them to continue working.

Thus neither language provides complete protection from the future, although C/C++ has a slight edge in this regard. Given the lack of synthesizable subsets, models written for one tool are to some degree captive to that tool in that you cannot expect to migrate a model into a different synthesis environment and expect consistent results or even that the model will be accepted. Neither language provides protection from tool advancement or provides interoperability today.

While Mentor felt that C++ was the right choice for algorithmic blocks, they also acknowledged the need to explicitly define concurrency, especially at the higher level, which is not directly supported by the C++ language. The dual strategy of supporting KPN for implicit concurrency and the insertion of channel components for explicit concurrency provides this capability. With this communications library in place high-level pipelining and communications architectures between blocks can be explored. Hierarchy will be discussed in Section 9.4.3.

9.3.1 Language Limitations for Synthesis

While the previous generation of behavioral synthesis tools had many restrictions on the constructs that would produce good quality of results (QoR), most of those restrictions are now gone from modern high-level synthesis tools. In fact it becomes easier to talk about exclusions than what is allowed. For Catapult, the single biggest restriction is that the code should be statically determinable. That means that all of its properties must be defined at compile time. This excludes functions such as malloc/free and new/delete since these expand or contract the state space of the system.

In a similar manner, pointers are supported so long as they point to statically allocated objects, which means that they can be converted into indexed arrays. Arithmetic is supported on pointers so long as it stays within the bounds of a single object, or put another way, a pointer cannot point to more than one object. There is an exception to this in that a pointer can be reassigned to point to an identical object such that, for example, ping-pong buffers can be supported. A converse of this limitation is that two pointers that are passed into the top level are assumed to point to different objects, or in other words, there cannot be a hidden dependency between them that would not be discernable by the synthesis tool.

There are also some other restrictions which are more good programming practices than actual restrictions on the language. An example of this is that any global is assumed to not be accessible from outside of the hardware. Without this restriction, it becomes impossible to know all of the dependencies within the system. Of course, the tool will not know if you choose to do this, but it would mean that you would have to be very careful with verification as there could be unexpected side effects.

One area of restrictions is related to the desire to provide consistent results independent of the machine it is being run on. Because of this, unions, void pointers, and typecasting of pointers are not supported since the results of these depend on the underlying memory architecture, be it big endian or little endian.

Finally, there are a few restrictions on functions. Recursion is not supported as this violates the statically determinable rule. Templates can be recursive so long as a recursion cap is defined.

As of this writing, Mentor has decided not to support floating point operations, as defined by IEEE-754-2008, within the Catapult product. This is an arbitrary decision based on the complexity of the hardware that would be generated, and the potential market size for this type of implementation. Having said that, Catapult is user extensible thanks to its usage of C++ as the input language and it is possible for a user to write a floating point library which would be fully supported in the current tool. Any user can extend the class libraries to provide support for certain operations and this has been done by some of their customers for the implementation of FFTs, neural networks, adaptive filters, and other application domains that do need to operate on floating point numbers. The general approach is to map the operations into 2's complement arithmetic for both the mantissa and the exponent. This is exactly the same process that Mentor used for the implementation of fixed-point and complex types. None of these library extensions require any extensions to the synthesis capabilities. In the future, Mentor may support a hardware friendly floating point library.

9.4 High-Level Synthesis Fundamentals

High-level synthesis is similar, in many ways, to other forms of synthesis, such as RTL synthesis, in that it contains three primary tasks: scheduling, allocation, and mapping. Scheduling is working out how many operations can fit into a single clock period, allocation is deciding how many resources can be afforded and mapping is the process by which operations are attached to the mapped resources. While allocation clearly has to happen before mapping, there is no prescribed order in which scheduling and allocation happen and in most cases it is likely to be an iterative process since the two are dependent on each other. It is thus not uncommon to find research programs [6] that attempt to perform both from within a single algorithm. High-level synthesis also needs to consider memories as an available resource for allocation and mapping and when to trade off memories for registers.

9.4.1 Schedule and Allocation Trade-offs

One of the common operations in high-level synthesis is that of loop management. Consider the simple loop:

```
for (i=0; i< 32; i++)
    q[i] = x[i] * y[i];
```

If we assume that the multiply operation can be performed in 1 clock cycle, and that memory operations take 1 clock cycle, then each iteration of the loop would take 4 clock cycles and the total loop would take 128 clocks. This is ignoring the time it may take to perform the increment and test operations for the loop control. In a trivial case of loop unrolling, we could try and replicate the loop 32 times and execute them in sequence. This would save the time and cost associated with the loop control structure and could also achieve some other efficiencies because the indexes are now known. But the savings are small.

Alternatively we could execute all of the iterations in parallel. Would the entire operation now take four clock cycles (assuming that the reads of the memories happen in parallel)? It would require almost 32 times as much area, the only savings being that we now no longer need to consume logic for the loop control. To achieve that kind of speedup would also require that we can access 32 words from each of the memories in parallel, which would be quite a feat. If we were to have dual port memories available to us, then we could do a loop unroll by two, so instead of the original code, it would become

```
for (i=0; i< 31; i+=2) {
        q[i] = x[i] * y[i];
        q[i+1] = x[i+1] * y[i+1];
        }
```

It should be noted that the user would not have to recode the algorithm as shown, as the effect of the recoding would be performed by the synthesis tool using the directives provided to it. While we have added some complexity, and the need for an additional incrementer, we will probably have obtained a two times speedup for about an increase in twice the area. Thus we can see that in this example, there is a direct trade-off between performance and area, but to obtain that performance the data bandwidth must also be increased by a similar amount.

We were previously expecting that the read operations would take place first, followed by the multiply, and then the write. What if we were to only allocate a single multiplier? After the first two memory locations have been read, those values can be multiplied together while the second set of values are being read, and then those multiplied together while the first result is being written. The total time for each iteration of the modified loop would now be 4 clock cycles, with the total loop taking 35 clocks, but we require only one multiplier and we do not need the extra complexity of the dual port memories since we are only reading one word at a time. The sharing of the multiplier would require some additional multiplexers, it would require a few extra registers to hold the read values and some additional control logic, but this is likely to be small considering the savings that have been achieved.

In addition, if these results were to be used in another computation, they could be written to local registers rather than main memory, with a different set of cost-performance trade-offs.

While this example is not meant to explain all of the optimizations that are possible with a tool such as Catapult, it is meant to show that many of these optimizations are complex trade-offs between area, speed, power, and less tangible aspects such as the complexity of the solution.

9.4.2 Synthesis at the Interface

So far we have talked about what the synthesis tool does to transform a module, but this is not enough. That module was part of a larger system and after synthesis it has to fit back into that system. That means the interfaces to the module have to be considered as well, and that cannot be done in isolation. When information is passed as parameters into a C or C++ module, it does not provide any timing information about those interfaces. For example, will the information be guaranteed to be stable on all inputs during the entire operation or is it only available for a short while after the module has been called?

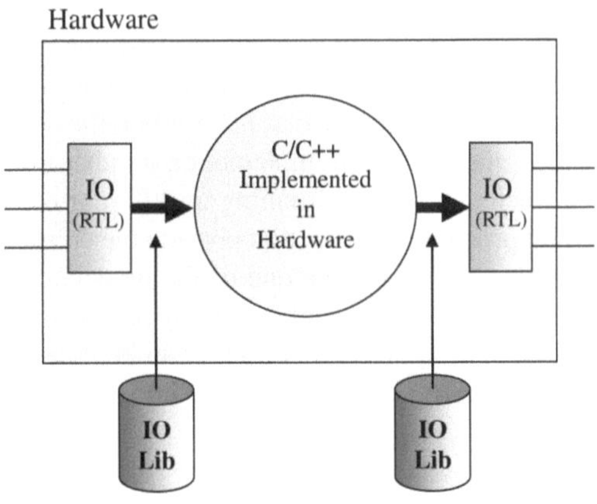

Fig. 9.9 Interface components

This problem is, in general, handled by a library of interface elements that represent protocols such as a wire, register, FIFO, memory interface, or a handshake protocol (see Figure 9.4). The type of interface is defined in the Catapult UI or by using pragmas in the design. In the case that a specific interface is required, such as a bus interface, this can be provided as a library module written in either VHDL or Verilog.

9.4.3 Hierarchy

Interfaces also become very important when we are dealing with a hierarchical design. In the example in Section 9.8, all of the blocks make up a single functional unit, so we do not need to consider interfacing between them. However, there are many cases where parts of a design need to execute concurrently and in these situations, the scheduling and communications between the blocks must be considered.

Consider a system that contains a video scaler. The top-level block diagram of this scaler can be seen in Figure 9.10.

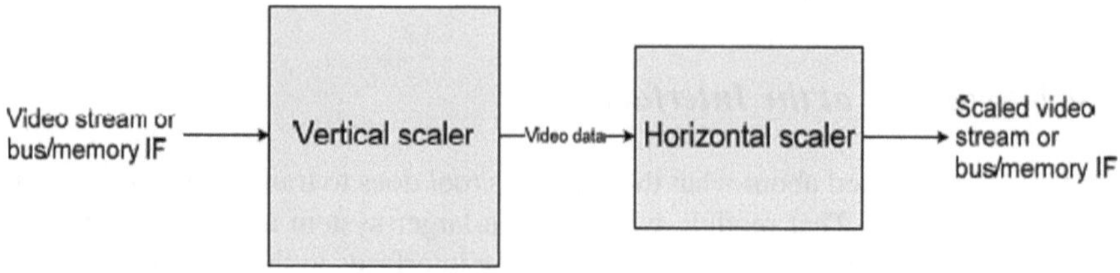

Fig. 9.10 Video scaler

In most cases, the scaler is required to run at 1 pixel per clock, which means that memory architecture is important, especially where information is passed between the vertical and the horizontal scaler. To help with this a library of communications primitives is required.

C++ can be enhanced with the addition of class libraries that provide additional capabilities, just as is done with SystemC. For communications, Mentor has built an *ac_channel* library that is based on the standard template library (STL) double-ended queues. These implement a number of streaming interfaces, including those between sub-systems running at different communications rates. This is based on Kahn process networks that provide the complete semantics of the communications and also happens to be one of the models of computation supported by SystemC. Sub-systems then communicate by performing read and write calls to the channel.

Different types of interfaces can be constructed depending on the communications patterns seen across the interface, some of which are shown in Figure 9.11.

9.4.4 Other Control

In addition to synthesizing the logic that is explicitly described in the C++ code, there are other hardware artifacts that are brought in by the synthesis process, such as the clock, reset, and possibly an enable operator. While the safest approach is to apply a reset to all of the registers in the block, area optimization may be achieved by only resetting the ones identified by the designer. Similarly for power optimization, clock gating may be desired such that all operations in the block can be stopped when no data is ready to be processed.

In a C environment, a process returns when the operation is complete. This implies a blocking operation, but this may not be what is required in all cases. Instead, it may be more convenient for the block to raise a finished flag when it has completed, allowing the caller to perform other tasks in parallel. It may thus

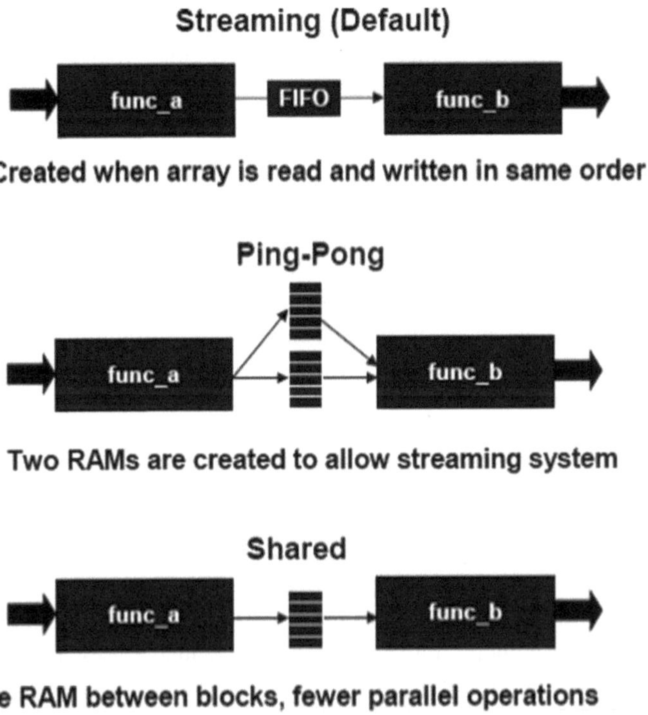

Fig. 9.11 Channel types

be desirable to provide block-level synchronization logic as well. The timing for a typical control system is illustrated in Figure 9.12.

In the timing diagram of Figure 9.12 two signals have been defined, namely `start` and `done`. After `reset` has been asserted, the `done` or completed signal is raised indicating that the block is ready to perform its intended task. It waits until `start` is set to high before executing the operation. After it has started it takes the `done` signal down and sets this back to high when it has completed, indicating that results are available and that it is ready to start the next operation.

It may also be beneficial to include an additional control signal that indicates the completion of a transaction. This has particular benefit when considering

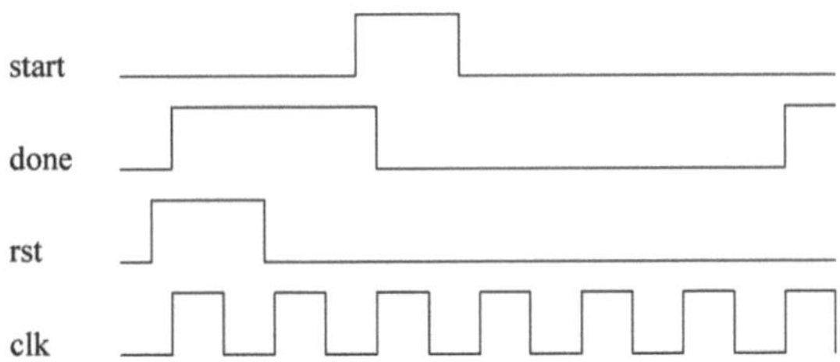

Fig. 9.12 Module control

verification as it may indicate the point at which the results from the synthesized design may be compared against those from the original high-level input. We will discuss verification further in Section 9.7.

9.4.5 Target Library

No description of the synthesis process would be complete without considering the target technology that the synthesized results are meant to be implemented in. These provide not only the speed that operations such as an ADD or memory access operation take, but also the elements that are available. For example, there is no use in attempting to use dual port memories, if the target library has none of these available. There may also be limits on the number of certain operators that are allocated. While this is more typical for FPGA synthesis, there may be other reasons why such limits are put in place, such as licensing restrictions.

There are two schools of thought on the use of technology libraries. One is that by integrating the behavioral exploration process with the RTL synthesis tool, characterization is not necessary since the .lib file can be read directly. This may be the solution where both high-level and RTL synthesis tools are coming from a single vendor. It can also mean that the scheduler may be able to be more aggressive since it can get more exact figures, rather than estimates in many cases. This however does not solve the problem of routing delays which are getting to be a larger proportion of the total delay, especially in FPGAs. These cannot be known until after place and route, and thus allowances must remain in the schedule for these.

If the tools come from different vendors, using the RTL synthesis tool to provide timing information would tie up two critical licenses. In this situation getting estimates for timing and area is a better trade-off. This can also result in the exploration being faster since it does not need to gather the low-level information, generating this instead only when good looking architectural candidates are identified. Catapult provides an automatic tool for doing this that uses the users' choice of downstream synthesis technology, such as Design Compiler [7] from Synopsys or FPGA synthesis tools from Altera or Xilinx. The characterization tool produces a range of points for various library elements, such as the fastest block or the one with the smallest area. Intermediate points are also stored for solutions that require some compromise between area and speed.

9.4.6 Data-Type Libraries for Synthesis

We have already discussed the need for fixed-point data types for hardware implementation in Section 9.3. In this section we will provide a very brief overview of one of the packages supplied by Mentor called Algorithmic C^{TM} Data types. This package is available as a free download at http://www.mentor.com/products/esl/high_level_synthesis/ac_datatypes.cfm. The library is implemented as a C++ class template and as of this writing has three types: ac_int<>, ac_fixed<>, and ac_complex< > with all of the necessary functions to convert between them and to perform functions on them. The fixed-point library supports arbitrary bit-widths. The library is written in C++ which means that no special compiler is needed to execute the models. The code written using ac_datatypes can be compiled and simulated using any standard off-the-shelf C++ compiler, such as Visual C++ or gcc.

For each type, several attributes can be defined, such as if it holds a signed or unsigned number, the position of the fractional point, and the rounding and overflow attributes of the variable. Figure 9.13 shows the effects of rounding modes as well as the implications associated with using them, such as the area impact.

Modes	Behavior n is integer, q is 2^{I-W}	Simulation/Synthesis cost
AC_TRN (**default**) (trunc towards $-\infty$)	nq (n+1)q	Delete bits, no cost except for /= (div assign) signed
AC_TRN_ZERO (trunc towards 0)	n < 0: nq (n+1)q n ≥ 0: nq (n+1)q	No cost for /=, or unsigned src For signed: incrementer, OR for deleted bits, AND with sign bit
AC_RND (round towards $+\infty$)	nq (n+1)q	Various forms differ only on the direction of the rounding for values that are exactly at half point. All require an incrementer, some require to OR deleted bits, some only require to look at the MSB of the deleted bits.
AC_RND_ZERO (round towards 0)	n < 0: nq (n+1)q n ≥ 0: nq (n+1)q	
AC_RND_INF (rounds towards $\pm\infty$)	n < 0: nq (n+1)q n ≥ 0: nq (n+1)q	
AC_RND_MIN_INF (round towards $-\infty$)	nq (n+1)q	
AC_RND_CONV (round towards even q multiples)	(2n-1)q 2nq (2n+1)q	

Fig. 9.13 Rounding (quantization) modes

Operators on the data types allow bit selection and slice read and write of arbitrary length.

It should be noted that SystemC also provides a fixed-point library. While Mentor normally attempts to use all available standards (SystemC is produced by OSCI and is standardized by the IEEE), they chose to implement their own because of performance and accuracy considerations. The Algorithmic C data types are arbitrary length and this allows the definition of semantics for all operators without regard to specific implementation limits. Special attention has been paid to these semantics to make sure that they are fully consistent between simulation and synthesis. In order to catch any violations of this, assertions have been added into the library that will flag illegal behaviors during simulation. If utilizing standards is more important to you, then Catapult does support the SystemC types, such as sc_int.h, sc_fixed, but the coding is not as efficient, and the result may suffer from the problems mentioned previously. The same is true with standard C/C++ data types. Again, coding will not be as efficient.

SystemC provides two data types for bit-accurate integers. The "fast" sc_int/sc_uint types are limited to 64-bit precision while the larger sc_bigint/sc_biguint can handle arbitrary precision but execute much slower. However, they do not always operate identically. An example cited in a paper at GSPx05 [8] states that while

```
(sc_int<6>) 1 > (sc_uint<6>) -1
```

returns false, the similar function using sc_bigint/sc_biguint returns true.

The second factor in deciding to implement a custom library was simulation speed. The ac_int type uses template specialization techniques that enable standard C++ compilers to generate optimized code that will execute much faster than the equivalent SystemC data types. Reference [2] reports that ac_int of bit widths in the range 1–32 can run 100× faster than the corresponding sc_bigint/sc_biguint data type and 3× faster than the corresponding sc_int/sc_uint data type. In addition to this, compilations speeds can also be 5× faster.

While the user has to specify bit-widths, this is not as onerous as it may sound [9]. In general it is only the bit-widths for the interface variables that need to be set or in other specific places. For example, if a variable is defined to be 13 bits wide and is the result of an assignment from an integer, then that integer will also be reduced to 13 bits, assuming that nothing else requires it to be larger. This automatic bit reduction not only minimizes the changes that are required in the source code but also makes it possible to write parameterizable designs as will be demonstrated in Section 9.8.

Consider the alternative. In Figure 9.14 we show the code for a simple 4-tap fir filter that utilizes the ac_fixed library. If we attempt to implement the same code in VHDL, for example, we may end up with something similar to that shown in Figure 9.15.

But is this a fair comparison? Not really, because the VHDL description defines one particular implementation of the FIR filter, rather than something that can be

```
#include "ac_fixed.h"
    typedef ac_fixed<16,16,true> mpy_t ;
    typedef ac_fixed<48,48,true> add_t ;
void fir(
    mpy_t *data,
    mpy_t coeff[4],
    add_t *result) {
    add_t acc = 0 ;
    static mpy_t taps[4] ;
loop:for (int i=0; i<4; i++) {
        if (i==3)
            taps[i] = *data ;
        else
            taps[i] = taps[i+1] ;
        acc += taps[i] * coeff[i] ;
    }
*result = acc ;
}
```

Fig. 9.14 C++ code for 4-tap FIR filter

```
library ieee ;
use ieee.std_logic_1164.all ;

package special_definitions is
    TYPE data_array IS ARRAY (0 TO 3) OF STD_LOGIC_VECTOR(15
DOWNTO 0) ;
    TYPE mpy_array IS ARRAY (0 TO 3) OF STD_LOGIC_VECTOR(31
DOWNTO 0) ;
end special_definitions;

LIBRARY IEEE;
USE IEEE.std_logic_1164.ALL ;
USE IEEE.std_logic_arith.ALL ;
USE IEEE.std_logic_signed.ALL ;
LIBRARY work ;
USE work.special_definitions.ALL ;

ENTITY fir IS
PORT (
    clk : IN std_logic ;
    a_rst_n : IN std_logic ;
    data : IN std_logic_vector(15 DOWNTO 0) ;
    coeff : IN data_array ;
    result : OUT std_logic_vector(47 DOWNTO 0)
) ;
END fir ;

ARCHITECTURE rtl OF fir IS
    SIGNAL taps : data_array ;
    SIGNAL mpy : mpy_array;
    SIGNAL add1 : STD_LOGIC_VECTOR(32 DOWNTO 0) ;
    SIGNAL add2 : STD_LOGIC_VECTOR(32 DOWNTO 0) ;
    SIGNAL final_add : STD_LOGIC_VECTOR(33 DOWNTO 0) ;
BEGIN

clocked : PROCESS(clk, a_rst_n)
BEGIN
    IF (a_rst_n = '0') THEN
        taps <= (OTHERS => (OTHERS => '0')) ;
        result <= (OTHERS => '0') ;
    ELSIF (clk'EVENT AND clk = '1') THEN
        result <= SXT(final_add,48) ;
        FOR i IN 0 TO 3 LOOP
            IF (i=3) THEN
                taps(i) <= data ;
            ELSE
                taps(i) <= taps(i+1) ;
            END IF ;
        END LOOP ;
    END IF ;
END PROCESS clocked ;

clocked1 : PROCESS(clk)
BEGIN
    IF (clk'EVENT AND clk = '1') THEN
        add1 <= SXT(mpy(0),33) + SXT(mpy(1),33) ;
        add2 <= SXT(mpy(2),33) + SXT(mpy(3),33) ;
        FOR i IN 0 TO 3 LOOP
            mpy(i) <= taps(i) * coeff(i) ;
        END LOOP ;
    END IF ;
END PROCESS clocked1 ;

final_add <= SXT(add1,34) + SXT(add2,34) ;

END rtl ;
```


Fig. 9.15 VHDL code for a 4-tap FIR filter

optimized in several ways. It uses four multipliers, it has no ability to define the interface so that it cannot be used in multiple design instances, it needs to manually handle precision extensions and perhaps most importantly, it has no handling for rounding or saturation, which could mean a bug, just like the [10] Ariane 5 disaster, waiting to happen.

9.4.7 Synthesis Tools

So far we have talked about the synthesis process in general and demonstrated how it is an interactive process, as the best trade-offs between latency, throughput, area, power, or other factors are considered. In order to make those trade-offs, information must be provided to the designer by the tool. This information educates and guides the decision-making process, enabling the designer to incrementally and iteratively close in on the best possible solution. Some of the most valuable charts are the Gantt chart and the datapath diagram.

9.4.7.1 The Gantt Chart

Perhaps one of the most important ways in which timing and schedule feedback are performed is through a Gantt chart. An example is shown in Figure 9.16. A Gantt chart shows the schedule that has been created, the slack that is available, and the dependency between certain events. This makes it possible to identify bottlenecks in a pipeline. This shows not only what can be performed during each clock cycle but also the total amount of time being taken by each of the levels of hierarchy in the design. In this particular case, almost all of the time is being taken by `LOOP SHIFT`.

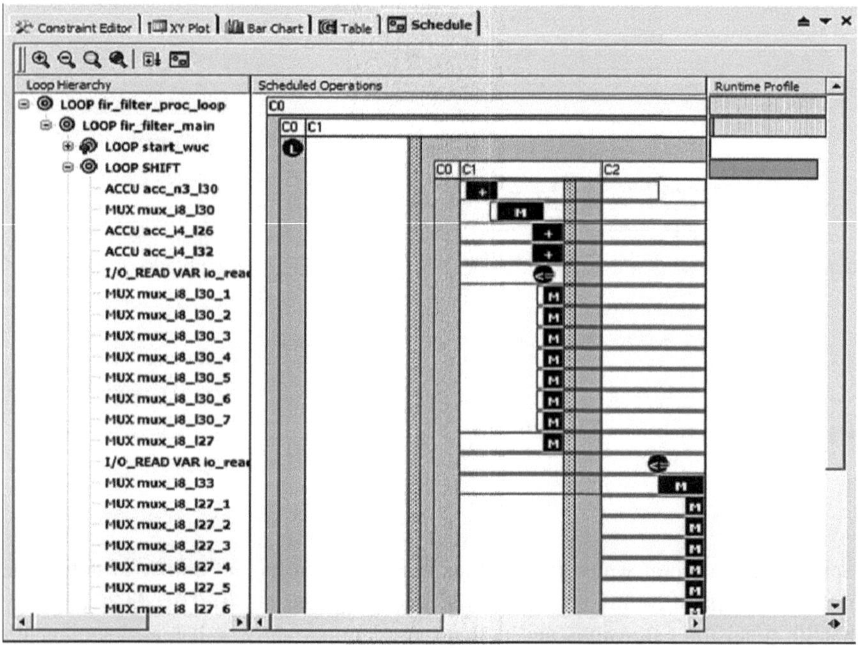

Fig. 9.16 Gantt chart

Within the Gantt chart, the dark box with a symbol inside it shows when the operation is scheduled to be performed. The larger outlined box that encloses the operator shows the slack associated with that operation. So, for example, while there is an ADD operation ACCU acc_n3_I30 in C1 of LOOP SHIFT, it could be moved into the next clock cycle if that would produce a better allocation. You may also note the dotted and grey areas toward the end of clock 1 and at other places. This is the time that is being set aside for setup delays on the registers and the slack that is desired to be kept for routing delays. It is also possible to show Gantt charts from the perspective of the elements, so that the designer can see when each element is being used.

In addition, the Gantt chart can be used to show dependency between operations. An example of this is shown in Figure 9.41. This can be very useful in understanding why certain decisions have been taken by the tool and give clues as to how it can be changed, if necessary.

9.4.7.2 Datapath Diagram

The datapath is the schematic equivalent of the micro-architecture that is currently being considered. It shows the major elements and how they are interconnected. In the example shown in Figure 9.17, the major operational blocks are shown along with the multiplexors to control the flow of data. However, the control circuitry for those multiplexors is omitted to help remove unnecessary clutter from the diagram.

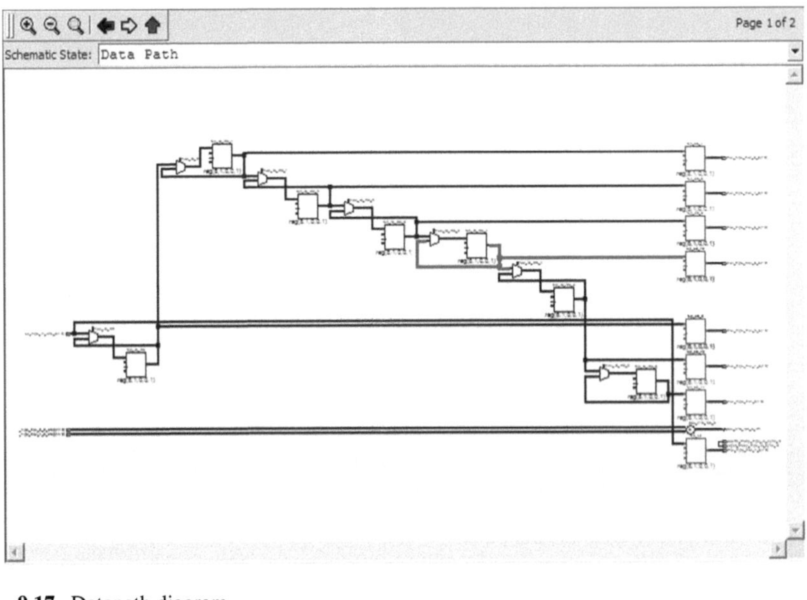

Fig. 9.17 Datapath diagram

9.4.7.3 Interactive Exploration

We have already discussed the fact that high-level synthesis does not have one correct solution that would fit all possible sets of requirements. It is equally unlikely that the first attempt by a designer will find the optimum solution. There are many parameters that can be adjusted and explored and it is more likely to be a number of optimization steps that lead to the solution that is finally accepted. As part of this process, it may be necessary to backtrack to a previous solution. This creates the need for extensive history being kept within the tool that remembers, not only the parameters used on each trial, but also the results so that they can be compared.

Figure 9.18 shows one such way of comparing results. Here, an *X–Y* plot is given showing the throughput cycles and area trade-offs of several solutions that were attempted. Many other attributes of the designs can also be plotted. In some cases, it is not necessary to have generated the RTL for solutions before trade-offs can be made. For example, quite a lot of performance data is available as soon as the design has been scheduled. This allows very fast iterations to be made without spending time generating the detailed results. Of course, area and speed figures for these will be less accurate as full generation of the control engines has not taken place. Pareto diagrams, such as this, are often found to show trade-offs in implementation options. Another example was seen in Chapter 6 for processor configurations.

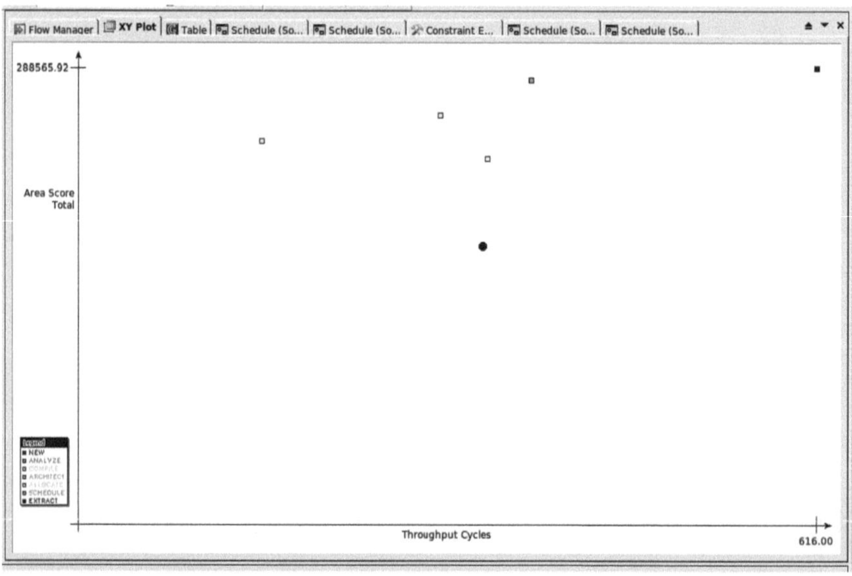

Fig. 9.18 *X–Y* plot of synthesis trial runs

9.5 Synthesis Domains

The first generation of behavioral synthesis tools created a lot of excitement, but this quickly faded when it was discovered that they had severe limitations on the type of input descriptions that they could handle. They were quite good at handling dataflow blocks, but that was about all. Engineers have long memories, especially when tied to bad experiences. From this kind of false start for an emerging technology, myths are born and these can take a long time to quell. One in particular is that high-level synthesis cannot handle control structures.

Most of the algorithms being designed today are complex adaptive algorithms that have a significant control aspect to them. This may include loops with variable iteration limits, prematurely exiting loops with constructs such as break and continue, and of course conditionals. While tools such as Catapult still specialize in data-dominant designs, these tools can handle all of the control aspects demanded from these modern algorithms. What may set solutions apart is the quality of results that may be obtained with specific examples. It is likely, although not shown, that some synthesis solutions will do better on certain classes of design, while others may excel on a different type of problem. This is a natural aspect of tool development.

An EDA company that is developing a synthesis tool will either target their existing customer base or a group of customers doing a similar type of design. This could be in fields such as encryption, video or audio processing, and wireless. All of the customers within this space are likely to have similar architectures, similar goals, and almost certainly working with the same set of algorithms. It makes sense for a tool company to optimize for this group of customers initially. As the tool matures, they will take on additional product segments and thus grow the optimization strategies that are present in the tool.

Tools from companies that have not targeted the area of development that you are in may well be able to accept the design and perform synthesis, but may not provide optimized quality of results. It is thus worth ensuring that the synthesis tool you are considering has had some exposure to the design domain that you are in and is able to demonstrate results on the kinds of algorithms that you deal with.

9.6 A Simple Example

While we have talked so far about high-level synthesis in general, many aspects of the tools and the languages are only fully understood by looking at examples. We will now look at one such example to demonstrate the difference between architecture and micro-architecture and the importance that this plays in the high-level synthesis flow.

9.6.1 Embedded Architecture

In Section 9.2 it was mentioned that high-level synthesis solutions, such as Catapult, cannot make architectural changes that are implicit in the source code. In a similar manner, people have often claimed that one of the advantages of migrating hardware into the C language is to have a single combined language from which we are free to make arbitrary hardware/software partitioning decisions. With a common language it is claimed that it will be possible to divide the input description and just substitute a different implementation strategy for various types of hardware or compile it for software. This is not correct as will be demonstrated in the following example. People who accept such statements are potentially missing out on finding the best implementation solutions.

In this section we will look at an example that shows how different architectures can affect the solution. The example is the Advanced Encryption Standard (AES) algorithm. The reason for choosing this is that it is a fairly simple algorithm that is available in source form from several web sites. The source used in this example came from http://www.hoozi.com/Articles/AES_Encryption.htm and is also mirrored on the book's web site http://electronicsystemlevel.com. Wikipedia [11] also provides a simple explanation to help understand the algorithm, parts of which are excerpted here.

The algorithm operates on a 4 × 4 array of bytes, called a state, and each state is subjected to a number of rounds, each round consisting of four transformation stages.

The complete algorithm can be described as

- First round
  - Add Round Key
- Subsequent rounds
  - Substitute bytes
  - Shift rows
  - Mix columns
  - Add round key
- Last round
  - Substitute bytes
  - Shift rows
  - Add round key

Each of the four stages is depicted in Figures 9.19, 9.20, 9.21, and 9.22.

In the "Add Round Key" step (Figure 9.19), each array of data is bitwise XORed with a subset of the encryption key derived from a key schedule that will not be described here. For a full description of this refer to the Wikipedia site [11].

Next a nonlinear function is performed that substitutes each byte for a corresponding byte in a table. This is called the Rijndael S-box as shown in Figure 9.20.

The third step mixes up the rows with the first row being left unchanged, the second row shifted by one, the third by two, and so on as shown in Figure 9.21.

The final step is a matrix multiply on each of the columns, shown in Figure 9.22.

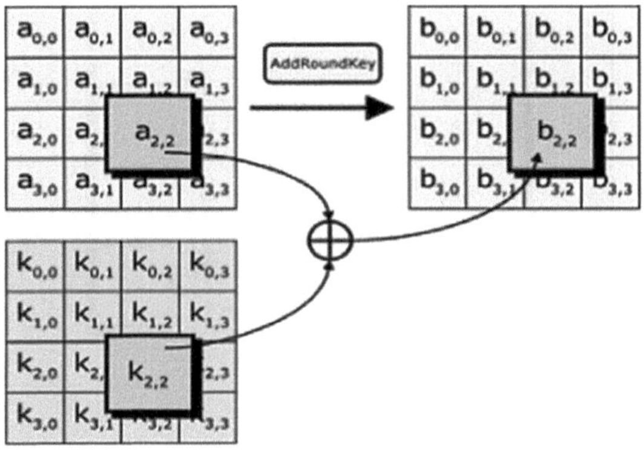

Fig. 9.19 Add round key

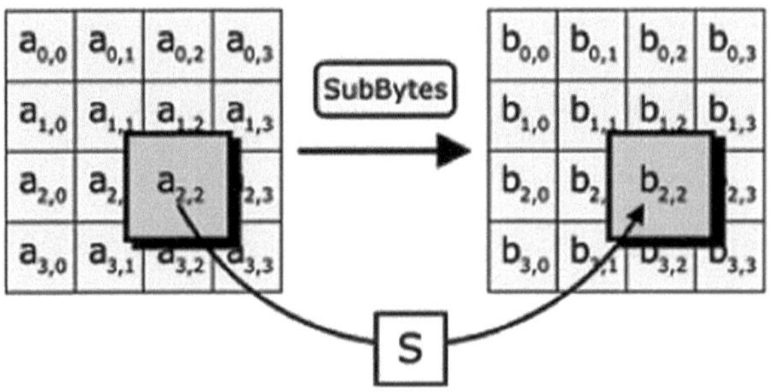

Fig. 9.20 Byte substitution

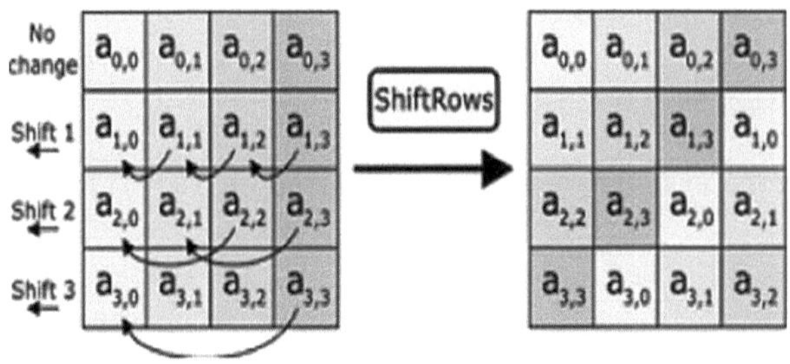

Fig. 9.21 Shift rows

Here is the code for each of the main computational elements as taken from [12].

```
// This function adds the round key to state.
// The round key is added to the state by an XOR function.
void AddRoundKey(int round)
{
    int i,j;
```

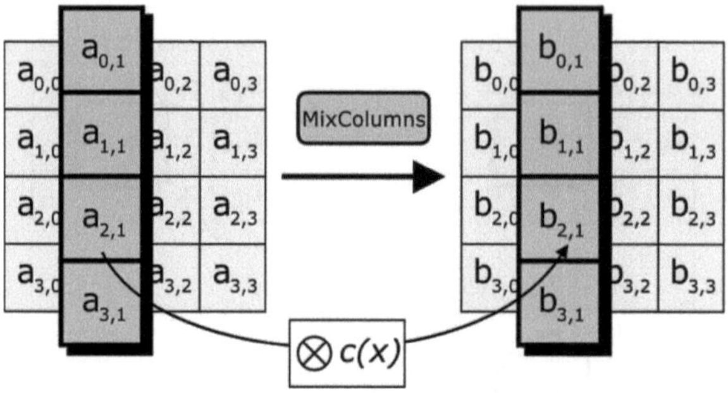

Fig. 9.22 Mix columns

```
        for(i=0;i<4;i++)
        {
            for(j=0;j<4;j++)
            {
                state[j][i] ^= RoundKey[round * Nb * 4 + i * Nb + j];
            }
        }
    }

    // The SubBytes Function Substitutes the values in the
    // state matrix with values in an S-box.
    void SubBytes()
    {
        int i,j;
        for(i=0;i<4;i++)
        {
            for(j=0;j<4;j++)
            {
                state[i][j] = getSBoxValue(state[i][j]);
            }
        }
    }

// The ShiftRows() function shifts the rows in the state to the left.
// Each row is shifted with different offset.
// Offset = Row number. So the first row is not shifted.
void ShiftRows()
{
    unsigned char temp;

    // Rotate first row 1 columns to left
    temp=state[1][0];
    state[1][0]=state[1][1];
```

```
    state[1][1]=state[1][2];
    state[1][2]=state[1][3];
    state[1][3]=temp;

    // Rotate second row 2 columns to left
    temp=state[2][0];
    state[2][0]=state[2][2];
    state[2][2]=temp;

    temp=state[2][1];
    state[2][1]=state[2][3];
    state[2][3]=temp;

    // Rotate third row 3 columns to left
    temp=state[3][0];
    state[3][0]=state[3][3];
    state[3][3]=state[3][2];
    state[3][2]=state[3][1];
    state[3][1]=temp;
}

// xtime is a macro that finds the product of {02} and the argument
   to xtime modulo {1b}
#define xtime(x)   ((x<<1) ^ (((x»7) & 1) * 0x1b))

// MixColumns function mixes the columns of the state matrix
// The method used may look complicated, but it is easy if you know
   the underlying theory.
void MixColumns()
{
    int i;
    unsigned char Tmp,Tm,t;
    for(i=0;i<4;i++)
    {
        t=state[0][i];
        Tmp = state[0][i] ^ state[1][i] ^ state[2][i] ^ state[3][i] ;
        Tm = state[0][i] ^ state[1][i] ; Tm = xtime(Tm); state[0][i]
          ^= Tm ^ Tmp ;
        Tm = state[1][i] ^ state[2][i] ; Tm = xtime(Tm); state[1][i]
          ^= Tm ^ Tmp ;
        Tm = state[2][i] ^ state[3][i] ; Tm = xtime(Tm); state[2][i]
          ^= Tm ^ Tmp ;
        Tm = state[3][i] ^ t ; Tm = xtime(Tm); state[3][i] ^= Tm ^
          Tmp ;
    }
}
```

And here is the main loop for the cipher.

```
// Cipher is the main function that encrypts the PlainText.
void Cipher()
{
    int i,j,round=0;

    //Copy the input PlainText to state array.
    for(i=0;i<4;i++)
    {
        for(j=0;j<4;j++)
        {
            state[j][i] = in[i*4 + j];
        }
    }
    // Add the First round key to the state before starting the
        rounds.
    AddRoundKey(0);

    // There will be Nr rounds.
    // The first Nr-1 rounds are identical.
    // These Nr-1 rounds are executed in the loop below.
    for(round=1;round<Nr;round++)
    {
        SubBytes();
        ShiftRows();
        MixColumns();
        AddRoundKey(round);
    }

    // The last round is given below.
    // The MixColumns function is not here in the last round.
    SubBytes();
    ShiftRows();
    AddRoundKey(Nr);

    // The encryption process is over.
    // Copy the state array to output array.
    for(i=0;i<4;i++)
    {
        for(j=0;j<4;j++)
        {
            out[i*4+j]=state[j][i];
        }
    }
}
```

Looking through the code we can see lots of nested loops indicating that it is an ideal candidate for implementation in hardware. In addition, it would appear that a nice pipeline can be created from the four stages. But let us stop here for a moment and consider our options.

If we were to synthesize hardware from this description, we may get an adequate solution, but it would be a big mistake. This is not actually the AES algorithm, but an implementation of it that has been architecturally transformed to be efficient for software. The culprit is the byte substitution step shown in Figure 9.20. If we dig a little deeper into the description of the algorithm on Wikipedia, we find this statement:

> The S-box used is derived from the multiplicative inverse over GF(2^8), known to have good non-linearity properties. To avoid attacks based on simple algebraic properties, the S-box is constructed by combining the inverse function with an invertible affine transformation.

In fact the whole notion of using a lookup table and value substitution for this is because the software cost to implement the actual function is very high. In one paper [13] it was stated that the calculation cost is between m and $3m$+2 where m is the order of the inverse function, namely 8, so between 8 and 26 cycles. It can clearly be seen that one memory access may actually be a good trade-off. On the other hand [14], the software implementation takes 256 bytes of storage for the tables, a high cost for a hardware implementation, especially given the logic required to access it. If we further think about each of the 16 bytes being performed in parallel, and then loop unrolling, 160 copies of the memory would be required, for a total of 40,960 bytes.

The actual function can be defined as

```
Out[i] =  In[i] XOR In[(i+4) mod 8]
                XOR In[(i+5) mod 8]
                XOR In[(i+6) mod 8]
                XOR In[(i+7) mod 8]
                XOR C(i)
```

where `In[i]` is the `i`th bit of the input byte and `C(i)` is the `i`th bit of a byte constant.

We can now see that this is a combinatorial function, which can be implemented quite efficiently in hardware and that the elimination of the memory access will actually make it possible to perform a lot more pipelining on the complete algorithm. One paper [15] claims to have realized a 64% area reduction by using this approach and also saw speed enhancement. It does not take a lot of searching on the Internet to find that there are other substantial architectural optimizations that can be made, many of which are dependent on the constraints of the environment in which they are to be used. For example, one paper [16] explores the architectural alternatives for key generation: the main options being to generate the keys on the fly, or to pre-calculate them.

Another interesting note on the Wikipedia web page is that the algorithm assumes a big-endian system and they state that attempting to run it on a little-endian machine requires considerable byte-swapping and reduced efficiency.

A final point to note is that while many implementations are referenced on the web site in many different languages, such as JAVA, C#, LISP, there are no algorithms provided that are optimized for hardware implementation.

With the major architectural options sorted out, the algorithm needs further preparations before being turned into hardware. The major one in this case is turning the data types into ones suitable for hardware implementation. For example, the loop counters and the number of rounds are all expressed as integers, which would result in 32-bit control systems. Given that the maximum count of these is 10 for a 128-bit key or 14 for a 256-bit key, these would easily fit into a 4-bit entity, saving huge amounts of logic. A more complete example of this algorithmic transformation process will be given in the example in Section 9.8.

What should be taken away from this small example is that significant architectural decisions are embedded into a description, even those written in C. No high-level synthesis tool can make this kind of transformation and never will be able to. It is the task of the designer to ensure that he is providing the right algorithm in a suitable way such that a high-level synthesis tool can create good hardware.

9.7 Tying It into a Verification Flow

When the synthesis process has been completed, it may still be necessary to verify the output that has been produced. It may seem strange that any verification would need to be performed to verify the synthesis process, as this would seem to imply an error in the tool. However, there are many cases where the semantics of a description written in C, C++, or SystemC are not totally clear when applied to hardware. Examples of this are actions taken with uninitialized variables, performing a shift using a negative number, or access to an array which is out of bounds. In addition, if aggressive bit-width reduction has been employed, it is possible that overflows may have occurred resulting in different behaviors. It is sometimes difficult to detect this kind of error without using simulation.

There are two primary ways in which verification can be performed, namely simulation and formal methods.

9.7.1 Verification with Simulation

If the synthesized block is to be verified using simulation, the goal is to reuse the testbench that was developed for the verification of the original high-level description from which the synthesis process started. This means it has to be possible to drop the synthesized block back into the original testbench. This process, using SystemC as the simulation environment, is depicted in Figure 9.23. Within Catapult, this is referred to as the SCVerify process.

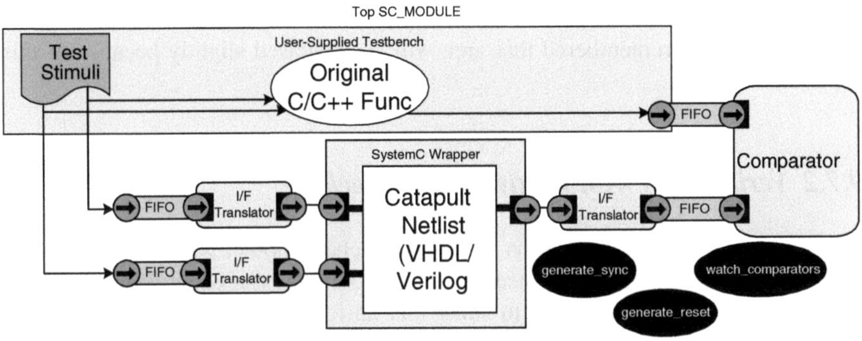

Fig. 9.23 Automatic verification with simulation

While the primary reason for doing this is to ensure that the synthesized result has the same functionality as the input description, it is also an important tool in calculating the actual latency of a synthesized design. This may be different from the initial results produced by Catapult because of a concept called data-dependent latency. A synthesis tool cannot know the actual data streams that may make up the stimulus stream, and in many cases, the data itself may determine the action within loops or conditional branches that may be taken. Thus Catapult will calculate the maximum latency that a design may produce, which may be pessimistic. We will see an actual example of this in the example in Section 9.8.

The transactors that surround the synthesized RTL results are necessary to convert the original function calls of the high-level input into the pin-level activity of the RTL. The FIFOs also deal with timing differences between the original and the generated hardware. Commercial synthesis systems, such as Catapult, generate all of the transactors, checkers, and integration code automatically as part of the synthesis process. They will also create all of the necessary scripts and make files necessary to perform the simulations, making the whole thing a push-button process. It should be noted that unlike a manual refinement process, what is generally referred to as a score board is not necessary. This is because Catapult guarantees that no data transactions will be reordered in the synthesis process, only retimed. This may not be true with other synthesis solutions.

In order to prepare for verification using simulation, a few changes may be necessary in the source files, usually within the testbench code. One of these changes enables what was the original "main" function, to become a testbench function. Return values are required in SystemC whereas they are not in ANSI C++, and the way in which command line arguments are handled is slightly different.

Finally, a switch in the synthesis flow can make verification a lot easier. This switch asks the synthesis tool to generate a "transaction done" signal which will allow the points in time when the synthesized results have completed processing a transaction to tell the testbench that this point corresponds to the end of a transaction in the C model. This way results can be compared easily. While the inclusion of this signal makes the design slightly larger, it does not affect scheduling and would be removed before the final synthesis runs are performed. However, when running this flow, it should be remembered that area will be increased slightly because of this capability.

9.7.2 Verification with Equivalence Checking

In addition to dynamic verification using testbenches, another way to perform verification is using formal equivalence checking. Combinatorial equivalence checking has been used for several years to show that an implementation at the gate level is logically the same as the RTL from which it was derived. This technology has almost removed the need to perform expensive gate-level simulation. Sequential equivalence checking is an emerging technology that can show that a behavioral model, written in C, C++, or SystemC is equivalent to an RTL model. It does this without the need for a testbench and can find any functional differences between the two models.

If sequential equivalence checking finds a difference, it will produce an example waveform which demonstrates the difference and this should help in understanding the problem and how to fix it.

Catapult creates all of the scripts and wrappers necessary to run the SLEC System-HLS [17] tool from Calypto Design Systems, Inc. [18] as shown in Figure 9.24. STARC's ASUKA II project reported success [19] using this combined flow although very few details were provided so it is difficult to tell if these were large designs or if any problems were actually uncovered.

In addition, ST reported at the Design Automation Conference in 2008 that the incorporation of this into their flow increased productivity by an additional 50% over and above the gains that came from using high-level synthesis. (see also Figure 9.52).

9.7.3 Verification Against Algorithmic Model

While the above two verification methods show that the results of implementation match the source, they do nothing to help show that the original source was correct. The creation of the C++ code which was fed into Catapult probably underwent some transformation from the original algorithmic description.

As an example, the original algorithm may have been described in floating point. While there are synthesis tools that can synthesize directly from this (Catapult can with an appropriate class library), most designers will probably want to perform a conversion of the algorithm into fixed point. This is a trade-off. On the one hand keeping everything in floating point may retain better algorithmic stability but has a high implementation cost. On the other hand fixed point provides more optimum hardware implementation and higher execution performance, but if the data is not

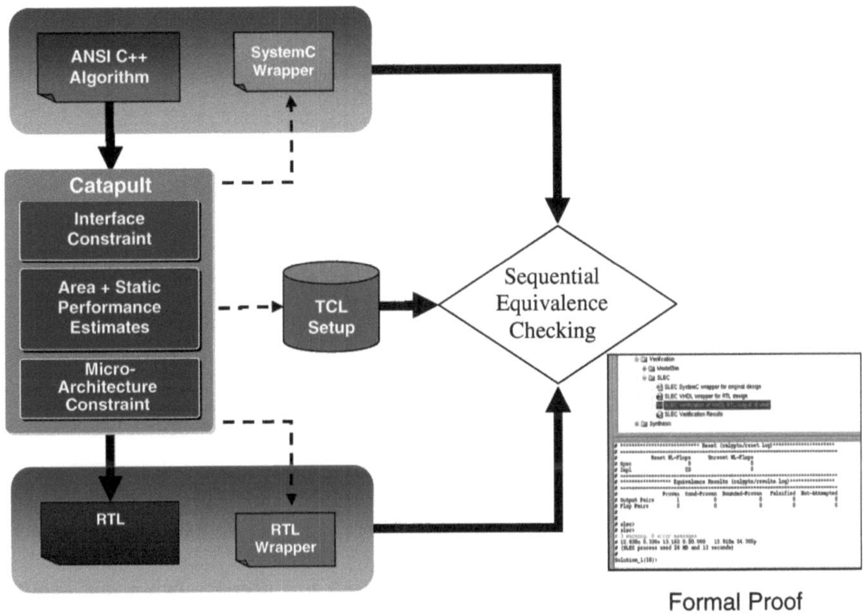

Fig. 9.24 Formal verification of synthesis results

sized right can cause noise or errors in the algorithm since it can become discontinuous. It should be remembered though that the IEEE floating point standard has a fixed size fractional part and that fixed point can have any arbitrary size, meaning that fixed point may provide greater accuracy than floating point. However, users can define their own custom floating point implementations based on Catapult fixed-point data types. An example of this is seen in Section 9.8. The goal is to find the minimum size for the data that produces acceptable results and yet remains true to the original algorithm within a defined margin of error.

It may thus be important to verify this modified specification directly against the original. Many companies use MATLAB/Simulink from The Mathworks [20] for their algorithmic modeling and verification. Catapult provides an integration with this environment that enables the C++ design to be measured against the results obtained from the original specification, such as signal-to-noise ratio, bit error rates. A wrapper is automatically created that makes it possible to import the Catapult C++ description into MATLAB/Simulink.

In Figure 9.25 the Catapult block icon was automatically created along with the wrapper enabling it to be simulated within the Simulink environment. In this particular testbench, three output waveforms are given: one for the pure FFT function, a second for the function as implemented by Catapult, and an error signal which is obtained by applying an inverse FFT function on the Catapult results.

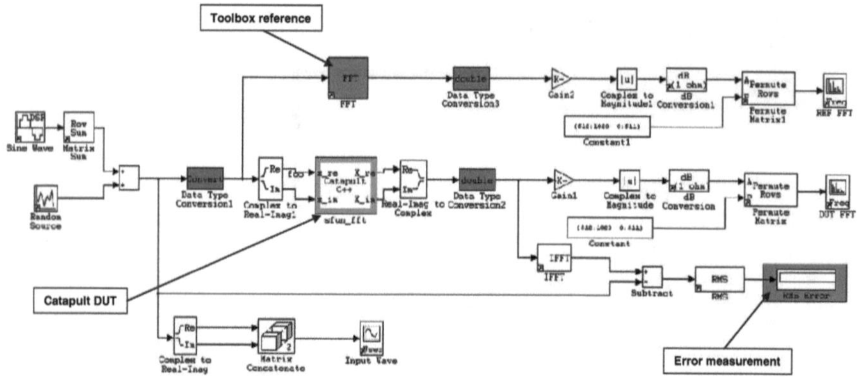

Fig. 9.25 FFT function verified in Simulink

9.7.4 Verifying Power

In high-level synthesis today, power is not yet seen as a primary factor that is optimized, and yet for many companies it can provide a competitive edge, or in the worst cases, produce a product that runs too hot. Part of the reason for this is that power optimization strategies are still very dynamic and thus the difficulties of automation are compounded. Only very recently were the verification strategies, such as VMM, extended to encompass current power optimization technologies [21].

While we can expect more complete solutions in the future, there are a few solutions that do exist today. The first is the ability to calculate the power that would be consumed by a particular synthesis solution and to annotate the synthesis results with power obtained by 3rd party tools such as Atrenta's SpyGlass Power [22] or Sequence's PowerTheater [23]. An example of this is shown in Figure 9.26. The second type of solution involves the creation of power-friendly RTL such that downstream RTL power optimization solutions can locate the maximum potential savings.

Power optimization is an area that is receiving increased attention these days, and while it is not yet possible for Catapult to perform power optimization at the architectural level, Catapult is a power friendly tool that will create RTL in such a way that a power analysis tool has an easy job of measuring the power consumption and flagging the blocks in the RTL netlist to help downstream RTL synthesis tools maximize power efficiency. One of the principle power optimization strategies is to employ clock gating, which reduces the dynamic power in registers and associated logic that are not changing. Typically, in an RTL description, a user would indicate which registers are clock gating candidates by given them an explicit enable signal. Catapult does this by examining registers in the design and extracts feedback paths from cones of logic and from this generates logic conditions that may be used as explicit enable signals for the registers. Thus maximum clock gating efficiency is enabled.

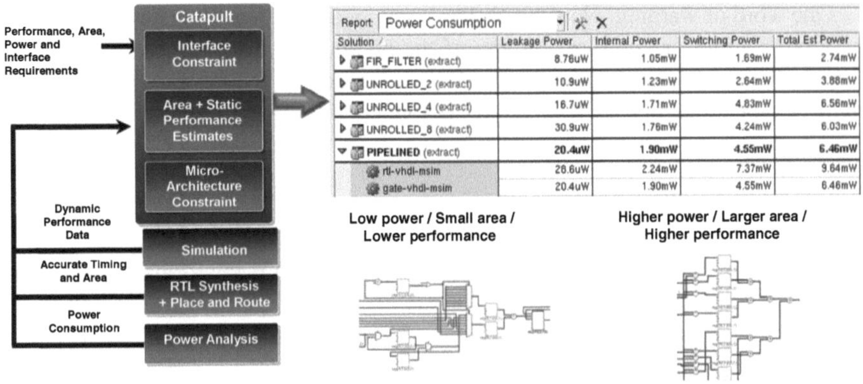

Fig. 9.26 Back annotation of power estimation

9.8 A More Complex Example

We have all seen the tutorials developed by EDA tool vendors. They have had lots of practice with those designs and the tool could have been optimized to produce fast and spectacular results for those designs. The demos are sleek and produce jaw-dropping results. How could anyone not want this tool? Rather than do the same here, it was decided to try something completely different. What is going to be shown here is the almost real-time work of a Mentor technical marketing engineer (TME) developing a model and synthesizing it all the way down to an implementation and performing RTL synthesis so that the accuracy of the high-level decisions can be seen. The application chosen is also a complex system and while he had some knowledge about the general problem area, this is a new application and thus required a new perspective to looking at this problem and the latest research in the area. Research in this area has been moving so rapidly, that his knowledge of the area emulates a typical senior design engineer, with familiarity with C++ coding, and the use of high-level synthesis.

As with a real design project, the objective here is not to find the optimal solution to the problem. Goals are set and the objective is to meet or exceed those goals within the time necessary to go to print with this book. This is thus not a benchmark and it is very likely that better solutions could have been found than the one shown here. The example is also one where there is no one right result. The design is being created to be parameterizable such that it could fit into a family of products, ranging from high performance to low cost. Many decisions, such as the allowable signal-to-noise ratio, are subjective and at the same time have a significant impact on implementation metrics.

The implementations highlighted are not meant to represent what a real company may find acceptable, but are used to demonstrate and match results reported in the latest technical papers that have been published in the area.

One word of warning. No analysis has been conducted regarding any patents that may be pending or approved on any of the approaches discussed in this section. Anyone wishing to attempt an implementation based on the information disclosed in this chapter should seek legal advice before continuing.

9.8.1 The Application

Frequency Division Multiplexing (FDM) is at the heart of many forms of communications, ranging from FM radio, digital subscriber line (DSL), and until recently, even the long distance telephone system. The key element they all share is the simultaneous transmission of information at multiple frequencies. Each signal has a unique carrier frequency which is modulated by the data being carried on that signal. FDM is limited by the amount of data that can be carried on each signal, so a logical extension of this technique is to use multiple carriers for the same signal and distribute the data amongst them. This is the essence of Orthogonal Frequency Division Multiplexing (OFDM) which uses a large number of carriers spread over a large frequency range at defined intervals.

Apart from increased data bandwidth, OFDM has many other advantages. Because the data can be spread across multiple carriers, the modulation frequency does not have to be as high. This in turn helps improve the attenuation characteristics and reduces interference. Due to these advantages, OFDM is being used in many systems today, such as WiFi, WiMAX, Digital Audio Broadcast DAB, digital TV, and mobile broadband in what is termed Long Term Evolution (LTE) of the mobile phone standards. This is sometimes referred to as fourth generation mobile networks.

Another significant advancement over recent years has been the adoption of multiple antenna systems at both the transmitter and the receiver. These are often referred to as multiple-input multiple-output (MIMO) technology. The principle advantage is that multi-path effects (caused by reflections) can be minimized and this increases both the range and the capacity of the communications. The downside to adding antennas is that it significantly increases system complexity. For a full description of this refer to [24].

These advances mean that much more complex signal processing algorithms are being adopted and these are sometimes beyond the capabilities of digital signal processors (DSPs). Designers are thus increasingly turning to dedicated hardware solutions. At the same time, power is becoming a primary consideration in design, whereas in the past this was a secondary factor. This is driven by the inability of battery technology to keep up with the increasing demands being placed on them.

Reference [24] identifies the main computation problem associated with this kind of radio system as being matrix inversion. The order of this matrix problem is defined by the number of antennas, and computation needs to be performed for every sub-carrier or group of adjacent sub-carriers. There are many possible solutions to this problem that utilize a plethora of different algorithms, such as Gauss–Jordan

Elimination, Cholesky factorization [25], matrix inversion Lemma [26], QR decomposition, and others [35] [36] [37]. The sheer number of algorithms for this indicates the degree of difficulty of the task and the importance of efficient solutions. While many of the solutions may be mathematically elegant, few of them are suitable for implementation in hardware.

9.8.2 The Flow

This example will follow a typical design and verification flow. The first stage of design requires some form of research, be it either original research or investigation of the techniques that have successfully been used by other designers. In this case, we are not attempting to develop a new algorithm, but want to reduce risk by selecting the ones that appear to have been successfully adopted by other people based on published results. This also gives us a way to determine the quality of the result that we achieve.

In order to produce a successful implementation, it not only takes a way to describe the design but also requires a way to verify that the design is correct and produces the desired results both in the functional sense and also in other non-functional attributes. In this case the noise levels produced by the solution are important. To do this a verification environment was established in MATLAB that not only acted as the reference model but also calculated the noise based on the results produced.

The design process itself contains several iterations that are associated with the architectural and micro-architectural optimization of the design. The high-level synthesis tool is associated with both of these loops, but in different ways. For architectural exploration, the tool identifies the bottlenecks and limitations of the design as coded. Recoding removes these. Thus high-level synthesis *assists* in the architectural exploration. At the same time, the high-level synthesis tool is directly manipulating the micro-architecture of the solution in an interactive manner with the designer. Each time that a transformation is made, verification can be repeated to ensure that no functionality has been broken.

The search for the best implementation is in reality always bounded. What is required is the best solution that can be found that meets all design constraints within a specified period of time. That dose of reality was also injected into this design example. While several sweeps were made of certain possible design parameters, there were many more that could have been made, and it is also possible that with further time, better modification could have been made to the code itself. However, finding *the* optimal design was not the criteria. What was desired was a design that was flexible enough to produce a range of solutions that could be built into a family of products. Each of those products may well have required a different solution. The design produced in this experiment allows very quick selection of the parameters and creation of an implementation through the setting of synthesis directives.

9.8.3 Design

Design is the most important part of a project, and this case is no different. We will be going through various aspects of the algorithm selection, as well as some of the changes that are needed to turn the algorithms into code that is amenable to high-level synthesis.

9.8.3.1 Step 1: Research Phase

The first step for a typical design engineer would probably be to conduct a significant amount of research into the approaches that had been used by other people. In our case, this included a collection of papers that addressed the problem domain and that talked about the algorithms that they had chosen for their analysis. While most of the papers found were from academia, the locations of the research and the size of the development teams indicated that significant funding was being provided by the major telecom companies around the world. The papers were separated into two primary groups:

(1) Those papers that provided actual results of a hardware implementation and
(2) Those that took a more academic approach at estimating the throughputs and sizes rather than actually performing an implementation.

While this latter group of papers may have presented advancements in the algorithms, the lack of results suggested that the risks associated with these would be higher and would not provide a reasonable goal against which the work could be compared.

Precedence was also given to papers that were cited by other papers, since this provides some level of endorsement by their peers. From the smaller pile of papers that remained, some were eliminated because they were special purpose solutions. For example, if an FPGA fabric with a hardcoded multiplier is the target, then this will heavily influence one particular solution. Another category that was excluded was those that made assumptions which were not realistic in a commercial setting, such as unacceptable signal-to-noise ratios.

The following group of papers were selected as the ones against which progress would be measured (see Table 9.1).

9.8.3.2 Step 2: Pick a Solution

At this point, it is likely that an initial approach to tackling the problem would be chosen. The primary selection criteria in this case were the level of success shown by previous implementations and the overall performance levels when compared to other solutions. The solution initially chosen was based on a systolic architecture [27].

However, as coding started, it was realized that a potential error had been made in the choice. While it would most certainly produce a high-performance solution, it

Table 9.1 Reference papers

Reference	Algorithm	Dimension	No. of bits	Frequency	Latency	Technology	Gates
Salmela	Gram–Schmidt	4×4 complex	16	160 MHz	139	130 nm	16.3 *k*
Lin	Givens rotation	4×4 complex	8	??	88	130 nm	27 *k*
Singh	Gram–Schmidt	4×4 real	16	277 MHz	67	180 nm	72 *k*
Luethi	Givens rotation	4×4 complex	??	125 MHz	65	250 nm	54 *k*
Percels	Givens rotation	4×4 complex	??	80 MHz	65	250 nm	33 *k*

would only be capable of implementing a single point on a trade-off curve between area and performance characteristics. This did not meet the requirement for a scalable solution, neither were there any results that could have been compared against. Also the research had revealed that the common wisdom is that a maximum of four antennas are worth considering, thus making the dimension scalability of the solution not as important as first thought. While it is possible that this may change in the future, most existing implementation use three antennas, so considering four antennas provides a considerable buffer for the future.

By selecting a different algorithm it would be possible, from a single high-level description, to target multiple solutions, some low cost, while others would be higher performance, and several implementation fabrics. The TME thus decided to pursue an algorithm using QR decomposition with modified Gram–Schmidt orthogonalization [28].

Several implementation data points using this algorithm as well as two using a different algorithm (Givens rotation) can be seen in Table 9.1. The results obtained will be compared against these in Section 9.8.7.

The latency of the solution is important as this defines the coherence window of the receiver. This in turn dictates how fast the receiver can be moving and still expect to get a coherent signal. So, for example, if the receiver were on a train traveling at 250 km/h, [29] Eilert calculates that the channel estimation computation must be completed in 1μs for a 4×4 MIMO system. Thus for a 500 MHz implementation the inversion must be performed in 150 clock cycles. Thus the primary optimization strategy for synthesis will be set to latency, except for a few cases where noted.

The top-level block diagram for the matrix inversion is shown in Figure 9.27.

The process consists of three steps: first QR decomposition, then computation of R^{-1}, and finally, the multiplication of matrix R^{-1} with Q^T.

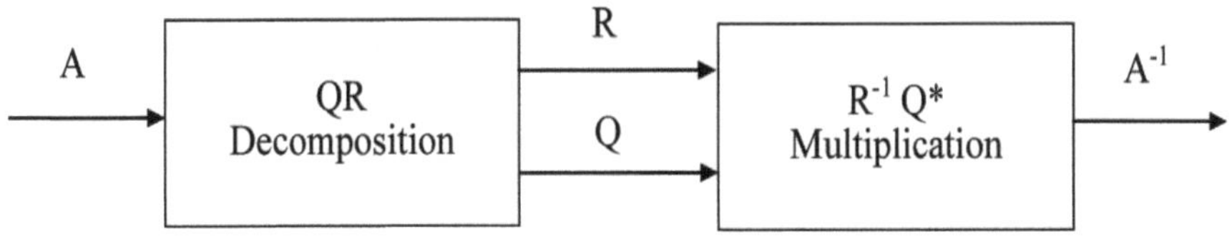

Fig. 9.27 Block diagram for matrix inversion

Given a matrix A, its QR decomposition is a matrix of the form

$$A = QR$$

where R is an upper triangulation matrix and Q is an orthogonal matrix, that is, one that satisfies

$$Q^T Q = I$$

where Q^T is the transposition of Q and I is the identity matrix.

The QR decomposition using the modified Gram–Schmidt algorithm is shown below:

```
for j = 1 : n;
    wj = aj
    for i = 1 : (j - 1),
        rij = (wj, qi)
          wj = wj - rijqi
    end
    qj = wj/||wj||2
    rjj = ||wj||2
end
```

In this pseudo code and that which follows, "*n*" represents the number of antennas in the system. Also for those less familiar with the math involved, "$||w_j||_2$" means that the norm of w_j is obtained, which means that it has a vector length of 1.

The computation of the R inverse algorithm is shown below:

```
for j = 1 : n
    for i = 1 : (j - 1)
        irij = ir(i, (1 : j - 1))* r((1 : j - 1),j)
    end
    ir(1:j-1),j = -ir(1:j-1),j/rjj
    irjj = 1/rjj
end
```

For readers wishing to understand the pseudo code further, consult one of the references provided in this section, and in particular [28] which is where these pseudo code snippets were taken from. This reference will also be one of the papers used for results comparison given that it is an implementation with results published.

9.8.3.3 Step 3: Code the Algorithm

The software industry has developed many methodologies, such as extreme programming and agile development methods, which call for the development of the

```
clear;

N = 4;

%% A = rand (N, N) + j * rand(N, N);

load "../dat/a_re.dat"
load "../dat/a_im.dat"

A = a_re + j * a_im;

[Q, R] = gram_schmidt_qr ( A );

%% Inverse computation via QR decomposition
R_inverse   = backsub_inverse ( R );
Q_hermitian = conj (transpose ( Q );
A_Inverse   = R_inverse * Q_hermitian;

%% calculate error
error_matrix = (A * A_inverse - eye(N, N);
error_vector = reshape (error_matrix, N*N, 1);
error_sumsq  = conj (transpose (error_vector)) * error_vector;
error_meansq = error_sumsq / (N * N);
error_rms    = sqrt (error_meansq);
error_rms_db = 20 * log10 (error_rms);

error_rms_db
```

Fig. 9.28 MATLAB testbench

test environment before the first line of code is laid down. That indeed was the case with this project.

A simple testbench was written in MATLAB and is shown in Figure 9.28. It basically takes a random N×N matrix A, computes Q and R using the MATLAB built-in function for Gram_Schmidt, and then performs the reverse function using the QR decomposition. The error is calculated and reported as both the root mean square (RMS) error and a decibel figure which represents the instability or noise created in the system.

The use of the inbuilt MATLAB functions shows that the algorithm for the inversion process is both correct and stable when using floating point arithmetic, as would be expected.

The next step is to code the QR decomposition algorithm in C++. You may have noticed in the MATLAB code that the random input vector was saved to a file. This value is fed into the top-level C++ testbench and the results of the QR decomposition are saved so that they can be fed back into the MATLAB model to calculate the errors. The MATLAB model thus continues to be the reference model against which the design is verified.

```
// tb.cpp

#include <mc_scverify.h>

#include "util.h"
#include "qrd.h"

CCS_MAIN (int argc, char *argv[] )
{
    //input
    cdat_t A[N_][N_];

    //output
    cdat_t Q[N_][N_];
    cdat_t R[N_][N_];
    cdat_t B[N_][N_];   // B = inv(A)

    load_matrix (A, "a_re.dat", "a_im.dat");
    disp_matrix (A, "Re(A)", "Im(A)");

    CCS_DESIGN (qrd)
    {
        A,
        Q,
        R,
        B
    };

    save_matrix (Q, "q_re.dat", "q_im.dat");
    disp_matrix (Q, "Re(Q)", "Im(Q)");

    save_matrix (R, "r_re.dat", "r_im.dat");
    disp_matrix (R, "Re(R)", "Im(R)");

    save_matrix (B, "b_re.dat", "b_im.dat");
    disp_matrix (B, "Re(B)", "Im(B)");

    CCS_RETURN(0);
}
```

Fig. 9.29 Top-level C++

In Figure 9.29 we see the saved matrix being loaded into A and then being fed into the design. The results are also saved in a similar fashion for post-processing of the errors in MATLAB.

The header file qrd.h is shown in part in Figure 9.30. Here we can see a number of interesting things that define the data types that will be used. First, the fixed-point sizes are defined. This is initially set at a total data width of 18 bits, with 12 bits for the integer part. In the definition of dat_t, a conditional is used to allow the variables to either be a double precision variable or use the fixed-point data type. In this case the fixed point is also set to truncate for its rounding behavior and to wrap for its overflow behavior. The complex data type cdat_t is then created using whatever definition of dat_t was defined.

```
#ifndef __QRD_H__
#define __QRD_H__

#include <ac_int.h>
#include <ac_fixed.h>
#include <ac_complex.h>

#define N_ 4

#define DAT_W  18
#define DAT_I  12

#ifdef OVERRIDE_WITH_DOUBLE
typedef double dat_t;
#else ac_fixed<DAT_W, DAT_I, true, AC_TRN, AC_WRAP> dat_t;
#endif

typedef ac_complex<dat_t> cdat_t;

cdat inner_product
{
    cdat_t V[N_],
    cdat_t W[N_]
};

void gram_schmidt_qr
{
    cdat_t A[N_][N_],
    cdat_t Q[N_][N_],
    cdat_t R[N_][N_]
};

 .
 .
 .

 #endif
```

Fig. 9.30 Partial header file

The plan is to execute a series of synthesis runs with varying values for the fixed-point sizes so that a plot can be obtained showing the trade-off between noise, latency, and area. A certain amount of digital noise will be quite acceptable, so long as it is small compared to the analog noise of the system, especially if it can save significant amounts of area and power.

The top-level design file is shown in Figure 9.31. The comments in this file correspond to the original lines in the MATLAB testbench in Figure 9.28.

```
#include "grd.h"

Using namespace std;

void qrd
{
    //inputs
    cdat_t A[N_][N_],

    //outputs
    cdat_t Q[N_][N_],
    cdat_t R[N_][N_],
    cdat_t A_inverse[N_][N_]
}
{
    cdat_t R_inverse[N_][N_];
    cdat_t Q_hermitian[N_][N_];

    // [Q, R] = gram_schmidt_qr (A);
    gram_schmidt_qr (A, Q, R);

    // R_inverse = backsub_inverse (R);
    backsub_inverse (R, R_inverse);

    // Q_hermitian = conj (transpose (Q));
    conj_transpose (Q, Q_hermitian);

    //A_inverse = R_inverse * Q_hermitian;
    matrix_product (R_inverse, Q_hermitian, A_inverse);
}
```

Fig. 9.31 Top-level C++ code

The are a total of four functions that have to be defined, two of which are shown in Figures 9.32 and 9.33. The other functions as well as these functions can be found on the companion web site for this book: http://ElectronicSystemLevel.com.

It should be noted that this is the first time that we have seen anything in the source listings that is synthesis specific. Until now, everything has been pure C++. The synthesis directive on line 44 is inserted because at this stage a synthesizable square root function has not been defined. This function is a core optimization possibility because it has been reported by [30] that it can be approximated for fixed-point arithmetic with very little overall impact on the performance. In fact it may be possible to implement it using only basic operators such as “add,” “mux,” and “shift”. A Synopsys DesignWare block also exists for this function that could be used. This implementation was found to be very efficient and in a real-life situation may well have been selected if licensing terms were acceptable. However, since this

```
#include "qrd.h"

void gram_scmidt_qr
{
    cdat_t A[N_][N_],
    cdat_t Q[N_][N_],
    cdat_t R[N_][N_],
}
{
L1: for (int j=0; j<N_; j++)
    {
        // W(1:N) = A(1:N, j);
        cdat_t W[N_];
L2:     for (int k=0; k<N_; k++);
            W[k] = A[k][j];

L3:     for (int i=0; i<N_; i++)
        {
          if (I >= j)
             R[i][j] = 0.0;
          else
          {
             // compute <W, Qi>:
             // R(I, j) = conj (transpose (Q(1:N, i))) * W(1:N);
             cdat_t sum(0, 0);
L4:          for (int n=0; n<N_; n++)
                sum += Q[n][i].conj() * W[n];
             R[i][j] = sum;

             // remove (<W, Qi> * Qi) from W;
             // W(1:N) = W(1:N) - Q([1:N], i) * R(I, j);
L5:          for (int n=0; n<N_; n++)
                W[n] -= Q[n][i] * sum;
          }

          // r = norm (W(1:N));
          cdat_t sum (0,0);
L6:       for (int i=0; I <N_; i++)
              sum += W[i].conj() * W[i];

          dat_t s = sum.r();

          // ir = 1/r = 1/sqrt(s);
#if (!defined (__SYNTHESIS__))
          dat_t ir = 1 / sqrt(s);
#else
          dat_t ir = x; // This will become the estimated function
#endif
          // ir = sqrt(s) = s * (1/sqrt(s));
          dat_t r = s * ir;

          // R(j, j) = r;
          R[j][j] = r;

          // Q(1:n, j) = W(1:N) * (1/r);
L7:       for (int i=0; i< N_; i++)
              Q[i][j] = W[i] * ir;

    }
}
```

Fig. 9.32 Gram_Schmidt function

```
function IR = backsub_inverse (R)

    N = length (R);

    IR = zeros(N, N);

    for j = 1 : N;
        for I = 1 : (j - 1);

            IR(i,j) = IR(I, i:j-1) * R(1:j-1, j);
        end
        IR(1:j-1, j) = -IR(1:j-1, j) / R(j, j);
        IR(j, j) = 1 / R(j, j);
    end
end
```

Fig. 9.33 Backsub inverse function

is a demonstration of high-level synthesis capabilities, it was decided to go with an algorithmic description and to synthesise the full results.

A compile-time switch is inserted so that either the core math function can be used or the approximation function can be used. The function that was inserted on line 47 at this stage is not a valid approximation function and will be discussed further in step 5.

With all of the functions coded in C++ it becomes possible to verify that the code actually works. First this was done using double precision numbers and it was found to give a 290 dB signal-to-noise ratio. This indicates that the code is functioning correctly with no significant accumulation of errors. This is the result that would be expected given the choice of algorithm.

Until a suitable function for the square root is inserted, it is not possible to find a valid signal-to-noise ratio for the fixed-point algorithm. This will be conducted later.

With most of the coding complete it is possible to run the code through Catapult to start investigating the algorithm as initially coded. This will identify any functional errors or coding styles that may limit synthesis.

9.8.3.4 Step 4: Taking Care with I/O Ports

There are a number of things in the algorithm, as first written, that are not very synthesis friendly. These can either cause the tool to miss optimization possibilities or cause it to create unnecessary logic. We will examine a few of these.

The first issue that we will investigate can be seen in the Gram_Schmidt function in Figure 9.32. On lines 5–7, the interface variables are defined, namely A, Q, and R. These are $N \times N$ arrays of complex numbers.

We want A to be an input and Q and R to be outputs. In Catapult, the interface direction is automatically computed based on the function. If we look at line 57, we see Q being written to and thus it is established as an output. However, on lines 27 and 33, Q is being read from. The synthesis algorithms are not able to discern that the only values being read are ones that were previously computed in the function. It has to assume that some values may have been those passed into the function and it will thus assume that Q is a bidirectional and insert the necessary multiplexers on the interface. This can be seen in the constraints window by looking at the port definitions, as shown in Figure 9.34.

Fig. 9.34 Port definitions

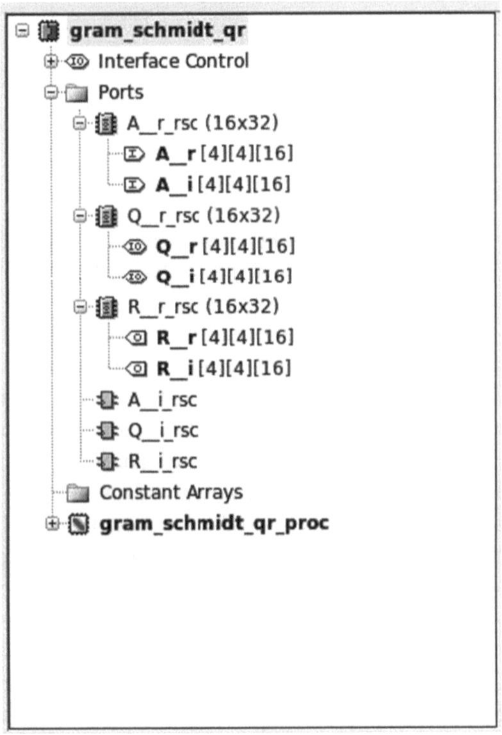

There is another problem with Q. When Q is being accessed, it is as a port and, given its size, would probably have been defined as a memory. This does not have to be the case especially if a hierarchic design is being created and as was discussed in Section 9.4.3. When Catapult attempts to schedule this code, it issues an error:

```
# Error: insufficient resources 'ram_sample-065 nm
    singleport_beh_dc.RAM(5,16,36,4,0,0,0,0,0,1,1,16,36,1)' to
    schedule 'gram_schmidt_qr_proc'. 2 are needed, but only 1
    instances are available (SCHD-4)
# Error: Design 'gram_schmidt_qr' could not schedule partition
    '/gram_schmidt_qr/gram_schmidt_qr_proc' - resource
  Competition
```

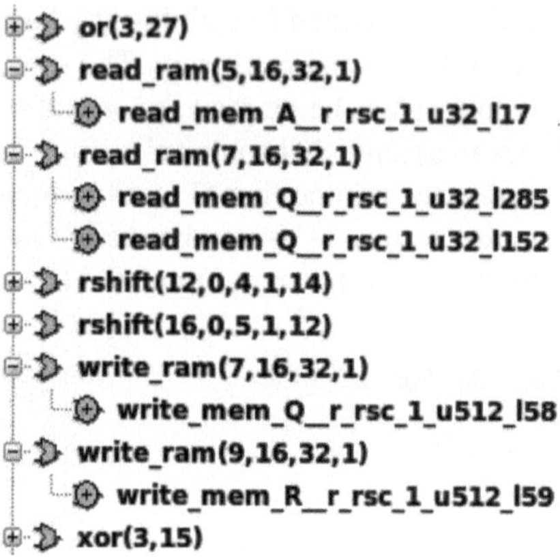

Fig. 9.35 Catapult constraints window

From the architectural constraints window shown in Figure 9.35 you can see that array Q is mapped to single-port RAM number 7 which is only a single memory port.

There is a lot of information contained here. Read_ram(7,16,32,1) tells us that this is memory element 7 and it is 16 deep and 32-bits wide. Immediately below that line, we see the two times that Q reads from this memory. Similarly the write to Q (a few lines down from the reads) is also mapped to ram number 7. We can thus see that there are two reads and one write to this memory. Catapult cannot access a single-port memory more than once in a clock cycle and thus to implement pipelining two ports would be required. It would be possible to specify that Q should be held in a two-port memory, but this is an expensive solution that would not be acceptable to most people as it would double the area for this component. A better way is to remove the bottleneck from the code itself.

The way to resolve both of these problems is to create a temporary array called Qt that will be used for intermediate results and by writing this out to Q only in one place, we can ensure that the interfaces are optimized. Thus lines 26–33 become

```
26 L4:          for (int n=0; n<N_; n++)
27                  sum += Qt[n][i].conj() * W[n];
28              R[i][j] = sum;
29
30              // remove (<W, Qi> * Qi) from W;
31              // W(1:N) = W(1:N) - Q([1:N], i) * R(I, j);
32 L5:          for (int n=0; n<N_; n++)
33                   W[n] -= Qt[n][i] * sum;
```

and lines 56 and 57 are replaced by

```
56 L7:        for (int i=0; i< N_; i++)
57              {
```

```
58                    cdat_t q_ij = W[i] * ir;
59                    Qt[i][j] = q_ij;
60                    Q[i][j] =  q_ij;
61                }
```

In addition, by creating a temporary variable Qt which is held in a register, multiple accesses become possible per clock meaning that restrictions on pipelining will have been removed and Catapult is now able to complete its scheduling successfully.

In the original code, there was a similar issue with the variable R which was written to three times. A similar set of code modification can be made here to streamline the pipelining opportunities that Catapult can exploit, but in addition an understanding of the algorithm can be used here to streamline the code. Not all elements in the matrices are actually used. In fact only the top triangle holds active data, the lower triangle being all 0.0s. This is called an upper triangular matrix. We can see this with the test on line 19 and setting the element to 0.0 on line 20. We can remove this potential write from here and instead migrate it into loop 7.

9.8.3.5 Step 5: Dealing with the Inverse Square Root

On line 47 of our original Gram_Schmidt function as shown in Figure 9.32, we saw the dummy function being added for the inverse square root function. While this is a common function in many multimedia and graphics functions, it is computationally intensive. Some microprocessors, such as the MIPS R10000/R12000, the IBM PowerPC 740/750, and many graphics engines, have dedicated instructions for this operation. In software, the function is implemented using division for the reciprocal function and the square root function also relies on the division operator or on convergence functions such as Newton–Raphson that use multipliers instead. Both functions would inject a long latency into the algorithm that would in turn limit the performance of the system. Reference [31] proposes a digit-recurrence algorithm that computes by iteration of carry-propagation-free additions, shifts, and multiplications by a single digit. Takagi noted that the implementation has a regular structure which makes it suitable for implementation in silicon and is amenable to pipelining.

Figure 9.36 shows the code for this function. Given the iterative nature of the algorithm, it makes sense to unroll a loop like this. We can see an example of a pragma being used to provide this synthesis directive just before the loop. It could have also been set as a constraint in the synthesis UI or script as well.

One unexpected benefit is that this algorithm provides both the square root and the inverse square root. Our algorithm actually requires both as seen on lines 47 and 49 of Figure 9.32. In order to make a higher level function pipelineable, everything that it contains also has to be pipelineable. Thus we can extract this snippet of the code and make it a function that returns both results. However, this algorithm does come with one side issue and that is the allowable range of values on its input. To overcome this, we have to normalize the values on the way in and then perform the opposite conversion after the function has been performed. This can be seen in Figure 9.37.

```
#include "rsqrt.h"

void rsqrt
(
    // input:
    fra_t  & x,

    // outputs:
    sqr_t  & sqr,
    rsq_t  & rsq
)
{
    fra_t L = 0.5;
    fra_t X = (x >> 1);
    w_t   W = w_t(1.0) - x;
    sqr_t P = x;
    rsq_t S = 1.0;

    #pragma hls_unroll yes
JN: for (int j = 0; j < NN; j++)
    {
        w2_t w2 = 2*W;

        if (w2 < -0.5)
        {
            // q = -1
            W = 2*W - X + 2*P;
            P = P - X;
            S = S - L;
        }
        else if (w2 > 0.25)
        {
            // q = +1
            W = 2*W - X - 2*P;
            P = P + X;
            S = S + L;
        }
        else
        {
            // q = 0
            W = 2*W;
        }

        L = L >> 1;
        X = X >> 1;
    }

    sqr = P;  // sqrt (x)
    rsq = S;  // 1 / sqrt (x)
}
```

Fig. 9.36 Square root function

```
#include "norm.h"

void norm
(
    // input:
    sum_t  & x,

    // outputs:
    sum_t  & r,    // sqrt(x)
    sum_t  & rr    // 1 / sqrt(x)
)
{
#if (defined (OVERRIDE_WITH_DOUBLE))

    r  = sqrt(x);
    rr = 1 / sqrt(x);
#else
    lop_t position = 0;
    uint1 polarity = SUM_F;
    uint1 valid    = 0;

    lod (x, position, valid);

    shf_t shift1 = position - (SUM_F - 1);

    if (position[0] == polarity)
        shift1 +=  1;

    fra_t fra = x >> shift1;

    sqr_t sqr = 0;
    rsq_t rsq = 0;

    rsqrt (fra, sqr, rsq);

    shf_t shift2 = shift1 >> 1;

    r  = sqr << shift2;  // sqrt(x)
    rr = rsq >> shift2;  // 1 / sqrt(x)
#endif
}
```

Fig. 9.37 Normalizing for SQRT

The *lod* function in Figure 9.37 locates the leading significant digit in the word and the routine for this can be found on the companion web site.

We now have the complete algorithm and it was verified that the complete system was still functioning. An initial pass through Catapult also revealed no further coding issues. This now means that the entire design is now ready for synthesis. The Gram_Schmidt function, with all of the modification, can be seen in Figure 9.38.

```
#include "qrd.h"

void gram_schmidt_qr
(
    cdat_t  A[N_][N_],
    cdat_t  Q[N_][N_],
    cdat_t  R[N_][N_]
)
{
    cdat_t  Qt[N_][N_];

L1: for (int j = 0; j < N_; j++)
    {
        cdat_t  W[N_];
        cdat_t  Rt[N_];

        // W(1:N) = A(1:N, j);
    L2: for (int i = 0; i < N_; i++)
            W[i] = A[i][j];

    L3: for (int i = 0; i < N_; i++)
        {
            cdat_t r_ij (0,0);

            // R(i, j) = conj (transpose (Q(1:N, i))) * W(1:N);
        L4: for (int n = 0; n < N_; n++)
                if (i < j) // upper triangle
                    r_ij += Qt[n][i].conj() * W[n];
                else
                    break;

            // W(1:N) = W(1:N) - Q(1:N, i) * R(i, j);
        L5: for (int n = 0; n < N_; n++)
                if (i < j) // upper triangle
                    W[n] -= Qt[n][i] * r_ij;
                else
                    break;

            Rt[i] = r_ij;
        }

        // r = norm (W(1:N));
        cdat_t sum (0, 0);
    L6: for (int i = 0; i < N_; i++)
        {
            cdat_t w_i = W[i];
            sum += w_i.conj() * w_i;
        }

        sum_t x = sum.r();
        sum_t r, rr;

        norm (x, r, rr);

        // Q(1:N, j) = W(1:N) / r;
    L7: for (int i = 0; i < N_; i++)
        {
            cdat_t q_ij = W[i] * rr;

            Q[i][j] = Qt[i][j] = q_ij;
            R[i][j] = (i < j) ? Rt[i] : (i == j) ? r : 0.0;
        }
    }
}
```

Fig. 9.38 Final form of Gram_Schmidt function

9.8.4 Verification

We had previously verified the functionality using the floating point arithmetic, but now we turn our attention to the fixed-point verification. Remember that all of the functions have been coded to accept variable widths, so that we can see exactly how the algorithm scales in terms of the signal-to-noise ratio vs. the number of bits of precision. The figure obtained for the floating point algorithm was 296 dB.

In Figure 9.39 we can see the signal-to-quantization noise ratio (SQNR) figures for a range of fixed-point sizes for the algorithm. At a bit-width of 8, with 4 bits for the integer and 4 for the fraction, no practical signal is available. Given the data that is being fed into the system, no more than 4 integer bits are necessary, so all additional bits are used for the fraction. It can be seen that the algorithm is quite nicely linear with each bit adding 6 dB in SQNR improvement.

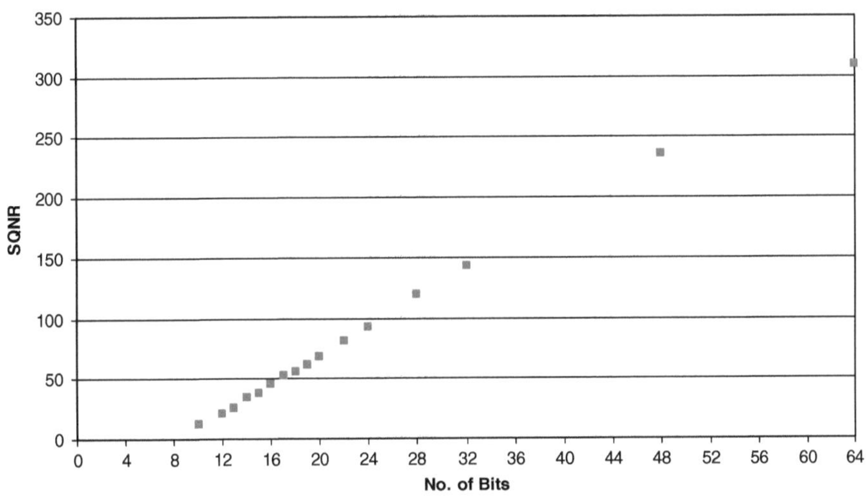

Fig. 9.39 SQNR vs. bit width

It can actually be seen that at 64 bits, the fixed-point results are better than for the floating point. This is because the floating point significand has less bits available because of the bits taken up by the exponent. This is a clear example of why floating point is not always the best choice, even when highly accurate results are required. In addition, the complexity of floating point operations is greater than the complexity of similar 64-bit fixed point. Floating point should only be used if the range of input numbers is wide.

For the purposes of this exercise, we chose a target of 45 dB SQNR because this is on the high end of what would be expected in a wireless link. Also, 16 bits are a typical half-word size, such that you can place two half-words together to get a 32-bit entity, which is the natural word size for most processors (CPU). While this is not important for this block, it may have implications for other parts of the system that could be operating in software. The actual target SQNR for a design could be higher or lower depending on the design specifications. The benefit of high-level synthesis is that the designer can explore all of the different possible scenarios, in this case bit-width vs. SQNR, and select the best one based on the target specification.

We can also look at the accuracy of the square root and inverse square root functions. For 16 bits of accuracy (4 integer and 12 fractional) the results are shown as the solid line in Figure 9.40. This corresponds to an SQNR of 99 dB for the square root and 98.7 dB for the inverse. To see how the error increases as the number of bits decreases, the * and o show the results for 10 bits. While the error appears to be quite small, this does represent additional noise created in the system. The SQNR at this level is 63.4 dB for the square root and 59.9 for the inverse. Given that this is just one component in the system, this probably sets the lower bound of acceptable accuracy.

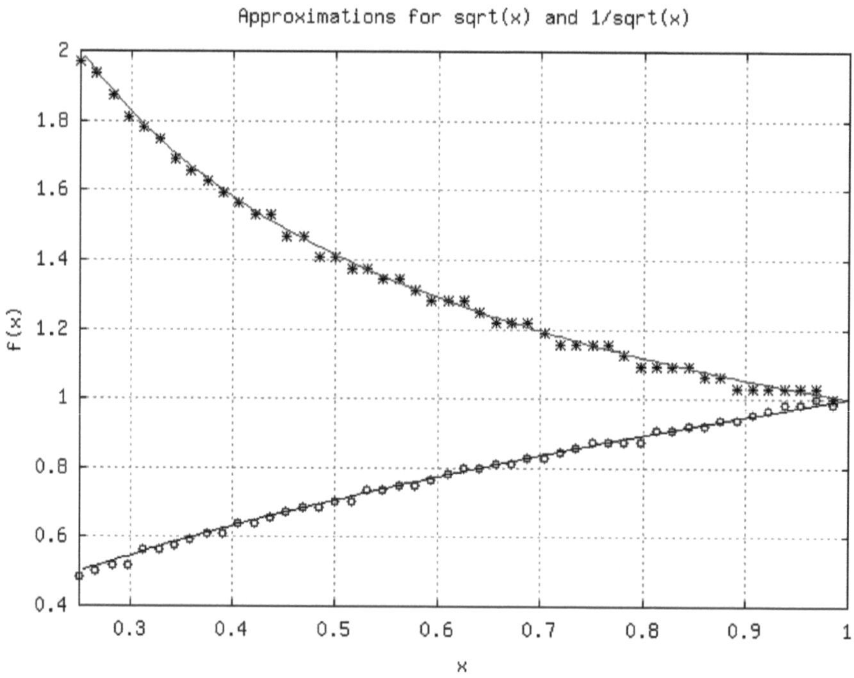

Fig. 9.40 SQRT accuracy

As a reminder, all of the synthesis results include the generation of the "transaction done" signal that was described in Section 9.7.1 as the SCVerify flow will be executed after every run to get accurate latency times.

9.8.5 *Synthesis*

When tackling a system, such as this, that is composed of several layers of hierarchy, the best approach is to start from the bottom and work up. This is because many high-level synthesis transformations, such as pipelining, require that to perform the function at that level, everything beneath them is also able to perform that function.

When you hit a problem, you need to investigate what produces the limitation and either set necessary constraints in the high-level synthesis tool that will enable you to overcome the problem or, if more serious, modify the code to remove the dependence.

9.8.5.1 Synthesizing the Square Root

The square root function is the lowest level function, and if you remember, we had already set a pragma in the code in Figure 9.37 to tell high-level synthesis to perform a loop unroll. As expected, performing synthesis on this function yields an adder tree, and all of the multiplication functions, which are by a single bit, are replaced with shift functions. Part of the schedule can be seen in Figure 9.41. You can also see from this diagram that some of the dependencies between the operations are shown. This can really help to understand what the tool is doing and to help uncover issues that may have prevented a better schedule from being found. In this case there was no problem and the complete function fits into 3 clock cycles when the target frequency is set to 160 MHz.

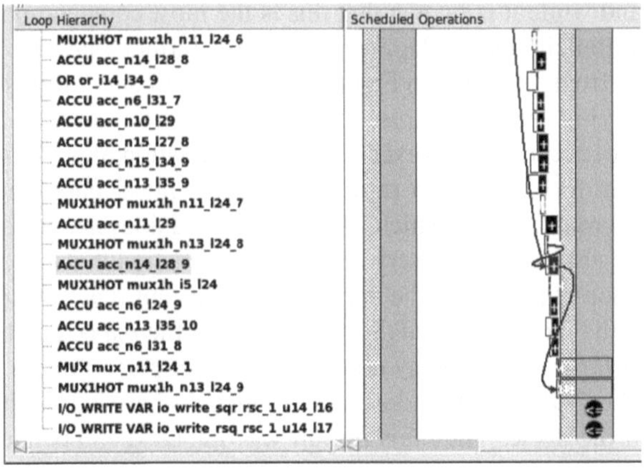

Fig. 9.41 SQRT schedule

Another way that some engineers assess the quality of the results is by looking at the schematics that are produced. These will quite often allow an engineer to see unexpected structures or congestion points in a block. In fact this very technique will be used to locate a problem a little further into the process. It is not, however, a good way to see details about the generated results as shown by the schematic

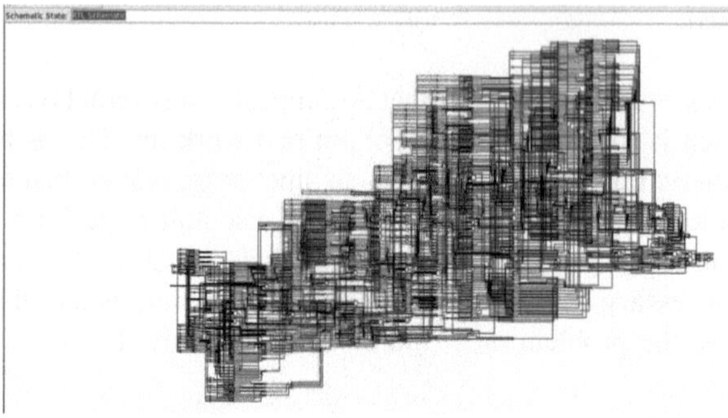

Fig. 9.42 Schematic of the SQRT function

shown in Figure 9.42. This is a relatively small block, and the results for the entire Gram_Schmidt algorithm take many sheets to display.

While a design of this size is easily created by a competent RTL designer, it is the ability to explore this algorithm that makes high-level synthesis such a powerful tool. Think of the time it would take to reproduce this with a different bit-width or in a different technology. High-level synthesis can do this in just a few seconds.

With good looking initial synthesis results for this, we moved one step up the chain, which is to incorporate the normalization code. Again there were no obvious problems and at this point we had a definite milestone that we could check against. The output of Catapult was fed through RTL synthesis and it showed that this function would take an area of 48 k μm^2. This is almost half of the total space claimed by Salmela [30], and while it is known that this is the most computationally intensive part of the algorithm, it raises a flag, which requires further analysis.

We have seen from the charts in Figures 9.39 and 9.40 that it is possible to create systems with varying degrees of noise immunity. Salmela showed an area of 16.3 k gates for 16 bits of accuracy. However, a closer examination of their implementation reveals the reason for their stellar result. They use a linear approximation for the reciprocal square root function which is very crude and does not perform well. This architecture does not balance the very coarse approximation in its `rsqrt()` unit to the 16-bit precision of the rest of the data path. This means that it would not be very practical where the data path needs to be tunable to the expected channel quality and needs to perform well in noisy conditions. In fact performing a mathematical calculation of the SQNR of this design showed it to have no ability to separate the signal from the noise. Thus this design was intended to operate either under very different conditions than those conceived for this design or the development of the system was not yet complete. In either case, it means that the results are not meaningful. (see also Section 9.8.7).

9.8.5.2 Manipulating Variables and Resources

In the Gram_Schmidt routine, we are constantly dealing with the real and imaginary parts of the variables. At the moment they are considered as two variables and reads

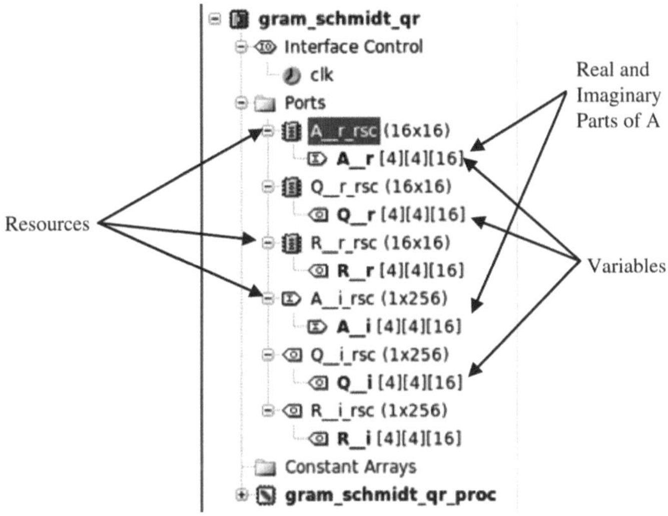

Fig. 9.43 Resources and variables

or writes of them would take place independently. However, they are never accessed in this manner, and so it makes sense to tell the synthesis program that they should be treated as if they were a single variable when accessing memory. We do this by mapping variables onto resources (Figure 9.43).

By default a resource is created for every variable, but by dragging and dropping, any variable can be made to share a resource. For example, in Figure 9.44, we can see the process where the imaginary part of A is being dragged and dropped onto the resource for the real part of A. This can be done for all of the variables.

As a second step they can be configured to be represented side by side in the resource, such that the memory operations will now fetch both pieces as shown in Figure 9.45.

The size of the memory has now changed from being a 16×16 to a 16×32.

9.8.5.3 Synthesizing the QR Block

We are now ready to tackle the Gram_Schmidt algorithm itself. The loops within the procedure can be seen in Figure 9.46. We can see the synthesis directives for each of the loops. Loops LO and JN, which are within the square root function, have been unrolled. Loops 2 – 7 have been pipelined. We can also see the notation (II=1) on these loops. This says that the initiation interval is 1 which means that we are feeding a new set of data in with every clock.

With the pipelining configuration as set in Figure 9.46 an initial result was obtained. It achieved a latency of 171 running at 160 MHz and an area of 305,271 μm^2. For this technology library, a gate occupies on average 5.1 μm^2 so this corresponds to about 59,857 gates. This clearly misses on just about every target level, so

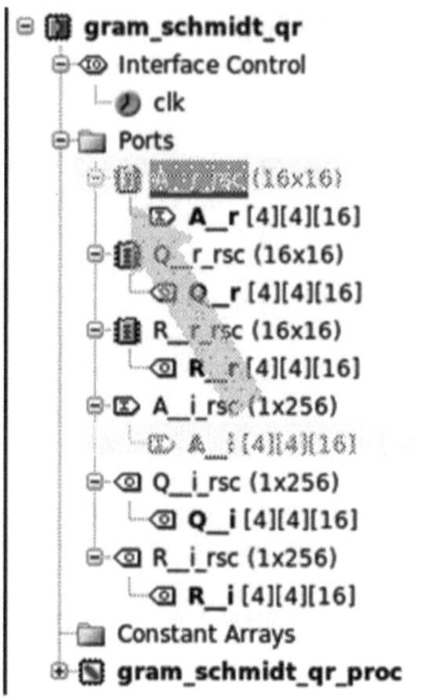

Fig. 9.44 Mapping variables to resources

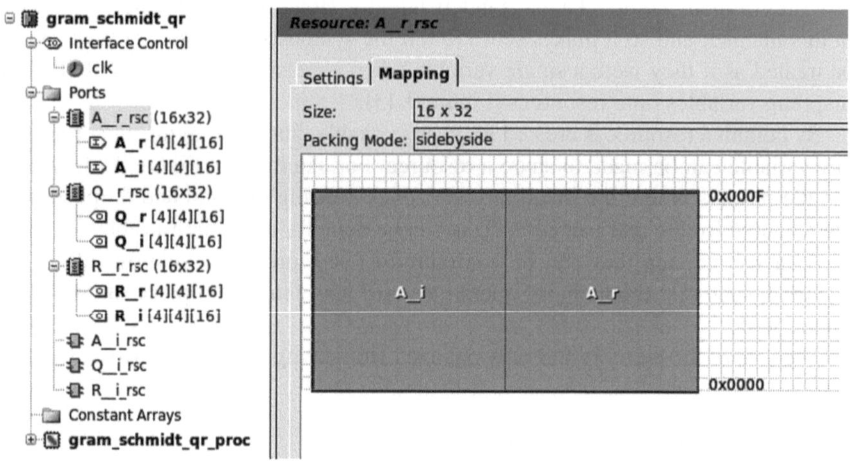

Fig. 9.45 Configuring variables

there is something that is constraining the tool, or we need to look at pipelining at the topmost level. However, after setting the top level to be pipelined, the tool failed to find a schedule. With one investigation, setting the pipelining interval to 2 instead of 1 initially managed to produce a schedule, but after full synthesis it could not find an acceptable schedule due to timing issues. While the slack time associated with each clock could have been reduced, this would be a dangerous approach to take as it could lead to problems further down through the implementation process. So another solution was sought.

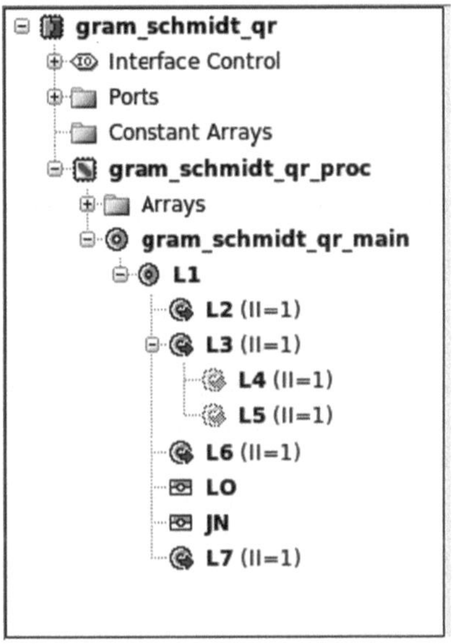

Fig. 9.46 Gram_Schmidt loops

The clues to the problem were found in the Gantt chart and the schematic, neither of which will be reproduced here because of their size. In the Gantt chart, it was seen that there were a lot of operations dependent on Qt and in the schematic large mux structures surrounded it. The resolution was to put Qt into a register file, rather than a memory. A further piece of information was provided to the synthesis tool related to memory precedence. Given that it is known that the memory reads and writes cannot interfere with one another because they are in separate iterations of the loop, the synthesis tool can relax the normal checks that it would make on these. These changes removed the bottlenecks across the loop iteration boundary making it much easier to schedule the block.

Providing synthesis directives such as memory precedences can be dangerous. If the information provided is wrong, it can result in the tool producing bad logic, and thus it is important that the verification be performed to ensure that the functionality remains true.

While it did not find a schedule at 160 MHz, it did find one at 100 MHz. It produced a latency of 94 and area of 242,243 μm^2 (47 k gates). While the performance is there, the area is still high. Setting the pipeline initiation interval to 2 should enable some sharing, and the design would then operate at 160 MHz, providing the same throughput with a latency of 184 but the area actually increased to 258,400 μm^2. This is clearly not a good direction to go in for this algorithm.

Removing the pipelining from the top level and setting the other loops to an initiation interval of 2 gave a latency of 222 at 160 MHz with an area of 175,560 μm^2 (34 k gates). This is a very competitive design for a true 16 bits of accuracy and shows that we are in the correct ballpark with regard to results. However, the latency is a little higher than was desired.

To find an extreme design point, the initiation interval was set to 1 and everything unrolled. The interfaces were set to wires rather than memories and at 160 MHz it gave a latency of 16 with an area of 3,982,906 μm^2 (780 k gates). While large, it would provide an astonishing throughput with a complete frame of data being consumed every cycle. While the need for a implementation like this is not obvious, it does demonstrate the range of solutions that can be found.

One encouraging data point was selected and was run through RTL synthesis. It fully met timing at 160 MHz, with a throughput of 130 cycles/factorization and a latency of 130 cycles. The final area is 200 μm^2 ~= 40 k gates in 0.13 μm CMOS. It used 16 bits of precision (4 integer and 12 fractional) and had a numerical noise level of 46 dB SQNR. Since this implementation meets all of the requirements, this would probably have been the end of the high-level synthesis process in a typical design flow.

Rather than continue with random data points, a set of design point sweeps was conducted and will be presented in the next section, followed by an analysis of how the results compare to the published literature. It should be noted that each of the synthesis runs takes between 60 and 80 seconds for most of the examples cited, so exploration can be very quick.

One last thing to note is that the distribution of area among the solutions can vary quite widely. For example, in Figure 9.47, the top piece of each chart shows the area dedicated to logic, the second chunk is adders and multipliers, third is multiplexers, and at the bottom are the registers. In some cases when a synthesis algorithm switches between different sets of options, unexpected changes can happen in total area and thus some exploration around the desired points is prudent to get the local minima as well as locating the global minima.

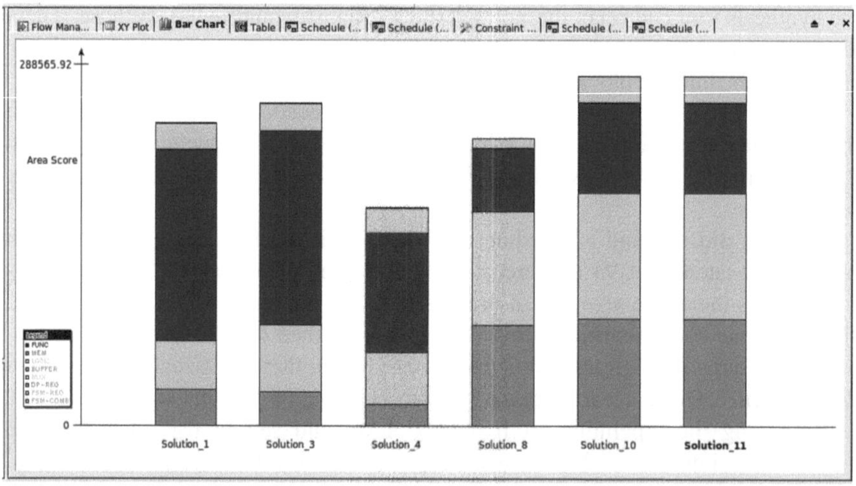

Fig. 9.47 Area distributions

9.8.6 Results

The first results chart seen in Figure 9.48 shows a range of solutions with different bit-widths. All solutions are at 160 MHz. Two lines are shown in this graph. The dashed line shows the results for the SQNR produced for different bit-widths. This data has already been seen before in Figure 9.39. The second, solid line shows the area achieved for each of these bit-widths. For the SQNR that we are looking for, we will be selecting the 16-bit fixed-point sizing.

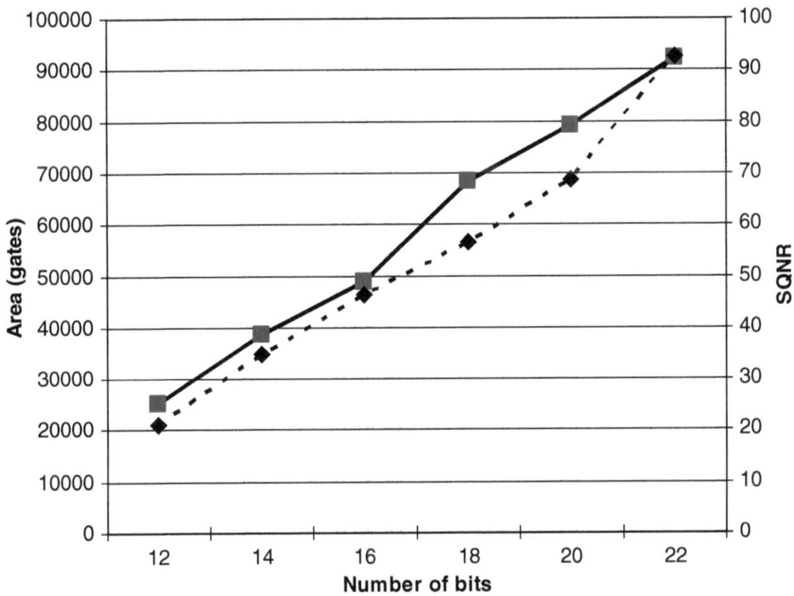

Fig. 9.48 Area vs. width

The next thing we will look at is the effect of frequency. With the design configured as it was, Catapult could not find a schedule for 200 MHz, so 180 MHz is the fastest that this design can run as configured. The results can be seen in Figure 9.49. You will notice that this curve is not smooth as might be expected. This is an example of the optimization algorithms within Catapult making decisions about sharing or the types and sizes of components to use. The solution it found at 100 MHz probably managed to optimize the usage of some component that was not available for the higher frequency solutions. It should be noted that the area figures reported here have not been run through RTL synthesis and optimization, so these are higher than would have been obtained with the extra optimization steps.

We have already talked about the performance impact that a reduction in the number of bits has, in terms of the signal-to-noise ratios. In Figure 9.50 we can see that relationship for the complete Gram–Schmitt algorithm. We can also look

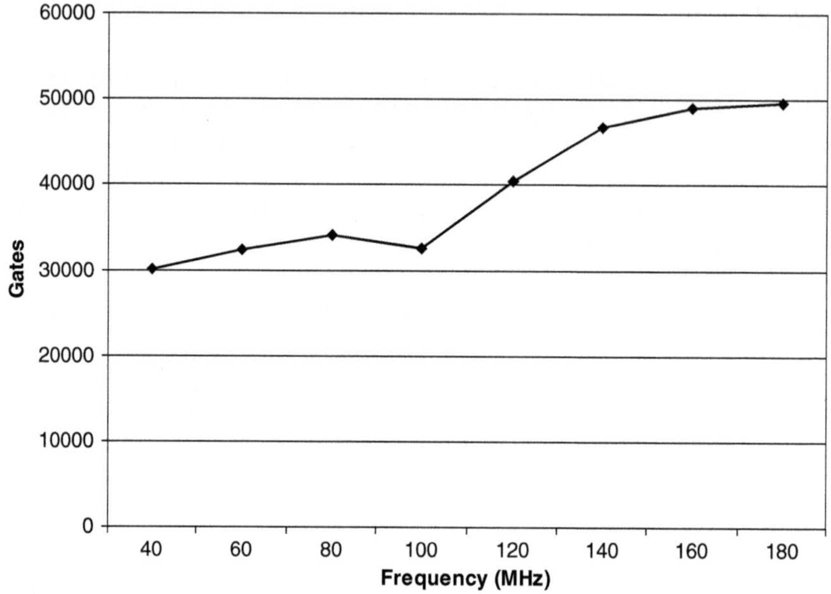

Fig. 9.49 Area vs. frequency

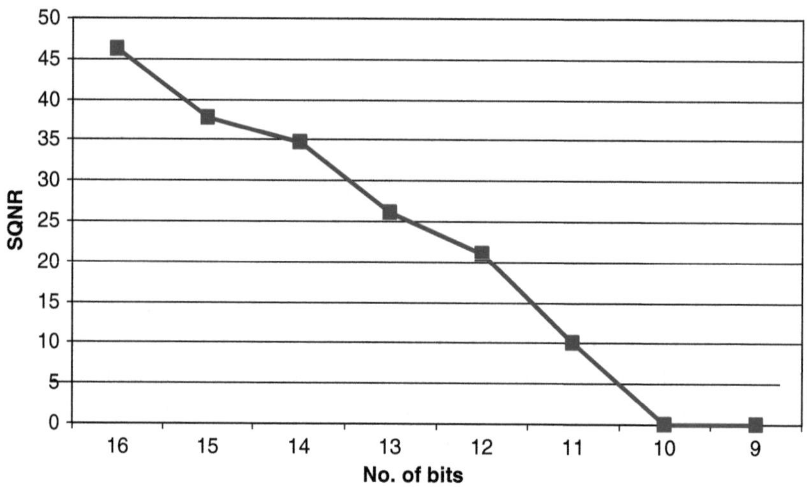

Fig. 9.50 SQNR vs. no. of bits

at the impact that the number of bits has on latency and area, since a reduction in the number of bits will have a significant impact on the complexity of the solution. This is shown in Figure 9.51. Here we see the latency drop from 130 to 122 as the number of bits decreases and the area drop from 40.5 k gates down to 14.9 k gates.

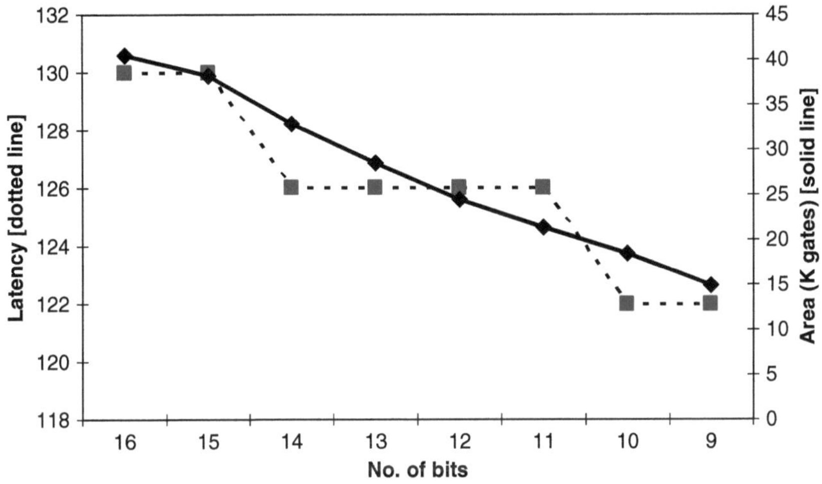

Fig. 9.51 Area and latency vs. no. of bits

9.8.7 Results Analysis

It should be remembered from the description at the beginning of this section that the result reported here is not a highly optimized solution. It is expected that given additional time, better results in terms of latency, area, power, or any other set of attributes would be attainable. It was shown that best reported results from the published papers on the subject could be attained by someone unfamiliar with the area, with the aid of a high-level synthesis tool. The reference results were shown in the table in Section 9.8.3.1.

We also discussed in Section 9.8.5 that the results obtained by Salmela are clearly incomplete as they were using a linear approximation for the square root function. This provides no ability to separate the signal from the noise and thus serves as no useful comparison.

Similarly, the 8-bit precision solution obtained by Lin may be too noisy for anything but the lowest cost systems. In our data sweeps, we did find a 12-bit solution, with a latency of 126 cycles at 160 MHz that would fit in 25 k gates. This is before RTL synthesis, so some further optimization is expected on this size.

Singh used real numbers rather than complex numbers which greatly simplify the design. In our design, the complex number is made up of two components representing the real and imaginary aspects of the signal, and the signal-to-noise ratio of this design is unknown. Even with these simplifications they did not manage to come close to the 40 k gates achieved here.

Luethi and Percels are very comparable to the results we obtained, but unfortunately they did not specify the data widths that they used or the resultant noise levels that they obtained, so we do not know if we have an apple-to-apple comparison, but the sizes are very similar. While Percels has half the latency, it is also operating at half the clock speed, which means that the total transformation time is identical.

What we have shown is that a designer, armed with a tool such as Catapult, can produce results that are very competitive with hand-crafted designs. But that is only half of the picture. We produced the entire design from start to finish in a few weeks (part time) and had a design that was fully scalable such that it can produce solutions that would fit in low-end or high-end products and everything in between. If the design needs to be mapped into a different technology or at a different clock speed, it would take about 2 minutes to produce a new design implementation.

This is a demonstration of the increased productivity that can come from providing engineers with this type of tool. If you wish to explore additional optimizations that have been made since the publication of this book, you should either contact Mentor or join in a discussion on the book's companion web site http:// ElectronicSystemLevel.com.

9.9 Successful Adoption

In a study performed by Toshiba [32], they were also looking to find the best architecture for a MIMO application. They had previously implemented both a QR decomposition approach and one based on the calculation of Eigen equations. They reasoned that in the future more and more antennas would be used, so they wanted to look at the implementation complexity of a four-antenna system which requires 4×4 complex number matrices.

Toshiba originally expected the Eigen equation approach to be the better choice since both algorithms had similar complexity when executed on a DSP, but the Eigen equation approach uses less division than QR decomposition. The use of high-level synthesis made it possible for them to explore both solutions and they did not have to spend months generating RTL for both solutions. Instead they spent just 5 days performing the study. At first, it looked as if they were right. The results they were obtaining showed Eigen equations to be superior. However, they continued to look for other micro-architectural solutions and they were surprised when they found a solution where QR decomposition produced smaller designs and operated at higher frequencies. This finished up being the best solution that they found.

Without the ability to explore the design space, this kind of evaluation would not be possible. It saved Toshiba 38% on their design area and 11.5% improvement in the frequency of operation. In a world of intense competition this can make the difference between success and failure. Probably more importantly though, it gave Toshiba the ability to explore, something they could not have performed in the past.

Assume for a moment that after reading this chapter you are convinced that your group or product would benefit from using high-level synthesis. How should you go about evaluating, piloting, planning, and getting the technology into the production flow? How do you convince your manager that it is the right thing to do? What can you expect the benefits to be? What are the pitfalls?

These are not easy questions to answer, and the benefits are likely to be different depending on your application and the attitude of your organization. As you will have seen from the examples in this chapter, it is not realistic to take a reference implementation that was targeted at software and to just run it through a high-level synthesis tool. This will not get you good results and will probably show you that hand-coded RTL is better. If this is the way that you look at a problem, then you probably will not have a great deal of success in using this technology. High-level synthesis is not a push-button solution, it is a tool that helps a designer become more productive, to explore an algorithm in search of the best architecture, the best micro-architecture and find a good solution that matches the desired requirements. If used correctly it will probably produce better solutions than a hand-coded algorithm. This is not because it has better optimization capabilities than an experienced engineer, but because the tool allows you to find better solutions, just as it did for Toshiba.

Education is a very necessary aspect to the adoption of new technologies. At the Design Automation Conference in 2008, STMicroelectronics [33] reported that they were seeing a 5 – 10× productivity improvement by using high-level synthesis. However, they also pointed out that managers would have to be a little patient because in the early phases of a new design, it would appear that productivity had been decreased, as shown in Figure 9.52.

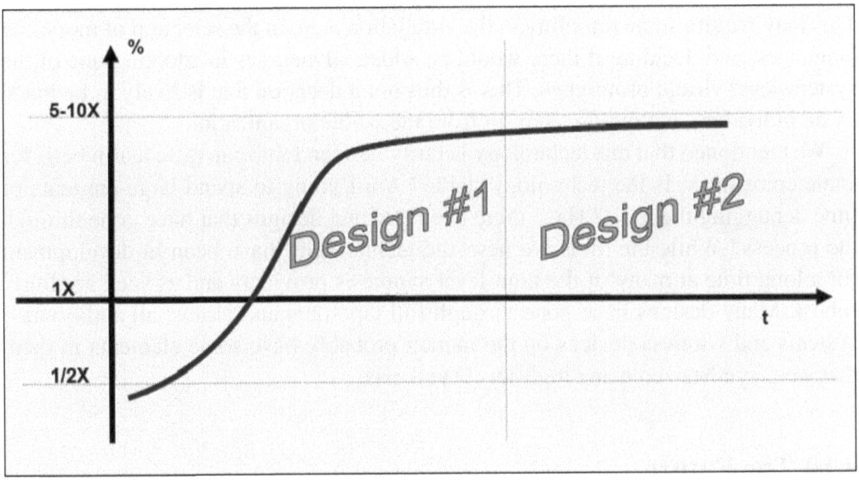

Fig. 9.52 Productivity gains from high-level synthesis

This initial decrease in productivity is the time taken to understand the algorithms, the potential architectures, and to learn about the tool, but that ground is quickly made up and gains are seen even in the first design. With all subsequent designs, the productivity improvements start at close to the maximum, although the authors feel that there is likely to be some continued improvement related to getting more familiar with the tool and the many options that are available within it.

ST also reported that when reusing high-level IP models, additional productivity gains resulted, demonstrating that behavioral IP reuse is just as valid, perhaps even more so, than RTL IP reuse. They also noted that this could open the door to future convergence of IP between divisions serving different markets which could have a huge financial impact on the company.

The writing of this chapter was the first exposure that the author had with this particular tool, and while tool training was attended and a lot of knowledge obtained through the development of the example in this chapter, I am surprised that ST feels that full productivity gains are established during the course of a single design. A tool such as Catapult has a large number of powerful options and settings and while many of them would be used on a large portion of designs, there are also many others that are quite specific to particular designs, situations, or problems. I also noticed the way in which the experience of the TME instinctively led to certain things being done, which I do not think would have been so obvious to someone less familiar with the tool. The author would thus expect that continued increases in productivity and in the quality of the solutions that could be found would increase as the designer builds their level of experience.

Along with the adoption of high-level synthesis, a methodology shift is also necessary in other parts of the organization to experience the full impact. For example, if the verification team is not geared up to deal with high-level design models, then they may not be able to fully utilize the high-simulation speeds that come with them. This may require some retooling in the simulation area, in the selection of modeling languages, and deciding if there would be added advantages in adopting one of the system-level virtual prototypes. This is thus not a decision that is likely to be made by an individual, but requires buy-in from the whole organization.

We mentioned that this technology is fairly new, and that can raise alarm bells for some companies. Is the technology stable? Am I going to spend large amounts of time debugging the tools? Have there been any real designs that have gone through the process? While the tools are new, the technologies have been in development for a long time at many of the high-level synthesis providers and as such are fairly robust. Many designs have gone through full tape-outs and almost all audio/video systems and wireless devices on the market probably have some elements in them that were synthesized using high-level synthesis.

9.10 The Future

While high-level synthesis has been dreamed about by academics for decades, successful commercial adoption has only happened over the past few years. During that time period rapid advancements have been made in terms of the language sets supported, the quality of the results produced, the flexibility in manipulating the micro-architectures, etc. What can we expect in the next few years? The obvious answer is the continued improvement in all of the areas mentioned, but there are new areas in which we can expect to see progress as well.

In the short term, I expect the biggest progress to be made in the size of the blocks that they can handle. While they are already quite significant, additional progress will come from component hierarchies where the tools can handle multiple blocks and at the same time optimize the scheduling, communications, and other overriding aspects between the blocks. While some capabilities for this already exist in the tools on the market, I expect to see a lot more of this being announced over the next couple of years.

Today's synthesis solutions tend to concentrate on the pieces of a system that have already been identified as going into hardware. To migrate high-level synthesis into the realm of true ESL synthesis, the tools must be able to analyze a complete system containing both hardware and software or a language suitable for describing the combined notions of both. Multiple implementation architectures would exist for the solution, which would be selected and then the functionality split between the major architectural blocks that have been instantiated. The principal difficulty with this is stitching everything back together again as this potentially requires modification of software architectural elements, such as operating systems and drivers. An early example of this was described in Chapter 7.

Another area of advancement will come in energy and power optimization. In the past, power was a second-class citizen when it came to optimization, but that is changing for many designs that rely on battery technology, and even for some designs that have access to unlimited power, but may have problems in getting rid of heat or just want to be environmentally friendly. While many of the high-level synthesis tools can analyze power-related issues and write out the results such that they are easy to optimize, they cannot as yet automatically find an architecture that would minimize the total energy consumed. Power optimization is related not only to minimizing active power but also in terms of multiple power domains, variable frequency, and many other complex and highly inter-related factors.

When RTL synthesis was maturing, it did not consider the back-end flow. It did not look at the results of layout and routing to find actual delays. This was not a big problem given that wire delays were small compared to gate delays. As fabrication technologies decreased, the interconnect delays began to rise, and they became a significant problem for the synthesis tools. Many companies reported issues in converging the timing in their designs. Eventually the RTL synthesis tools became more aware of the back end and could take these timing issues into account. Today, high-level synthesis tools output RTL. They place a timing buffer into their scheduling calculations to take into account unexpected back-end delays. The size of this buffer is determinable by the user. This can lead to overdesign or possibly similar timing convergence problems, even though the technology libraries are better qualified than they were in the initial troubled days with RTL. While there are currently hand-offs between C-to-RTL synthesis and RTL-to-gate synthesis tools, we can expect to see even tighter integration between the C and the RTL similar to what we are now used to with RTL and back-end tools.

9.11 Summary

In this chapter we have demonstrated the state of the art in high-level synthesis technology. What has been demonstrated is that this technology is ready for adoption today and is capable of contributing to the design, implementation, and verification of complex systems. High-level synthesis supports the new ESL methodologies that enable incremental implementation and verification at each level of design process.

The technology is advancing rapidly, with tools adding greater language flexibility, improving quality of results both in initial execution and in incremental optimization of the solution with the ability to quickly explore multiple implementation micro-architectures.

High-level synthesis has demonstrated a very high return on investment (ROI) with only a short learning curve that has to be overcome before engineers can expect to become highly productive. The main benefits are increased productivity, higher quality designs, and can often achieve better results than hand-coded RTL, principally due to their ability to explore a greater number of solutions. In addition, it makes it very easy to develop a family of solutions from a single abstract model, further cutting down verification times and increasing quality.

While Mentor's CatapultC tool has been used for these demonstrations, it should be noted that there are other tools available in the marketplace that may be capable of providing similar results. This is thus not an endorsement of any tool and people interested in this technology should talk to their EDA suppliers to see which one will best fit with their requirements. Other high-level synthesis tools are available from Bluespec, Forte, Synfora, and others. Cadence recently entered this product category [34] which should heat up the competition and ensure that the rate of innovation remains high since there are now two of the major EDA companies competing to be the market-dominant supplier. In October 2009, Synopsys introduced a new high-level synthesis tool, Synphony, that uses MATLAB m-code as the input. This means that three of the major EDA companies are now competing in the HLS area.

It is often said that in order to shift a market, there has to be a 10× gain to make it worthwhile. High-level synthesis can easily deliver on that gain. The gain comes not only from the speed with which a high-quality solution can be obtained but in the reduced verification times that it enables, and in many cases will provide a much better result than would have been possible by coding at the RT level due to the opportunities enabled by architectural exploration.

RTL engineers do not have to be scared of this technology. While it is true that most gate-level designers did not make good RTL designers, the migration to behavioral is different. It requires knowledge of RTL implementation structures, but now you do not need to implement them, just decide which ones would produce the best results. At the same time, a software engineer is unlikely to take their job away, as these tools utilize the experience that RTL developers have built up over time in order to drive the tool to producing good results. High-level synthesis is a productivity tool for a hardware developer; it is not a replacement for them. It enables a good engineer to become a better engineer.

References

1. Philippe Coussy, Adam Morawiec (Eds.). High-Level Synthesis: From Algorithm to Digital Circuit, Springer, 2008.
2. Philippe Coussy, Adam Morawiec (Eds.). High-Level Synthesis: From Algorithm to Digital Circuit, Springer – Chapter 3, 2008.

3. Brian Bailey, Grant Martin, Andrew Piziali. ESL Design and Verification: A Prescription for Electronic System Level Methodology, Elsevier Morgan Kaufmann, 2007.
4. Gary Smith EDA website. http://www.garysmitheda.com
5. Brian Bailey, Grant Martin, Thomas Anderson (Eds.). Taxonomies for the Development and Verification of digital systems, Springer, 2005
6. Richard Cloutier, Donald Thomas. "The combination of Scheduling, Allocation, and Mapping in a Single Algorithm". Proceedings of the 27th ACM/IEEE Conference on Design Automation. 1991
7. Synopsys website for Design Compiler. http://www.synopsys.com/Tools/Implementation/RTLSynthesis/Pages/default.aspx
8. Andres Takach, Peter Gutberlet, Simon Walters. Fast Bit-Accurate C++ Datatypes for Functional System Verification and Synthesis. GSPx 2005
9. Andres Takach, Bryan Bowyer, Thomas Bollaert. C Based Hardware Design for Wireless Applications. DATE 2005
10. Ariane 5 Flight 501. http://en.wikipedia.org/wiki/Ariane_5_Flight_501
11. http://en.wikipedia.org/wiki/Advanced_Encryption_Standard
12. Advanced Encryption Standard (AES) Implementation in C/C++ with comments. http://www.hoozi.com/Articles/AES_Encryption.htm
13. Nicolas Sklavos, Odysseas Koufopavlou. Architectures and VLSI Implementations of the AES-Proposal Rijndael. IEEE Transactions on Computers, Vol. 51, No. 12, December 2002 http://www.nsklavos.gr /Papers/%5BJ01%5D.pdf
14. David Canright. A Very Compact S-box for AES. Cryptographic Hardware and Embedded System Sept. 2005
15. Tsung-Fu Lin et al. A High Throughput Low Cost AES Cipher chip. Proceedings of 3rd IEEE Asia-Pacific Conference on ASICs. 2002
16. Frank Gürkaynak et al. A 2 Gb/s Balanced AES Crypto-Chip Implementation. GLSVLSI 2004.
17. http://www.calypto.com/slecsystemhls.php
18. http://www.calypto.com
19. Press release: Mentor Graphics and Calypto Design Systems Announce Customer-Proven Electronic System Level Synthesis and Verification Flow Featuring Catapult C Synthesis and SLEC Sequential Equivalence Checker. Jan 14 2008
20. The Mathworks website. http://www.mathworks.com/
21. Srikanth Jadcherla, Janick Bergeron, Yoshio Inoue, David Flynn. Verification Methodology Manual for Low Power, Synopsys, 2009
22. Atrenta website. http://atrenta.com/
23. Sequence website. http://www.sequencedesign.com/solutions/powertheater.php
24. Alexander Maltsev et al. Triangular Systolic Array with Reduced Latency for QR-decomposition of Complex Matrices. ISCAS 2006
25. Adrian Burian, Jarmo Takala, Mikko Ylinen. A fixed point implementation of matrix inversion using Cholesky decomposition. Proceedings of IEEE International Midwest Symposium on Circuits and Systems. Dec 2003
26. Isabelle LaRoche, Sebastien Roy. An efficient regular matrix inversion circuit architecture for MIMO processing. Proceedings of IEEE International Symposium on Circuits and Systems. May 2006
27. Abbas Bigdeli, Morteza Biglari-Abhari, Zoran Salcic, Yat Tin Lai. A New Pipelined Systolic Array-Based Architecture for Matrix Inversion in FPGAs with Kalman Filter Case Study. EURASIP Journal on Applied Signal Processing, 2006
28. Chitranjan K. Sing, Sushma H. Prasad, Poras T. Balsara. "VLSI Architecture for Matrix Inversion Using Modified Gram-Schmidt Based QR Decomposition". in Proceedings of International Conference on VLSI Design, pp. 836–841. Jan 2007
29. Johan Eilert, Di Wu, Dake Liu. Efficient Complex Matrix Inversion for MIMO Software Defined Radio. IEEE International Symposium on Circuits and Systems. 2007

30. Perttu Salmela, Adrian Burian, Harri Sorokin, Jarmo Takala. Complex-Valued QR Decomposition Implementation for MIMO Receivers. in Proceedings of ICASSP. Apr 2008
31. A Hardware Algorithm for Computing Reciprocal Square Root. Proceedings of the 15th IEEE Symposium on Computer Arithmetic. 2001.
32. Toshiba Information Systems (Japan) Corporation. Advanced Algorithmic Evaluation for Imaging, Communication and Audio Applications – Eigenvalue Decomposition Using The Catapult® C Algorithmic Synthesis Methodology. April 2008
33. STMicroelectronics presentation at Design Automation Conference 2008. http://communities.mentor.com/mgcx/docs/DOC-1110
34. Cadence community blog entry for C to silicon launch. http://www.cadence.com/Community/blogs/sd/archive/2008/07/14/c-to-silicon-launch.aspx
35. Kuang-Hao Lin, Robert C. Chang, Chier-Lin Huang, Feng-Chi Chen, Shih-Chun Lin. "Implementation of QR Decomposition for MIMO-OFDM Detection Systems". in Proceedings of ICECS. Sep 2008
36. Peter Luethi, Andreas Burg, David Perels, Norbert Felber, Wolfgang Fichtner. "VLSI Implementation of a High-Speed Iterative MMSE QR Decomposition". in Proceedings of ICECS. 2007.
37. David Perels, Simon Haene, Peter Luethi, Andreas Burg, Norbert Felber, Wolfgang Fichtner, Helmut Bolcskei. "ASIC Implementation of a MIMO-OFDM Transceiver for 192Mbps WLANs". in Proceedings of ICECS. 2005.

Acronyms

The following list defines the acronyms used throughout the book:

AES	Advanced Encryption Standard
AMBA	Advanced Microcontroller Bus Architecture
AHB	Advanced High-performance Bus
AMP	Asymmetric Multi-Processing
API	Application Programming Interface
ASIC	Application-Specific Integrated Circuit
ASIP	Application-Specific Instruction set Processor
AT	Approximately Timed
AV	Algorithmic View or Application View
AVM	Advanced Verification Methodology
AXI	Advanced eXtensible Interface
BCA	Bus Cycle Accurate
BFM	Bus Functional Model
CAD	Computer Aided Design
CDT	C/C++ Development Toolkit
DASC	Design Automation Standards Committee
DCT	Discrete Cosine Transform
DDF	Dynamic Dataflow
DLL	Dynamic Linked Library
DMA	Direct Memory Access
DMI	Direct Memory Interface
DPI	Direct Programming Interface
DSL	Digital Subscriber Line
DSP	Digital Signal Processor
DTD	Document Type Definition
ECIX	Electronic Component Information eXchange
EDA	Electronic Design Automation
EDK	Embedded Development Kit
EHCI	Enhanced Host Controller Interface
ESL	Electronic System level
*e*RM	***E*** reuse Methodology

B. Bailey, G. Martin, *ESL Models and their Application*, Embedded Systems,
DOI 10.1007/978-1-4419-0965-7, © Springer Science+Business Media, LLC 2010

FDM	Frequency Division Multiplexing
FIFO	First In First Out
FIR	Finite Impulse Response
FLIX	FLexible Instruction eXtensions
FPGA	Field Programmable Gate Array
FPU	Floating Point Unit
FSM	Finite State Machine
GP	Generic Payload
GUI	Graphical User Interface
HAL	Hardware Abstraction Layer
HDL	Hardware Description Language
HVL	Hardware Verification Language
HW/SW	Hardware/Software
ICE	In-Circuit Emulator
IDCT	Inverse Discrete Cosine Transform
IDE	Integrated Development Environment
IJG	Independent JPEG Group
IP	Intellectual Property
IPCM	Incisive® Plan-to-Closure Methodology
ISA	Instruction Set Architecture
ISR	Interrupt Service Routine
ISS	Instruction Set Simulator
ITF	Interpret–Translate–Feedback
JIT	Just In Time
JPEG	Joint Photographic Experts Group
KPN	Kahn Process Network
LAU	Least Addressable Unit
LT	Loosely Timed
LTE	Long Term Evolution
MAC	Multiply Accumulate
MIMO	Multiple-Input Multiple-Output
MOC	Model of Computation. Sometimes written MoC
NOC	Network-on-Chip. Sometimes written NoC
NRE	Non-Recurring Engineering
OCP	Open Core Protocol
OCP-IP	Open Core Protocol International Partnership
OFDM	Orthogonal Frequency Division Multiplexing
OS	Operating System
OSCI	Open SystemC Initiative
OPB	On-chip Peripheral Bus or Open Peripheral Bus
OTG	On-The-Go
OVM	Open Verification Methodology
PBD	Platform-Based Design
PLB	Processor Local Bus
PLD	Programmable Logic Device

PLI	Program Language Interface
PSL	Property Specification Language
PV	Programmer's View
PV+T or PVT	Programmer's View plus Timing
RAM	Random Access Memory
RASSP	Rapid prototyping of Application Specific Signal Processors
RDL	Register Description Language
RMS	Root Mean Square
ROI	Return-on-Investment. Sometimes called RoI
RTL	Register Transfer Level
RTOS	Real-Time Operating System
SAHE	Software Accessible Hardware Elements
SCE-MI	Standard Co-Emulation – Modeling Interface
SCV	SystemC Verification library
SDK	Software Development Kit
SDL	Specification and Description Language
SEC	Sequential Equivalence Checking
SIMD	Single Instruction Multiple Data
SLVP	System-Level Virtual Prototype, synonymous with functional prototype or virtual system prototype
SMP	Symmetric Multi-Processing
SNR	Signal-to-Noise Ratio
SoC	System on Chip. Sometimes called SOC.
SPIRIT	Structure for Packaging, Integrating and Re-using IP with Toolsflows consortium
SQNR	Signal-to-Quantization Noise Ratio
STL	Standard Template Library
SVA	SystemVerilog Assertions
TIE	Tensilica Instruction Extension
TL	Transaction Level
TLI	Transaction Level Interface
TLM	Transaction Level Modeling
TLP	Transaction Level Platform
TME	Technical Marketing Engineer
URM	Universal Reuse Methodology
USB	Universal Serial Bus
UML	Universal Modeling Language
VCT DWG	Virtual Component Transfer Development Working Group
VCX	Virtual Component eXchange
VHDL	VHSIC Hardware Description Language
VHPI	VHDL Procedural Interface
VIP	Verification IP
VLIW	Very Long Instruction Word
VRE	Virtual platform Runtime Environment
VSIA	Virtual Socket Interface Alliance

VMM	Verification Methodology Manual
VSP	Virtual System Prototype, synonymous with functional prototype or system-level virtual prototype
VV	Verification View
XCL	Xilinx Cache Link
XML	eXtensible Markup Language
XPG	Xtensa Processor Generator

Index

MIX
Papier aus verantwortungsvollen Quellen
Paper from responsible sources
FSC® C105338

If you have any concerns about our products,
you can contact us on
ProductSafety@springernature.com

In case Publisher is established outside the EU,
the EU authorized representative is:
Springer Nature Customer Service Center GmbH
Europaplatz 3, 69115 Heidelberg, Germany

Printed by Libri Plureos GmbH
in Hamburg, Germany